Lecture Notes in Computer Science 14588

Founding Editors

Gerhard Goos
Juris Hartmanis

The series Lecture Notes in Computer Science (LNCS), including its subseries Lecture Notes in Artificial Intelligence (LNAI) and Lecture Notes in Bioinformatics (LNBI), has established itself as a medium for the publication of new developments in computer science and information technology research, teaching, and education.

LNCS enjoys close cooperation with the computer science R & D community, the series counts many renowned academics among its volume editors and paper authors, and collaborates with prestigious societies. Its mission is to serve this international community by providing an invaluable service, mainly focused on the publication of conference and workshop proceedings and postproceedings. LNCS commenced publication in 1973.

Daniel Mendez · Ana Moreira

Editors

Requirements Engineering: Foundation for Software Quality

30th International Working Conference, REFSQ 2024
Winterthur, Switzerland, April 8–11, 2024
Proceedings

 Springer

Editors
Daniel Mendez 🆔
Blekinge Institute of Technology
Karlskrona, Sweden

fortiss GmbH
Munich, Germany

Ana Moreira 🆔
NOVA University Lisbon
Lisbon, Portugal

NOVA LINCS
Costa da Caparica, Portugal

ISSN 0302-9743   ISSN 1611-3349 (electronic)
Lecture Notes in Computer Science
ISBN 978-3-031-57326-2   ISBN 978-3-031-57327-9 (eBook)
https://doi.org/10.1007/978-3-031-57327-9

Preface

This volume contains the papers presented at REFSQ 2024, the *30th International Working Conference on Requirements Engineering: Foundation for Software Quality*, held on April 8–11, 2024 in Winterthur, Switzerland.

The REFSQ series was established in 1994, at first as a workshop series, and since 2010 as a "working conference", dedicating ample time for presentations, substantial discussions, and community building. We consider it to be among the major international scientific events in Requirements Engineering, and the only one to be permanently located in Europe and fostering also a special connection to European industry.

To celebrate this connection and the growing trend for practically relevant contributions, the special theme for REFSQ 2024 was consequently **Out of the Lab, into the Wild!** We intended to foster discussion around the following questions:

- What are domain-specific problems and challenges relevant to Requirements Engineering in practice?
- What are solutions (methodologies and technologies) that guide professionals when handling requirements in real-world environments?
- What different interpretations and manifestations can Requirements Engineering take in practical settings?
- How can we effectively demonstrate that the methods, techniques, and tools we develop address practical problems?
- How can we foster fruitful academia-industry collaborations?
- What are insights, experiences, and lessons learned that emerge from our investigations of how RE is applied in practice?
- How can we ensure reproducibility and transparency in our research artifacts?

We were very happy to observe that the challenge was promptly taken up by the research community, with many submissions focusing on exactly those issues. Several of those contributions were accepted for presentation at the conference, and are now part of this volume.

In response to the Call for Papers, we received 59 submissions, out of which, after screening, 53 were followed by independent peer-review by at least three program committee members. Based on the review results and profound online discussions among the program committee members, 15 papers were accepted and 7 were accepted on the condition that certain improvements would be made (those underwent an additional check by a PC member before final acceptance). Overall, 22 papers were finally accepted for publication, and are now collected in this volume. In particular, we accepted papers in the following categories:

- Scientific Evaluation papers investigating real-world problems and evaluating existing artifacts implemented in real-world settings.
- Technical Design papers describing the design of new artifacts.

- Experience report papers providing retrospective reports on experiences in applying RE techniques in practice, or addressing RE problems in real-world contexts.
- Vision papers reasoning on where research in the field should be heading.
- Research Preview papers describing well-defined research ideas at an early stage of investigation which may not be fully developed yet.

The acceptance rate of the research track contributions was 28.5% for full papers (Scientific Evaluation and Technical Design papers) and 42% for short papers (Experience Reports, Vision, and Research Preview papers) leading, in total and covering all categories, to an acceptance rate of 37% (22/59).

As in previous years, the conference was organized as a three-day symposium (Tuesday to Thursday), with one day devoted to industrial presentations (in a single track), and two days of academic presentations (in two parallel tracks). In addition to paper presentations and related discussions, the program included three keynote talks by Lorenz Hilty, Irina Koitz, and Awais Rashid; a Poster & Tools session organised by Oscar Dieste and Nelly Condori-Fernandez; an Education and Training track, organized by Stan Bühne, Jennifer Hehn, and Birgit Benzenstadler; an Industry track organized by Martin Glinz and Rainer Grau; and awards to recognize the best contributions in various categories.

Prior to the symposium with its various tracks, the conference featured the Doctoral Symposium and several workshops:

- RE4AI: 5th International Workshop on Requirements Engineering for Artificial Intelligence, organized by Renata Guizzardi, Khan Mohammad Habibullah, Anna Perini, and Angelo Susi.
- NLP4RE: 7th Workshop on Natural Language Processing for Requirements Engineering, organized by Sallam Abualhaija, Chetan Arora, Davide Dell'Anna, Alessio Ferrari, and Sepideh Ghanavati.
- AgileRE: the AgileRE Workshop, organized by Fabiano Dalpiaz and Jan-Philipp Steghöfer.
- VIVA RE: Virtues and Values in Requirements Engineering 2024 Workshop, organized by Alexander Rachmann and Jens Gulden.
- Green Digital Design - How can modern RE of digital products contribute to more sustainability?, organized by Andrea Müller, Martina Beck, and Dominik Birkmeier.
- CreaRE: 11th International Workshop on Creativity in Requirements Engineering, organized by Andrea Herrmann, Maya Daneva, Patrick Mennig, and Kurt Schneider.

The proceedings of the co-located workshops as well as the tracks for Posters & Tools and the doctoral symposium are published in a separate volume via CEUR.

We would like to thank all members of the Requirements Engineering community who prepared a contribution for REFSQ 2024: there would be no progress in our community (and no proceedings!) without the devotion that so many brilliant researchers dedicated to the field. We would also like to thank members of the Program Committee and additional reviewers for their invaluable contribution to the selection process. Special thanks are due to all our colleagues who served in various distinguished roles to make the organization of REFSQ 2024 possible. Your help in assembling and organising this rich program has been invaluable:

- The REFSQ Steering Committee has provided excellent support and guidance throughout the process;
- The previous PC chairs were always available to share their experiences and discuss new ideas;
- The Local Organizers Marcela Ruiz and Norbert Seyff were fantastic organizers and hosts;
- The Steering Committee Chair, Anna Perini, and Vice-Chair, Fabiano Dalpiaz, and the Background Organization Chair, Xavier Franch, constantly gave collegial support.
- The Social Media and Publicity Chairs, Anne Hess, Oliver Karras, and Gunter Mussbacher, always acted proactively to disseminate information and engage with the community through various channels.
- The Open Science Chairs, Sallam Abualhaija, Alessio Ferrari, and Davide Fucci, helped to provide support to authors in making their research publicly accessible.
- Last but not least: The webchair, Anton Luckhardt, and the proceedings chair, Julian Frattini, showed endless patience and dedication. It must have sometimes felt to them like herding cats.

Thank you!

February 2024

Daniel Mendez
Ana Moreira

Organization

Program Committee Chairs

Daniel Mendez	Blekinge Institute of Technology, Sweden, and fortiss GmbH, Germany
Ana Moreira	NOVA University of Lisbon and NOVA LINCS, Portugal

Local Organization Chairs

Marcela Ruiz	Zurich University of Applied Science, Switzerland
Norbert Seyff	University of Applied Sciences and Arts Northwestern Switzerland, Switzerland

Industry Track Chairs

Martin Glinz	University of Zurich, Switzerland
Rainer Grau	Juropera GmbH, Switzerland

Workshop Chairs

Jennifer Horkoff	Chalmers University of Technology and the University of Gothenburg, Sweden
Thorsten Weyer	Technische Hochschule Mittelhessen—University of Applied Sciences, Germany

Education and Training Chairs

Stan Bühne	IREB, Germany
Jennifer Hehn	Bern University of Applied Science, Switzerland
Birgit Penzenstadler	Chalmers University of Technology, Sweden

Posters and Tools Chairs

Oscar Dieste   Universidad Politécnica de Madrid, Spain
Nelly Condori-Fernandez   Universidad de Santiago de Compostela, Spain

Doctoral Symposium Chairs

Maya Daneva   University of Twente, The Netherlands
Michael Unterkalmsteiner   Blekinge Institute of Technology, Sweden

Open Science Chairs

Sallam Abualhaija   University of Luxembourg, Luxembourg
Alessio Ferrari   CNR-ISTI, Italy
Davide Fucci   Blekinge Institute of Technology, Sweden

Social Media and Publicity Chairs

Anne Hess   Fraunhofer IESE, Germany
Oliver Karras   Leibniz Information Centre for Science and
                                        Technology, Germany
Gunter Mussbacher   McGill University, Canada

Web Chair

Anton Luckhardt   fortiss GmbH, Germany

Proceedings Chair

Julian Frattini   Blekinge Institute of Technology, Sweden

Finance Chair

Samuel Fricker   Fachhochschule Nordwestschweiz, Switzerland

REFSQ Series Organization

Steering Committee

Anna Perini (Chair)	Fondazione Bruno Kessler, Italy
Fabiano Dalpiaz (Vice-chair)	University of Utrecht, The Netherlands
Xavier Franch (Chair of BO)	Universitat Politècnica de Catalunya, Spain
Alessio Ferrari	Consiglio Nazionale delle Ricerche, Italy
Birgit Penzenstadler	Chalmers Tekniska Högskola, Sweden and Lappeenranta-Lahti University of Technology, Finland
Paola Spoletini	Kennesaw State University, USA
Nazim Madhavji	Western University, Canada
Vincenzo Gervasi	University of Pisa, Italy
Andrea Vogelsang	University of Cologne, Germany
Daniel Mendez	Blekinge Institute of Technology, Sweden, and fortiss, Germany
Ana Moreira	NOVA University of Lisbon and NOVA LINCS, Portugal

Background Organization

Xavier Franch	Universitat Politècnica de Catalunya, Spain
Quim Motger	Universitat Politècnica de Catalunya, Spain

Program Committee

João Araújo	Universidade NOVA de Lisboa, Portugal
Chetan Arora	Monash University, Australia
Fatma Başak Aydemir	Boğaziçi University, Turkey
Dan Berry	University of Waterloo, Canada
Stefanie Betz	Furtwangen University, Germany
Mitra Boakei Hosseini	University of Texas at San Antonio, USA
Travis Breaux	Carnegie Mellon University, USA
Jaelson Castro	Universidade Federal de Pernambuco, Brazil
Fabiano Dalpiaz	Utrecht University, The Netherlands
Maya Daneva	University of Twente, The Netherlands
Oscar Dieste	Universidad Politécnica de Madrid, Spain
Jörg Dörr	Fraunhofer IESE, Germany
Alessio Ferrari	ISTI-CNR, Italy

Jannik Fischbach	Netlight Consulting GmbH, Germany, and fortiss GmbH, Germany
Xavier Franch	Universitat Politècnica de Catalunya, Spain
Julian Frattini	Blekinge Institute of Technology, Sweden
Matthias Galster	University of Canterbury, New Zealand
Vincenzo Gervasi	University of Pisa, Italy
Martin Glinz	University of Zurich, Switzerland
Miguel Goulão	NOVA LINCS, Universidade Nova de Lisboa, Portugal
Eduard C. Groen	Fraunhofer IESE, Germany
Iris Groher	Johannes Kepler University Linz, Austria
Alicia Grubb	Smith College, USA
Paul Grünbacher	Johannes Kepler University Linz, Austria
Andrea Herrmann	Herrmann & Ehrlich, Germany
Hans-Martin Heyn	Chalmers University of Technology, Sweden, and the University of Gothenburg, Sweden
Emilio Insfran	Universitat Politècnica de València, Spain
Marcos Kalinowski	Pontifical Catholic University of Rio de Janeiro, Brazil
Erik Kamsties	University of Applied Sciences and Arts Dortmund, Germany
Oliver Karras	TIB - Leibniz Information Centre for Science and Technology, Germany
Eric Knauss	Chalmers University of Technology, Sweden, and the University of Gothenburg, Sweden
Matthias Koch	Fraunhofer IESE, Germany
Sylwia Kopczyńska	Poznan University of Technology, Poland
Kim Lauenroth	University of Applied Science and Arts Dortmund, Germany
Emmanuel Letier	University College London, UK
Nazim Madhavji	University of Western Ontario, Canada
Anastasia Mavridou	KBR/NASA Ames Research Center, USA
Daniel Mendez	Blekinge Institute of Technology, Sweden, and fortiss, Germany
Luisa Mich	University of Trento, Italy
Sabine Molenaar	Utrecht University, The Netherlands
Lloyd Montgomery	University of Hamburg, Germany
Ana Moreira	NOVA LINCS and NOVA University of Lisbon, Portugal
John Mylopoulos	University of Ottawa, Canada
Maleknaz Nayebi	York University, Canada
Nan Niu	University of Cincinnati, USA
Marc Oriol Hilari	Universitat Politècnica de Catalunya, Spain

Barbara Paech	Universität Heidelberg, Germany
Elda Paja	IT University of Copenhagen, Denmark
Liliana Pasquale	University College Dublin, Ireland
Oscar Pastor	Universidad Politécnica de Valencia, Spain
Nitish Patkar	University of Applied Sciences and Arts Northwestern Switzerland, Switzerland
Anna Perini	Fondazione Bruno Kessler, Italy
Bjorn Regnell	Lund University, Sweden
Mehrdad Sabetzadeh	University of Ottawa, Canada
Huma Samin	Durham University, UK
Simon Andre Scherr	Fraunhofer IESE, Germany
Klaus Schmid	University of Hildesheim, Germany
Kurt Schneider	Leibniz Universität Hannover, Germany
Laura Semini	University of Pisa, Italy
Paola Spoletini	Kennesaw State University, USA
Jan-Philipp Steghöfer	XITASO GmbH IT & Software Solutions, Germany
Angelo Susi	Fondazione Bruno Kessler, Italy
Sira Vegas	Universidad Politécnica de Madrid, Spain
Colin C. Venter	University of Hudderfield, UK
Andreas Vogelsang	University of Cologne, Germany
Liping Zhao	University of Manchester, UK
Didar Zowghi	CSIRO's Data61, Australia

Supporters

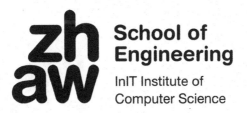

fortiss

Contents

Artificial Intelligence for Requirements Engineering

Natural Language Processing for Requirements Engineering

Requirements Engineering for Artificial Intelligence

Crowd-Based Requirements Engineering

Emerging Topics and Challenges in Requirements Engineering

Quality Models for Requirements Engineering

How Explainable Is Your System?
Towards a Quality Model
for Explainability

Hannah Deters$^{(\boxtimes)}$ (ID), Jakob Droste(ID), Martin Obaidi(ID), and Kurt Schneider(ID)

Software Engineering Group, Leibniz University Hannover, Hannover, Germany
{hannah.deters,jakob.droste,martin.obaidi,
kurt.schneider}@inf.uni-hannover.de

Abstract. **[Context and motivation]** Explainability is a software quality aspect that is gaining relevance in the field of requirements engineering. The complexity of modern software systems is steadily growing. Thus, understanding how these systems function becomes increasingly difficult. At the same time, stakeholders rely on these systems in an expanding number of crucial areas, such as medicine and finance. **[Question/problem]** While a lot of research focuses on how to make AI algorithms explainable, there is a lack of fundamental research on explainability in requirements engineering. For instance, there has been little research on the elicitation and verification of explainability requirements. **[Principal ideas/results]** Quality models provide means and measures to specify and evaluate quality requirements. As a solid foundation for our quality model, we first conducted a literature review. Based on the results, we then designed a user-centered quality model for explainability. We identified ten different aspects of explainability and offer criteria and metrics to measure them. **[Contribution]** Our quality model provides metrics that enable software engineers to check whether specified explainability requirements have been met. By identifying different aspects of explainability, we offer a view from different angles that consider different goals of explanations. Thus, we provide a foundation that will improve the management and verification of explainability requirements.

Keywords: Explainability · Requirements Engineering · Quality Models · Metrics · Literature Studies

1 Introduction

With the ongoing rise of Artificial Intelligence (AI) throughout society, end-users of all kinds make contact with increasingly opaque [2] and complex software systems [20]. Even systems without AI components are becoming increasingly complex these days. Thus, the need for explanations is no longer limited to AI systems [13]. As a consequence, understanding how a software operates can become difficult for its stakeholders [19]. This lack of understanding might lead to distrust and ultimately to the user rejecting the system [19]. Research in the field of Explainable AI (XAI) has identified explainability as a way to address

D. Mendez and A. Moreira (Eds.): REFSQ 2024, LNCS 14588, pp. 3–19, 2024.
https://doi.org/10.1007/978-3-031-57327-9_1

these issues [10,12]. Recent research in requirements engineering has established explainability as a non-functional requirement (NFR) [7,19].

In many cases, explanations are used to influence related NFRs like understandability and transparency. As such, explainability is often considered to be a means to an end [8,11]. The explainability of a system depends on the context of use [7] and the users themselves [13,28]. An explanation that is useful to a certain stakeholder in a specific use case might not be helpful for another stakeholder in the same use case. Similarly, the same stakeholder might prefer different kinds of explanations in different contexts [7]. The effectiveness of explanations also depends on the goals of the explainer [11]. Short and concise explanations might increase usability, while long and detailed explanations provide more transparency. In order to specify requirements for explainable systems, it is essential to understand what makes a system explainable in the first place. To this end, different aspects of explainability, such as understandability or transparency, must be considered. Furthermore, there is a need for metrics that can effectively measure these aspects, in order to evaluate the degree of explainability of a system [29].

The goal of this paper was to *develop a comprehensive user-centered quality model for explainability that highlights different aspects of explainable systems and includes metrics to measure explainability with respect to these aspects.*

In this paper, we present a quality model for measurable explainability. Building on an existing Systematic Literature Review on explainable systems [7], we conducted a Snowball Search (SBS) in the direction of explainability metrics and measurement. Among the 1025 examined works, we found 84 papers which were concerned with measuring explainability. Within these 84 papers, we identified 36 distinct criteria and 35 metrics for explainability, which we organized in 10 aspects. These ten aspects are *Understandability, Transparency, Effectiveness, Efficiency, Satisfaction, Correctness, Suitability, Trustability, Persuasiveness* and *Scrutability*. Depending on the explainer's goal, different aspects are then chosen to specify and validate explainability requirements. Our model provides the means to specify explainability requirements with respect to different aspects of explainability and enables requirements engineering for explainable systems.

We structure this work as follows: Section 2 discusses the background and related work of this paper. The research design is laid out in Sect. 3. Our results and discussion are found in Sects. 4 and 5. We close this paper with the conclusion and plans for future work in Sect. 6.

2   Background and Related Work

2.1   Explainability

Explainability is the quality of a software system to be explained to its stakeholders [9]. A large part of past explainability research has focused on XAI [10,12]. However, the topic has recently gained more traction in requirements and software engineering, independent of AI [7,8]. In requirements engineering, explainability is considered to be a non-functional requirement and quality aspect [7,19].

Explainability is often said to be related to other quality aspects such as transparency [14], understandability [19] and end-user trust [5,14,19]. As such, explainability may be considered a "means to an end" [8,11].

Not every type of explanation is appropriate in every context and for every stakeholder [7,28]. In user-centered design, personalization can be used to provide appropriate explanations for any specific stakeholder, according to their needs [26,30]. Therefore, different systems need to incorporate different explanations for their users. Using a quality model supports the communication about these needs in the development of explainable systems.

2.2 Quality Models and Metrics

In software engineering, quality models are used to enable the objective measurement of software quality aspects [25]. Depending on what quality aspect is measured, different quality models might be more or less applicable [3]. In some cases, these models are specifically built around certain quality aspects, e.g., security models are concerned with security [22].

Quality models may contain metrics that can be used to scale and measure quality aspects [4]. They are a means to describe to what degree a software system fulfills certain criteria, i.e., by assigning a numerical value [17], and are needed during different steps of the requirements engineering process [11]. In the early stages, metrics can help requirements engineers to formulate quality requirements [17]. During evaluation and testing, metrics are used to determine whether or not a given system fulfills its quality requirements [16,17]. Furthermore, metrics can be used to identify defects in a software [16,23].

2.3 Related Work

Chazette et al. [7] conducted the SLR that our work builds upon. In their work, they established a definition for explainable systems and examined the relation between explainability and other NFRs such as understandability, transparency and many others. In our work, we focus on the different aspects of explainability itself and on the different ways to measure and evaluate explainability.

Deters et al. [11] examined the effectiveness of goal-oriented metrics for explainability. Their findings suggest that when choosing the metrics to evaluate an explainable system, it is important to consider the goals of the system designer and provider. For instance, a system that aims to be as understandable as possible needs different explanations than a system that is tuned towards efficiency. However, they only considered four aspects of explainability and provided heuristics for evaluation. In contrast, this work highlights the need for compiled metrics relating to different aspects of explainability.

Speith and Langer [29] discussed the separation of evaluation methods for explainability approaches. They divided the approaches according to the XAI processes they target: evaluation methods for explanatory information, for understanding and for desiderata. Furthermore, they distinguished between objective and human-centered evaluation methods. While we consider all of these categories to be relevant to this work, we focus on explainability for end-users, whereas Speith and Langer [29] discussed explainability in general. Hence, we deliberately omit aspects such as debuggability from our quality model and instead provide more in-depth information on aspects that are relevant for end-users.

Tintarev et al. [31] conducted a literature search in 2011 to compile criteria and metrics for the explainability of recommender systems. Firstly, their work is therefore limited to recommender systems. Secondly, a lot of research has been done on explainability since then. This is also reflected in the fact that 80% of the papers in our dataset were published after 2011. There are several other papers that examine what aspects of a system increase explainability and thus compile criteria for *good* explainability [6,21,32]. However, these papers refer to specific types of systems and focus on XAI. Notably, this work is not limited to XAI systems and explicitly not to the evaluation of different XAI algorithms. Instead, we enable the evaluation of explainability from the users' point of view. By compiling different aspects of explainability, we illustrate its versatility as an NFR and offer a suitable evaluation method depending on the goal of the explanation.

3   Research Design

Our quality model will be composed of existing criteria and metrics found in literature. To this end, we conducted a literature review by performing a snowball search with one forward and one backward snowballing step.

3.1   Research Questions

RQ1: *What explainability criteria have already been defined in literature?*
RQ2: *What metrics are used in literature to measure explainability?*
RQ3: *How can the criteria and metrics be combined and categorized into aspects?*

The main component of our quality model are criteria that describe the quality aspect (RQ1). In order to assess whether the criteria are met, metrics are needed that measure the criteria (RQ2). In addition, aspects of explainability are identified in order to organize the criteria and metrics (RQ3).

3.2   Methodology

The snowball search was based on four relevant papers that discuss criteria and metrics for explainability [15,24,27,31]. Based on this, one forward and one backward snowballing step was performed. We used the Google Scholar database, as it is not restricted to specific publishers. We identified these four papers using a pre-existing SLR on explainability by Chazette et al. [7]. Following Kitchenham's [18] recommendation, we established inclusion and exclusion criteria to select papers. Only papers that fulfill one of the inclusion criteria but none of the exclusion criteria were considered (Table 1).

Table 1. Inclusion and exclusion criteria

	Inclusion Criteria
I1	The paper specifies what constitutes good explainability. (RQ1)
I2	The paper specifies how explainability can be measured. (RQ2)
I3	The paper evaluates the explainability of a given software system. (RQ2)
	Exclusion Criteria
E1	The paper was not peer-reviewed
E2	The paper is not available with the university's resources
E3	The paper is not written in English or German

The papers were examined in 3 phases (see Fig. 1). In phase one and two, papers were selected according to the inclusion and exclusion criteria. In the first selection phase, we read the title and abstract. In the first selection phase, we read the title and the abstract. Only papers that clearly did not satisfy the inclusion or exclusion criteria were rejected. Papers that we were unsure about were included at this stage. In the second selection phase, we scanned the remaining papers for relevant information. Papers for which we did not find any relevant information related to the inclusion criteria were excluded. After the second selection phase, we identified a set of 84 papers that fit the inclusion and exclusion criteria. In the third phase, all 84 papers were then read in depth and all relevant information was extracted.

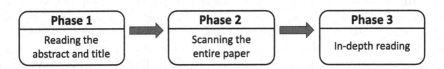

Fig. 1. Phases during literature review

Startset. We identified the start set using an SLR by Chazette et al. [7] which included 229 papers. After reading the abstract and title, there were 33 papers that did not fail the inclusion or exclusion criteria. After scanning these 33 papers, another 18 papers were sorted out, leaving fifteen papers that contained relevant information on criteria and metrics for explainability. From these fifteen papers, four relevant papers were identified [15, 24, 27, 31], which formed the start set for forward and backward snowballing.

Phase 1	Phase 2	Phase 3
229 paper	33 paper	15 paper

Forward Snowballing. We used the Google Scholar database for forward snowballing as it is not restricted to specific publishers. The four selected papers were cited 769 times in total (by 4/25/2022). After sorting out duplicates in the set itself and with the start set (preprocessing) 665 papers remained. In the first selection phase, 583 of these papers did not meet the inclusion and exclusion criteria, leaving 82 papers. In the second selection phase, a total of 45 papers were found to be relevant for our research.

Preprocessing	Phase 1	Phase 2	Phase 3
769 paper	665 paper	82 paper	45 paper

Backward Snowballing. A total of 404 literature references were made in the four selected papers. Many of these papers were already included in the start set or the forward snowballing set, resulting in 131 *new* papers that we examined. Of these 131 papers, 84 were rejected in the first selection phase, leaving 47 papers to be considered in the second selection phase. In the second selection phase, a total of 24 were finally found to be relevant to our research.

Preprocessing	Phase 1	Phase 2	Phase 3
404 paper	131 paper	47 paper	24 paper

Overall Process. From a total of 1024 analyzed papers, we were able to identify 84 papers that contained relevant information on criteria and metrics for explainability. We stopped the literature search after the first forward and backward snowballing step. For one, almost two-thirds of the papers were already identified as duplicates in the first backward snowballing step, which indicated that we had already covered a large proportion of available papers. Furthermore, the 84 papers we found formed a sufficient basis for creating a quality model.

3.3　Designing the Quality Model

The quality model was created within four steps. In the first step, ten aspects describing explainability were compiled on the basis of the 84 retrieved papers. To this end, we noted which aspects of explainability were mentioned in the paper while reading, and we combined similar aspects (for example, user efficiency and time efficiency). Aspects that are very domain-specific or could not be generalized and only appeared in one of the papers were not included in this step. In the second step, all the criteria and metrics collected from the papers were each assigned to one of the aspects. Criteria or metrics that did not fit into an aspect were collected in an additional list. For most papers, it was possible to deduce to which of the aspects the criteria and metrics belonged. If the papers did not provide any information on this, we assigned them to the best of our judgment.

In the third step, we then checked based on the additional list whether another aspect should be added. This was not the case. The final step was to analyze which criterion can be measured by which metric. Again, for most of the metrics, it was specified which criteria they measure. If the paper did not indicate which criteria the metric assesses, we reflected and assigned criteria ourselves.

4   Results

4.1   Aspects of Explainability

We were able to identify ten aspects of explainability during the SLR. The aspects are described in Table 2. Depending on the goal that an explanation is supposed to fulfill, different aspects need to be considered when evaluating the explanation. For example, if explanations are intended to enable faster use, efficiency criteria and metrics can be applied. On the other hand, if explanations are intended to ensure that the system delivers the best possible result, the focus should be on the aspect of effectiveness.

Table 2. Definition of explainability aspects

Criterion	Description
C1 Understandability	The explanations are easy to understand by the addressee
C2 Transparency	The explanations provide sufficient insight into how the system works
C3 Effectiveness	The explanations help the addressee to use the system *better*. Make better decisions, use functions that fit best
C4 Efficiency	The explanations help the addressee to use the system faster. Make decisions faster, execute actions faster
C5 Satisfaction	The explanations increase the comfort of use and enjoyment
C6 Correctness	The explanations are truthful
C7 Suitability	The explanations are suited to the context, the user and the goal of the use
C8 Trustability	The explanations help to have confidence in the system
C9 Persuasiveness	The explanations convince the addressee to use, try or buy some system related item
C10 Scrutability	The explanations help to correct the system if necessary

4.2   Quality Model

We developed a quality model based on a total of 84 papers. We identified 36 criteria and 35 metrics for explainability. As our quality model is user-centered, most metrics require the involvement of users. For each aspect, the criteria and metrics are illustrated in a directed acyclic graph. The structure of these graphs can be seen in Fig. 2. The top box displays the aspect, e.g. *Understandability*. The boxes below the aspect contain criteria that contribute to *good* explainability with regard to that aspect. The dotted boxes below the criteria contain metrics that measure the respective criteria. The dotted lines between the metrics and criteria indicate which criterion can be measured by which metric. In this example, *MetricA* measures both *CriterionA* and *B*. *CriterionB* can also be measured by *MetricB*.

Fig. 2. Scheme of the Quality Model

Each aspect is based on numerous papers. The criteria and metrics are also based on at least one, usually several papers. We provide supplementary material indicating from which papers each aspect, criterion and metric originates [1].

4.3   Criteria and Metrics

Understandability is the first aspect of explainability. It focuses on whether the explanations are easy to understand for the addressee. The related criteria and metrics are listed in Fig. 3. Criterion C1.1 states that the explanations should convey information with little mental effort. Explanation should be kept simple and thus be written in natural language and contain few elements - e.g. few sentences (C1.2). The form of presentation also affects the understandability of an explanation. (C1.3). For example, some information can be better conveyed using illustrations, whereas some information can be better conveyed using text. Furthermore, the parts of the explanation (e.g. sentences in a text) should be logically coherent (C1.4).

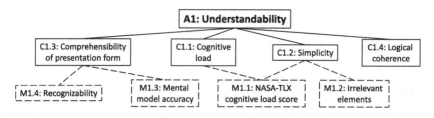

Fig. 3. Criteria and Metrics for Understandability

The NASA-TLX Cognitive Load Score is a standardized questionnaire which can be used to measure the cognitive load (M1.1). Furthermore, multiple choice questions can be used to examine whether the users have processed the information correctly (M1.3). The recognizability (M1.4) can be assessed to check if the user understands the representation. This metric can be applied by showing users various explanations (e.g. heat maps), some of which do not match the situation in the software. The user must then judge whether the explanation fits the situation or not. Metric M1.2 suggest asking users to remove elements (e.g. sentences) from the explanation that were not necessary for their understanding. This way, it can be checked whether the explanation was as simple as possible.

Transparency focuses on whether the explanation provides sufficient insight into how the system works (see Fig. 4). Explanations should align the mental model of the user to the system model (C2.1). To ensure transparency, the explanation should provide justifications why certain outcomes were obtained (C2.2). Furthermore, the explanation should convey which elements (e.g. input values) have a particularly large influence on the outcomes (C2.3). A high proportion of explainable components is also desirable (C2.4).

The mental model accuracy can be obtained by asking the users to explain how the system works (M2.3). The users can also be asked to predict the outcome (M2.1). Asking the users to predict whether the behavior of the system changes in specific situations can assess the situation awareness (M2.4). Another possibility is to let users predict which elements (e.g. input values) are most relevant (M2.2).

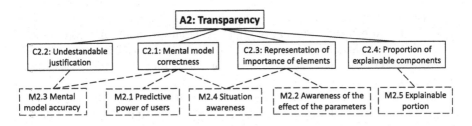

Fig. 4. Criteria and Metrics for Transparency

Effectiveness focuses on whether the explanation help the addressee to use the system better (e.g. make better decisions, use functions that fit the best). Criteria and metrics are listed in Fig. 5. Explanation should enable the user to perform actions to improve the overall situation (C3.1 Actionability). Moreover, explanations should help the user to assess the situation in order to intervene if necessary (C3.2). Effective explanations allow users to make the optimal decision to generate the best possible outcome for them (C3.3).

A common metric used to measure effectiveness was metric M3.1, the *Absence of a Difference*. It tests whether users would make a different decision if all possible information was available to them compared to if only the explanation was available to them. For example, a recommender system for streaming services makes suggestions which films a user might also like. An explanation can now illustrate what the suggestion is based on. If users make the same decision when they only see the explanation compared to when they see a trailer of the movie, the explanation seems to be very effective. Another way to measure the effectiveness is to check if the user decides for a more satisfactory choice when the explanation is available to them (M3.2). To check whether the user is enabled to use functions that fit the intended goal, the number of functions identified by the user can be counted (M3.3). Finally, it can also be measured how successfully the user fulfills their tasks (M3.4).

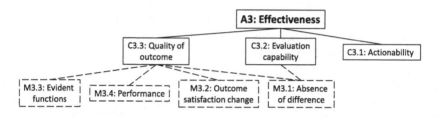

Fig. 5. Criteria and Metrics for Effectiveness

Efficiency focuses on whether an explanation helps the addressee to use the system faster. (see Fig. 6). Explanations should help the user to make decisions faster or execute actions faster (C4.1). Therefore, the user should be granted quick access to information (C4.2). Navigating to an online manual, for example, would not be an efficient explanation. The explanation should not contain any information unnecessary for the user (C4.3). If the explanations are generated while the system is used, the generation of the explanation must take little time (C4.4).

The measurement of time is an often mentioned way to assess efficiency (M4.1). This includes recording the time while completing a task or the time needed to answer a question about the information. Another mentioned possibility is to count the number of interactions the user needs to complete a task (M4.2).

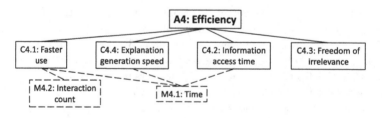

Fig. 6. Criteria and Metrics for Efficiency

Satisfaction focuses on whether the explanations increase the comfort of use and the enjoyment. Criteria and metrics are listed in Fig. 7. As satisfaction is subjective, the focus lies on the perception of the users. Explanations should increase the perceived usefulness (C5.1), the perceived quality (C5.2) and the subjective enjoyment (C5.3). To ensure that the explanations do not lower satisfaction with the system, the explanations themselves have to be easy to use (C5.4). Moreover, the explanations should be simple - e.g. containing few elements and being short (C4.5).

The most frequently mentioned metric for satisfaction is the assessment of subjective perceptions using questionnaires and likert scales (M5.1). Participants are asked to indicate their satisfaction with the overall system or with the explanations themselves. It can also be assessed how inclined users are to use the product again (M5.2). The usability of the explanations themselves can be measured with established usability tests (M5.3). Counting certain expressive actions (e.g. displayed explanations, frustrations of the user, usability issues) can also indicate satisfaction with the system (M5.4).

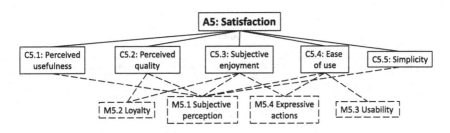

Fig. 7. Criteria and Metrics for Satisfaction

Correctness focuses on whether the explanations are truthful (see Fig. 8). The explanation must not try to mislead the users (M6.1). Furthermore, they should be sound (M6.2) and complete (M6.3). Explanations for similar situations (inputs) should also be similar, especially when using different underlying (ML) models (M6.4).

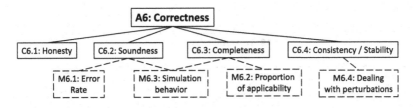

Fig. 8. Criteria and Metrics for Correctness

To measure how often an incorrect explanation is generated, model-specific metrics exist (M6.1). Completeness can be assessed by counting the number of instances to which the explanation applies (M6.2). Another mentioned method is to simulate scenarios with inputs and target outputs and compare target with actual (M6.3). Furthermore, it can be assessed how highly sensitive the explanation is to insignificant disturbances (M6.4).

Suitability focuses on whether the explanations are appropriate for the context, the user and the goal of use (see Fig. 9). Based on the context, the explanations should be triggered at a suitable timing and focus on unexpected events (C7.1). Explanations should also be suited to the user - especially to their prior knowledge and personal preferences (C7.2). Furthermore, users should only receive information they need to achieve their goal (C7.3).

To measure suitability it was suggested to ask the user what further information he would have liked to receive (M7.1). Another proposed method it to examine the user's prior knowledge and/or character traits with the help of a pre-study, and to evaluate ratings on this basis (M7.2, M7.3). We did not find any metrics in literature to measure suitability for the context.

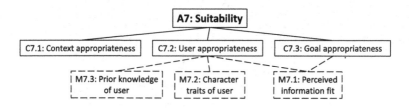

Fig. 9. Criteria and Metrics for Suitability

Trustability focuses on whether the explanation helps users to have confidence in the system (see Fig. 10). Explanations can increase trust in the system by increasing the user's perceived competence (C8.1). Furthermore, users tend to trust the system more when the explanations are consistent with expectations (C8.2). According to literature, revealing limitations of the system can also increase trust in reliable situations (C8.3). Moreover, similar scenarios should evoke similar explanations to increase confidence (C8.4).

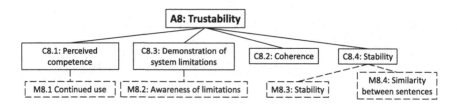

Fig. 10. Criteria and Metrics for Trustability

Frequent use of a system/explanation is a good indicator that the system/explanation is trusted. Thus, it can be counted how often the system in general or the explanations are used to assess the perceived competence. (M8.1). Asking the user to judge when and how to rely on the system (M8.2) was also mentioned as metric. The stability can be assessed by calculating the similarity between sentences (M8.4). Adding gaussian noise to the input and measuring how much the explanation changes can also assess the stability (M8.4).

Persuasiveness focuses on whether the explanation convinces the user to use, try or buy a system related item (see Fig. 11a). Explanations can convince the user to buy or use the product recommended by the system by increasing the perceived accuracy of the system (C9.1). For example, if a recommender system suggests a movie and the explanation convinces the user that the suggestions are very suitable, the user is more likely to buy or watch the movie. Moreover, a higher acceptance of the system's output means that users are more likely to be persuaded by what the system says (C9.2).

Counting how many users are willing to buy a product (M9.1), how many users follow a recommendation (M9.4) or how often a user changes his mind after seeing an explanation (M9.3) can assess persuasiveness. Another mentioned metric is asking the user how confident they are that the proposed product is suitable. (M9.2).

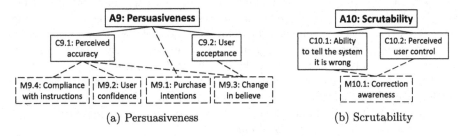

(a) Persuasiveness   (b) Scrutability

Fig. 11. Criteria and Metrics for Persuasiveness and Scrutability

Scrutability focuses on whether the explanation helps to correct the system if necessary (see Fig. 11b). An explanation should enable the user to correct the system if an error occurs (C10.1). Furthermore. the explanation should give the user the feeling of being able to control the system more themselves (C10.2).

To measure scrutability it can be checked if the users understand that and how they can correct the system - e.g. with thinking out loud or interviews (M10.1).

5 Discussion and Threats to Validity

5.1 Answering the Research Questions

Based on the results and the findings of Sect. 4, we answer our three research questions as follows:

RQ1: We found 36 criteria indicating *good* explainability in the literature. A concise overview of these criteria is provided in Sect. 4.

RQ2: A total of 35 metrics, accompanied by explicit execution methods for measuring the identified criteria, are available based on our analysis. An overview of these methods is also provided in Sect. 4.

RQ3: The criteria and metrics are categorized into ten aspects, resulting in a quality model. The ten primary aspects are: understandability, transparency, effectiveness, efficiency, satisfaction, correctness, suitability, trustability, persuasiveness and scrutability. The literature highlights the close relationship between these criteria and metrics, and the intended objective of explainability.

5.2   Discussion of the Results

The established quality model has the potential to close a gap in the area of explainability in requirements engineering. The model is based on 36 criteria and the associated 35 metrics identified from 84 papers in our conducted snowball search. This enables requirements engineers to specify explainability requirements and to ensure that they are fulfilled. The variety of aspects and criteria also allows them to have an overview of the different explainability goals, and to categorize and prioritize them accordingly.

The quality model is user-centered, meaning that all criteria and metrics are oriented towards the end user. Further aspects of explainability should be considered when looking at other stakeholders. For example, when considering developers, the aspect of debuggability plays a major role, which was also mentioned in some papers. As we designed a user-centered quality model, most of the metrics are only applicable with the involvement of end users.

Identifying different aspects of explainability enables requirements engineers to address explainability in a more goal-oriented manner. Depending on the goal to be achieved through explanations, different aspects of explainability can be used to define the appropriate requirements.

5.3   Limitations and Threats to Validity

In the following, we present threats to validity according to our study. We categorize the threats according to Wohlin et al. [33] as construct, internal, external, and conclusion validity.

Construct Validity. Since the starting set consisted of pre-filtered papers, it is possible that this already introduced a bias. In addition, the subsequent evaluation of the literature was made by a single person, which may allow subjectivity. Applying the snowball principle repeatedly could have led to the discovery of other potentially relevant papers. However, we believe that the papers included in our SBS are sufficient to answer the research questions.

Internal Validity. During our SBS, we utilized a set of criteria to determine the inclusion or exclusion of papers. For instance, papers lacking peer review were excluded. To enhance the internal validity of our process, we derived our inclusion and exclusion criteria from the guidelines established by Kitchenham et al. [18]. If uncertain, we added the paper and revisited this choice during a subsequent phase of our process. In addition, the extraction of the criteria and metrics from the papers was done by one of the authors. It is therefore possible that criteria or metrics were omitted. The compilation of the directed graphs also poses a potential threat to the internal validity of our work. It is possible that some criteria and metrics could be assigned to a different aspect. To address this, we plan to validate the quality model with other experts.

Conclusion Validity. For mitigation, explicit inclusion and exclusion criteria were defined to make the selection as objective as possible. The papers from the start set from the paper by Chazette et al. [7] were also selected with a defined procedure such that the bias was also kept as small as possible here. Overall, however, it should be noted that it is likely that not all existing papers relevant to this topic were included. In addition, 50% of the papers in the start set referred to recommender systems. It is therefore possible that our quality model is biased towards recommender systems. In the overall dataset, about a quarter of the papers were related to recommender systems.

External Validity. Our quality model is based on literature, and we did not perform a hands-on validation of our results. Thus, we cannot claim the applicability of the quality model in practice. However, as we only considered peer-reviewed papers, we are confident that the criteria and metrics are reasonable.

6 Conclusion and Future Work

In this paper, we established a quality model for user-centered explainability consisting of ten graphs reflecting ten aspects of explainability. We were able to identify 36 criteria and 35 metrics in literature, which we assigned to the ten explainability aspects. We used directed acyclic graphs to visualize which metrics can be used to measure the criteria.

The quality model enables requirements engineers to specify explainability requirements and check whether they have been fulfilled. The identified aspects also enable requirements engineers to consider explainability from different perspectives. Depending on the goal that the explanations are meant to achieve, required aspects of explainability can be determined.

The quality model is solely based on criteria and metrics found in the literature. An important next step is validating and expanding the quality model by consulting experts. We plan to conduct interviews with experts to validate the categorization and to find further criteria and metrics that can be used to assess explainability. The next step will be to validate the quality model in practice in order to draw conclusions about its applicability and usefulness. Once the quality model is validated, the process of eliciting explainability requirements requires further scrutiny. In particular, we want to investigate how to avoid biases in the elicitation process, and how explainability needs can be identified during runtime, by observing user behavior and physiological triggers.

Acknowledgments. This work was funded by the Deutsche Forschungsgemeinschaft (DFG, German Research Foundation) under Grant No.: 470146331, project softXplain (2022-2025).

References

1. Supplementary material. https://doi.org/10.5281/zenodo.10640067
2. Adadi, A., Berrada, M.: Peeking inside the black-box: a survey on explainable artificial intelligence (XAI). IEEE Access **6**, 52138–52160 (2018)
3. AL-Badareen, A.B., Selamat, M.H., Jabar, M.A., Din, J., Turaev, S.: Software quality models: a comparative study. In: Mohamad Zain, J., Wan Mohd, W.M., El-Qawasmeh, E. (eds.) ICSECS 2011. CCIS, vol. 179, pp. 46–55. Springer, Heidelberg (2011). https://doi.org/10.1007/978-3-642-22170-5_4
4. Al-Qutaish, R.E.: Quality models in software engineering literature: an analytical and comparative study. J. Am. Sci. **6**(3), 166–175 (2010)
5. Brunotte, W., Specht, A., Chazette, L., Schneider, K.: Privacy explanations - a means to end-user trust. J. Syst. Softw. **195**, 111545 (2023). https://doi.org/10.1016/j.jss.2022.111545. https://www.sciencedirect.com/science/article/pii/S0164121222002217
6. Caro-Martínez, M., Jiménez-Díaz, G., Recio-García, J.A.: Conceptual modeling of explainable recommender systems: an ontological formalization to guide their design and development. J. Artif. Intell. Res. **71**, 557–589 (2021). https://doi.org/10.1613/jair.1.12789
7. Chazette, L., Brunotte, W., Speith, T.: Exploring explainability: a definition, a model, and a knowledge catalogue. In: 2021 IEEE 29th International Requirements Engineering Conference (RE), pp. 197–208. IEEE (2021)
8. Chazette, L., Klös, V., Herzog, F., Schneider, K.: Requirements on explanations: a quality framework for explainability. In: 2022 IEEE 30th International Requirements Engineering Conference (RE), pp. 140–152. IEEE (2022)
9. Chazette, L., Schneider, K.: Explainability as a non-functional requirement: challenges and recommendations. Requirements Eng. **25**(4), 493–514 (2020)
10. Das, A., Rad, P.: Opportunities and challenges in explainable artificial intelligence (XAI): a survey. CoRR abs/2006.11371 (2020). https://arxiv.org/abs/2006.11371
11. Deters, H., Droste, J., Schneider, K.: A means to what end? Evaluating the explainability of software systems using goal-oriented heuristics. In: Proceedings of the 27th International Conference on Evaluation and Assessment in Software Engineering, pp. 329–338 (2023)
12. Došilović, F.K., Brčić, M., Hlupić, N.: Explainable artificial intelligence: a survey. In: 2018 41st International Convention on Information and Communication Technology, Electronics and Microelectronics (MIPRO), pp. 0210–0215. IEEE (2018)
13. Droste, J., Deters, H., Puglisi, J., Klünder, J.: Designing end-user personas for explainability requirements using mixed methods research. In: 2023 IEEE 31st International Requirements Engineering Conference Workshops (REW), pp. 129–135. IEEE (2023)
14. Fox, M., Long, D., Magazzeni, D.: Explainable planning. arXiv preprint arXiv:1709.10256 (2017)
15. Hoffman, R.R., Klein, G., Mueller, S.T.: Explaining explanation for "explainable AI". Proc. Hum. Fact. Ergon. Soc. Annu. Meet. **62**(1), 197–201 (2018). https://doi.org/10.1177/1541931218621047
16. Honglei, T., Wei, S., Yanan, Z.: The research on software metrics and software complexity metrics. In: 2009 International Forum on Computer Science-Technology and Applications, vol. 1, pp. 131–136. IEEE (2009)
17. Institute of Electrical and Electronics Engineers: IEEE standard for a software quality metrics methodology. IEEE Std 1061-1992, pp. 1–96 (1993). https://doi.org/10.1109/IEEESTD.1993.115124

18. Kitchenham, B.: Procedures for performing systematic reviews. Keele, UK, Keele University **33**(2004), 1–26 (2004)
19. Köhl, M.A., Baum, K., Langer, M., Oster, D., Speith, T., Bohlender, D.: Explainability as a non-functional requirement. In: 2019 IEEE 27th International Requirements Engineering Conference (RE), pp. 363–368. IEEE (2019)
20. Kuhrmann, M., Tell, P., Klünder, J., Hebig, R., Licorish, S., MacDonell, S.: HELENA stage 2 results. ResearchGate (2018)
21. Langer, M., et al.: What do we want from explainable artificial intelligence (XAI)? - A stakeholder perspective on XAI and a conceptual model guiding interdisciplinary XAI research. Artif. Intell. **296** (2021). https://doi.org/10.1016/j.artint.2021.103473. https://www.sciencedirect.com/science/article/pii/S0004370221000242
22. McLean, J.: Security models. Encycl. Softw. Eng. **2**, 1136–1145 (1994)
23. Mladenova, T.: Software quality metrics-research, analysis and recommendation. In: 2020 International Conference Automatics and Informatics (ICAI), pp. 1–5. IEEE (2020)
24. Nunes, I., Jannach, D.: A systematic review and taxonomy of explanations in decision support and recommender systems. User Model. User-Adap. Inter. **27**(3), 393–444 (2017)
25. Samadhiya, D., Wang, S.H., Chen, D.: Quality models: role and value in software engineering. In: 2010 2nd International Conference on Software Technology and Engineering, vol. 1, pp. V1–320. IEEE (2010)
26. Schneider, J., Handali, J.: Personalized explanation in machine learning. CoRR abs/1901.00770 (2019)
27. Sokol, K., Flach, P.: Explainability fact sheets: a framework for systematic assessment of explainable approaches. In: Proceedings of the 2020 Conference on Fairness, Accountability, and Transparency, FAT* 2020, pp. 56–67. Association for Computing Machinery, New York, NY, USA (2020). https://doi.org/10.1145/3351095.3372870
28. Sokol, K., Flach, P.: One explanation does not fit all: the promise of interactive explanations for machine learning transparency. KI-Künstliche Intelligenz **34**(2), 235–250 (2020)
29. Speith, T., Langer, M.: A new perspective on evaluation methods for explainable artificial intelligence (XAI). In: 2023 IEEE 31st International Requirements Engineering Conference Workshops (REW), pp. 325–331. IEEE (2023)
30. Tintarev, N., Masthoff, J.: Effective explanations of recommendations: user-centered design. In: Proceedings of the 2007 ACM Conference on Recommender Systems, pp. 153–156 (2007)
31. Tintarev, N., Masthoff, J.: Designing and evaluating explanations for recommender systems. In: Ricci, F., Rokach, L., Shapira, B., Kantor, P.B. (eds.) Recommender Systems Handbook, pp. 479–510. Springer, Boston, MA (2011). https://doi.org/10.1007/978-0-387-85820-3_15
32. Vilone, G., Longo, L.: Notions of explainability and evaluation approaches for explainable artificial intelligence. Inf. Fusion **76**, 89–106 (2021). https://doi.org/10.1016/j.inffus.2021.05.009. https://www.sciencedirect.com/science/article/pii/S1566253521001093
33. Wohlin, C., Runeson, P., Höst, M., Ohlsson, M.C., Regnell, B., Wesslén, A.: Experimentation in Software Engineering. Springer, Heidelberg (2012). https://doi.org/10.1007/978-3-642-29044-2

Identifying Relevant Factors of Requirements Quality: An Industrial Case Study

Julian Frattini[✉][iD]

Blekinge Institute of Technology, Valhallavägen 1, 371 41 Karlskrona, Sweden
julian.frattini@bth.se

Abstract. [**Context and Motivation**]: The quality of requirements specifications impacts subsequent, dependent software engineering activities. Requirements quality defects like ambiguous statements can result in incomplete or wrong features and even lead to budget overrun or project failure. [**Problem**]: Attempts at measuring the impact of requirements quality have been held back by the vast amount of interacting factors. Requirements quality research lacks an understanding of which factors are relevant in practice. [**Principal Ideas and Results**]: We conduct a case study considering data from both interview transcripts and issue reports to identify relevant factors of requirements quality. The results include 17 factors and 11 interaction effects relevant to the case company. [**Contribution**]: The results contribute empirical evidence that (1) strengthens existing requirements engineering theories and (2) advances industry-relevant requirements quality research.

Keywords: Requirements quality · Case study · Interview

1 Introduction

Requirements specifications constitute the input to many subsequent software engineering activities and artifacts. Requirements specifications are used to design architecture, develop code, or derive test cases. Hence, the quality of requirements specifications impacts the software engineering process [18,27].

The requirements quality research domain aims to aid practitioners in understanding and managing the quality of their requirements specifications by detecting and removing requirements quality defects [20]. However, empirical evidence in this research domain remains scarce [16]. This hampers the adoption of research results in practice [7,12]. Recent systematic studies of the requirements quality research domain have identified a lack of industrial relevance as a main factor holding the field back [20].

The aim of this study is to contribute empirical evidence about the impact of requirements quality defects by identifying factors of requirements quality that are relevant in practice. To this end, an industrial case study at Ericsson has been conducted. The gathered evidence both strengthens existing requirements engineering theories and steers future research efforts toward solving practically

© The Author(s), under exclusive license to Springer Nature Switzerland AG 2024
D. Mendez and A. Moreira (Eds.): REFSQ 2024, LNCS 14588, pp. 20–36, 2024.
https://doi.org/10.1007/978-3-031-57327-9_2

relevant problems. While the results cannot be generalized due to the employed research method, we encourage replication in different industrial contexts with the disclosure of our replication package.[1]

The remainder of this article is structured as follows: Section 2 introduces related work on requirements quality research. Section 3 describes the applied method and Sect. 4 reports the obtained results. These results are discussed in Sect. 5 before concluding in Sect. 6.

2   Background

2.1   Requirements Quality

Organizations use requirements specifications in several subsequent software engineering activities. Non-functional requirements influence a system's architecture, functional requirements determine the input and expected output of the system's features, and all requirements are ultimately translated into test cases to assert whether the developed system meets the customers' expectations. However, quality defects in requirements specifications like missing or ambiguous requirements impede this reuse [20, 27].

Two factors aggravate the impact of requirements quality defects on subsequent activities. Firstly, requirements specifications are predominantly written in natural language [13] (NL). The inherent complexity and ambiguity of NL promotes quality defects. Secondly, the cost to remove a quality defect scales the longer it remains undetected [3]. Clarifying and rewriting an ambiguous requirement takes significantly less time than re-implementing a wrong feature that was built based on a misunderstood requirement.

Consequently, many organizations attempt to detect and remove relevant requirements quality defects as early as possible [20]. A popular frame for this is the requirements quality factor.[2] A quality factor is a normative metric that maps a requirements specification to a level of quality [16]. For example, the quality factor *passive voice* associates the use of passive voice in a requirements sentence with bad quality due to the potential omission of the subject of the sentence [23]. Requirements quality research abounds with quality factors and automatic tools to detect violations against them [16].

2.2   Requirements Quality Theory

Despite their usability, requirements quality factors suffer from shortcomings. Most significantly, the majority of them lack empirical evidence for their relevance [15, 16, 20], i.e., they are purely normative. Most publications empirically investigate the performance of a tool automatically detecting a violation against a quality factor, but only the fewest empirically investigate whether this violation

---

[1] Available at https://zenodo.org/doi/10.5281/zenodo.10149474.

[2] In the remainder of the paper, a requirements quality factor is referred to as a *quality factor* as the scope of this article remains focused on requirements specifications.

does have an actual impact and is, therefore, even worth detecting and mitigating. This undermines the practical relevance of requirements quality research, and research results are rarely adopted in practice [7,12].

In response, previous research proposed a harmonized *requirements quality theory* (RQT). This theory frames requirements quality as the impact that properties of requirements specifications have on the properties of dependent activities in a given context [9]. Consequently, the RQT does not consider a violation against a quality factor as harmful in itself, but only if this violation has an impact on the activities that use the requirement as an input.

Figure 1 visualizes a reduced version of the RQT [15]. The RQT consists of three groups of concepts: artifact concepts, the context concept, and activity concepts. The concept of an *entity* represents all types of requirements artifacts like use cases or sentences. *Quality factors* are properties of these entities [16], like the length of a specification, the completeness of a use case, or the voice (active or passive) of a single sentence. *Activities* represent every software engineering process that uses a requirements entity as an input [9]. This includes activities like implementing or generating test cases but also more implicit processes like understanding an entity and estimating its effort. *Attributes* are properties of these activities and include metrics like the duration or correctness of an activity. *Context factors* represent properties of the process and involved stakeholders [21], like the domain knowledge of involved engineers, the used process model, or the distribution of the organization. Finally, the *impact* represents the relationship between the quality factors, context factors, and attributes.

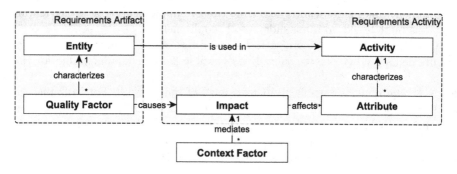

Fig. 1. Groups and concepts of the reduced Requirements Quality Theory [15]

Not only does the RQT guide the framing of requirements quality, but it also enables operationalization in practice [15]. By measuring quality factors, context factors, and attributes, all input and output variables to the impact become quantified. Once all variables are quantified, a statistical model trained on historical data can estimate the probability that a certain quality factor in a given context will affect the attribute of an activity [15]. This quantification was not attempted yet [8] but advocated for in requirements quality research roadmaps [7,15] since it allows to (1) empirically assess and compare the criticality of quality factors, and (2) predict how a requirements specification will impact dependent activities. This prediction model would meet the initially mentioned need of organizations to reliably detect quality defects that impact the software engineering process.

2.3 Gap

Requirements quality research faces two major gaps. Firstly, requirements quality research lacks empirical evidence for the relevance of requirements quality factors [15,16,20]. This entails the risk that requirements quality research does not focus on problems relevant to practice.

Secondly, and by extension, the RQT is difficult to operationalize without empirical evidence about the relevance of quality factors. Previous research has already identified 206 mostly normative requirements quality factors [16]. Measuring all of them is not feasible, given their amount and complexity. Empirical evidence about the relevance of quality factors will aid in prioritizing and selecting factors to consider in statistical models for impact estimation.

3 Method

This study aims to address the gaps outlined in Sect. 2.3 by contributing empirical evidence about relevant factors of requirements quality. The study addresses the following research questions:

- RQ1: Which factors of requirements quality do engineers that process requirements perceive to be relevant?
- RQ2: Which factors of requirements quality do engineers that process requirements report in issues?

The study contrasts two perspectives: relevant factors of requirements quality as *perceived* by engineers using requirements (RQ1) and as *reported* in issues (RQ2). The study employs case study research to obtain insights on the necessary level of detail at the expense of generalizability [24]. The contemporary software engineering phenomenon [28] that is subject of the study is the impact of requirements quality defects as described through its factors. The case study research method lends itself to the investigation since the boundary between the phenomenon and the context is unclear [28].

The method follows Runeson et al. [24] and is reported according to the guidelines by Höst et al. [17]. Figure 2 visualizes the steps of the process. Verbatim examples shown in the figure and throughout this section (in quotation marks) are artificial as the raw data is confidential.

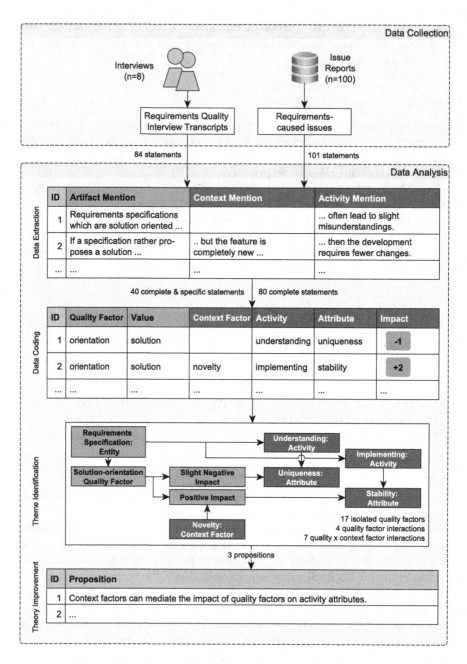

Fig. 2. Overview of the data collection and analysis method

3.1   Data Collection

Ericsson, the case company providing the data, is a large, globally distributed software development organization. Ericsson follows an agile development approach but completes a feature-level requirements specification for each new feature or change of an existing one before committing to subsequent phases like design, implementation, and testing. These requirements specifications use unconstrained natural language to specify one or more requirements related to a specific feature or change request. The study evaluates two different types of data to triangulate and strengthen the results [28]. The upper section of Fig. 2 visualizes the two data collection approaches.

Interview Data. To understand the factors that engineers processing requirements perceive to be relevant, interviews were conducted (RQ1). A contact at the case company provided a sample of eight software engineers from one team which are responsible for processing the requirements specification and developing a solution specification. These eight software engineers represent more than half of the population of their respective role in the team. The interview participants had an average of 3.5 years of experience in their role, 7.5 years with the organization, and 15.3 years as software engineers.

A protocol aided conducting the semi-structured interviews based on previous requirements quality research. In particular, we structured the protocol around eleven high-level attributes of requirements quality [20], including, for example, ambiguity, completeness, and correctness. For each attribute, the interview participants were asked whether they experienced any issues of this type when processing requirements. If yes, they were prompted to elaborate on the issue using the RQT as a frame [15]. Otherwise, the interview moved to the next theme. A slide deck introduced and visualized the RQT for the interview participants to aid the conversation. The author of this paper conducted all eight interviews, each taking up to one hour.

The recorded interview sessions were automatically transcribed using Descript.[3] Afterwards, the author manually checked all transcripts and ensured that the automatic transcription matched the recording. The replication package contains all supplementary material, including the interview guidelines. The transcripts contain confidential information and cannot be shared.

Issue Data. To understand which factors of requirements quality cause an impact that ends up in a report, issues from an issue tracker were analyzed (RQ2). The contact at the case company provided access to Ericsson's database of issue reports. This database contains issues for one specific system raised both during the internal development process and from external customers. Every issue denotes the development phases in which it was detected and in which the root cause of this issue has been introduced. Domain experts procure and document the latter information. For this study, the available issues were filtered to

---

[3] https://www.descript.com/.

obtain only those that have been introduced during the requirements engineering phase, resulting in 100 component-level issue reports from January 2021 until September 2023.

3.2 Data Analysis

To analyze the large body of textual data, thematic synthesis as proposed by Cruzes and Dybå [6] was employed and reported according to their guideline. The lower section of Fig. 2 visualizes the data analysis steps.

Data Extraction. The first step comprised of extracting relevant data from the textual corpus. Each defect perceived by an interview participant constitutes one statement about requirements quality. The eight interview transcripts produced 84 statements about perceived requirements quality. The 100 issues contained 101 statements about reported requirements quality. The three groups of the RQT (artifact, context, and activity) provided a frame for the data extraction. For each of the three groups, all relevant mentions from a statement were extracted. For example, the statement "If a specification rather proposes a solution, but the feature is completely new, then the development requires fewer changes." contains one mention of each group: "a specification rather proposes a solution" describes a property of the artifact, "the feature is completely new" describes the context, and "the development requires fewer changes" represents the activity impacted by the quality and context factor.

Data Coding. Each mention received codes for all the concepts contained within the respective RQT group. An artifact mention can contain a number of quality factors with a value associated with each of them. A context mention can contain a number of context factors. An activity mention can contain a number of activities, each associated with one attribute and an impact value. All codes and coding instructions were documented in coding guidelines.

Coding the artifact and context mentions followed a deductive approach [19]: quality factors [16,20] and context factors [21] identified during previous research constituted the available codes. Coding the activity mentions followed an inductive approach [26] since literature is still lacking in this regard [7,15]. Figure 3 visualizes the activity and attribute codes generated inductively from the data. Descriptions of all activities and attributes can be found in the coding guideline contained in the replication package.

The artifact mention "if a specification rather proposes a solution" received the quality factor code *orientation* and the value *solution*, since it describes a *solution-oriented* requirements specification. The context factor mention "but the feature is completely new" was coded *novelty*. Finally, the activity mention "then the development requires less changes" received the following codes: the *stability* attribute of the *implementing* activity experiences a *positive* impact (+2). We coded the strength of the impact on a discrete seven-point scale. Positive (+2), negative (-2), and no impact (±0) were used as the default

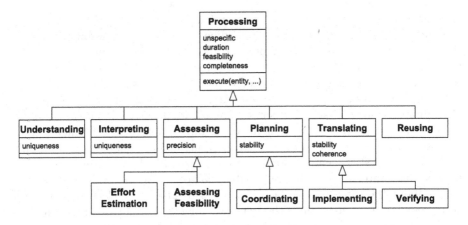

Fig. 3. Model of activities (as classes) and attributes (as their attributes)

codes depending on the direction of the impact. Particularly strong (+3 and
-3) or weak (+1 and -1) codes were only used when explicitly mentioned by
the interview participant, e.g., if an impact was "very critical" or just "slight."

Independent coders were involved to ensure the reliability of the subjective
coding task. For the coding of the interview transcripts, a senior researcher from
the same department as the author, who is also familiar with the case company, performed the coding task independently using the coding guideline on
a random sample of 10 statements ($10/84 \approx 12\%$). The inter-rater agreement
of codes achieved a percentage score of 83.8% and a Cohen's Kappa of 71.8%.
However, Cohen's Kappa is known to be unreliable for uneven marginal distri-
butions [10]. Still, the more robust S-Score [1] resulted in an agreement of 82.3%.
All scores represent a substantial agreement and support the reliability of the
coding process.

For the coding of the issue reports, a senior engineer from the case company
was involved. In a session between the author and the involved engineer, ques-
tionable codes were reviewed and adjusted. This process confirmed the applicability of the chosen codes as well as that 21 of the selected issues were not
caused by requirements quality defects but unrelated circumstances. These 21
statements were not considered in future phases of the data analysis.

Theme Identification. In this step, individual codes were aggregated into
"more meaningful and parsimonious units" [6]. The theme identification imposed
two further conditions on the coded statements: (C1) A statement must contain
at least one quality factor and at least one activity, and (C2) a statement must
not contain any unspecific activity (activity code: *processing*, root of Fig. 3)
or an unspecific attribute (attribute code: *unspecific*). This ruled out 44 vague
interview statements of the form "<quality factor> is *bad*" (C1, missing activity)
or "<quality factor> is *bad* for implementing" (C2, unspecific attribute).

The final analysis considered 40 interview and 80 issue statements. The amount of information per mention differed between the two data sources. While the interview encouraged participants to elaborate on all concepts of the RQT, the issue data lacked the same level of control [24]. Issue statements always contained the activity concepts since they explicitly report the effect of an issue, but the level of information on the root cause in requirements engineering was often limited, and the issues did not contain any information about the context.

The analysis of the more granular and complete interview statements split the data into three groups: statements containing a single quality factor and no context factors, statements containing multiple quality factors but no context factors, and statements containing context factors. Within each group, statements about the same quality factors were aggregated to collect all impacted activities and attributes. Since the issue statements contained codes on higher levels and no context factors, they were aggregated into one matrix showing the distribution of quality factors impacting activity attributes.

Model Creation. The final inferential step of the guideline by Cruzes et al. is the description of higher-order themes, a taxonomy, model, or a theory [6]. Because this study was already grounded in the theoretical foundation provided by the RQT [15] and none of the encountered data challenged this theory, developing a new model or theory was not deemed constructive. Instead, this study evolved the existing RQT by deriving propositions from the identified themes. These propositions enriched the theory with empirical insights and contributed falsifiable hypotheses for further research.

Trustworthiness Assessment. The final overall step of the guideline by Cruzes et al. is to assess the trustworthiness of the synthesis [6]. As these concerns align with threats to validity, they are addressed in Sect. 5.2.

4   Results

4.1   Interview Data

Impact of Single Quality Factors. The interview data contains information about 17 unique quality factors with an impact on at least one subsequent activity. For brevity, Table 1 lists only the four quality factors that were contained in at least two statements. Note that these constitute the most prevalent but not necessarily the most influential quality factors. Each cell of the *impact* column in Table 1 lists the perceived direction and strength of the impact that the quality factor has on the attribute of the activity in this cell. For example, two statements of the interview data described a negative impact of solution-oriented requirements on a unique understanding of that requirement. One statement stressed that this impact is major (-3), the other one did not (-2). The replication package contains the remaining quality factors and their impact.[4]

---

[4] Available at https://github.com/JulianFrattini/rqi-relf/blob/main/src/analytics/results.md.

Table 1. Perceived impact of single quality factors on subsequent activities

Quality Factor	Activity	Attribute	Impact
Solution-oriented	Understanding	Uniqueness	-3 -2
	Verifying	Completeness	-2 -2
	Effort Estimation	Traceability	+2
	Translating	Stability	+2
	Feasibility Assessment	Precision	+2
	Planning	Stability	+2
Non-atomic	Translating	Duration	-2 -2
	Planning	Stability	-2
Non-concise	Understanding	Uniqueness	-1
		Duration	-2
Too dense	Understanding	Duration	-2
	Interpreting	Uniqueness	-2
	Verifying	Duration	-2

The table shows that the interview participants perceived the four most often mentioned quality factors to impact a variety of activities and their attributes. The most frequently mentioned quality factor is *solution-orientation*, i.e., requirements that impose on the solution space rather than elaborating on the problem space [11]. This quality factor is perceived to cause misunderstandings (i.e., causing the *understanding* activity to be not *unique*) and lack of coverage when deriving test cases (i.e., causing the *verifying* activity to be not *complete*). The table also shows that some quality factors have a mixed impact on different activities. For example, a solution-oriented requirements specification is also perceived to aid effort estimation, translating, and planning.

Interaction of Multiple Quality Factors. The interview data contains information about four unique interactions between quality factors. Table 2 lists the four interaction effects. The interview participants reported that redundant requirements that were also not connected through horizontal trace links (i.e., links between requirements) caused incoherent implementations. Furthermore, non-functional requirements were reported to be susceptible to ambiguous understanding when providing to little details. Additionally, requirements specifications that were yet immature but also already committed to were quicker to implement due to the applied time pressure (+2), but implementation became much less feasible (-3). Finally, the precision of feasibility assessment suffered from jargonic and dense requirements.

Interaction with Context Factors. The interview data contains information about seven unique interactions between quality factors and context factors.

Table 2. Interaction between two quality factors

Quality Factor 1	Quality Factor 2	Activity	Attribute	Impact
Redundant	Missing horizontal trace links	Implementing	Coherence	-2
Too little details	Non-functional requirement	Understanding	Uniqueness	-2
Immature	Committed	Implementing	Duration	+2
			Feasibility	-3
Jargonic	Too dense	Assessing Feasibility	Precision	-2

The two most prominently perceived interaction effects involve the quality factor *solution-orientation* and *density*. Figure 4 visualizes one statement describing the interaction between the quality factor *solution-orientation* and context factor *involvement* on the *uniqueness* attribute of the *understanding* activity. If the stakeholder responsible for processing the requirement was also involved in writing it, the impact on the understandability is mitigated. The remaining interaction effects are detailed in our replication package but follow a similar pattern: Context factors like *involvement, experience*, and *supplementary communication* can mitigate the negative impact of quality factors. Additionally, solution-oriented requirements exhibit an even stronger positive impact on several activities like feasibility assessment and effort estimation when the context of the requirement is new.

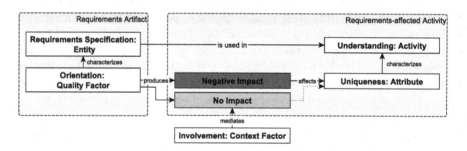

Fig. 4. Interaction effect between *solution-orientation* and *involvement* on the *understanding* activity

4.2   Issue Reports

The issue data only allows to infer general relationships between quality factors and activities. Table 3 lists the number of statements per constellation of quality factor, activity, and attribute. The most prominent impact of requirements quality that results in a reported issue is *completeness*, i.e., caused by a missing requirement. This results mostly in *incorrect* (i.e., bugs) or *incomplete* implementations (i.e., missing features). In rare cases, the *understanding, interpreting*, and *verifying* activity are reported to be impacted. Behind completeness, the most often reported impact is *consistency* and *ambiguity*.

Table 3. Requirements impact as recorded in the issue reports (Comp. = Completeness, Corr. = Correctness, Cons. = Consistency)

| Activity | Understanding | Interpreting | Implementing | | | Verifying | |
Attribute	Unique	Unique	Comp.	Corr.	Cons.	Coverage	Feasible
Quality Factor							
Completeness	1	0	13	34	3	1	1
Consistency	1	0	0	6	2	2	0
Ambiguity	1	2	2	1	2	0	0
Correctness	1	0	0	2	1	0	0
Feasibility	0	0	0	1	0	1	0
Relevance	0	0	1	0	0	0	0

4.3 Propositions

The triangulation of interview and issue analysis results allows to derive the following propositions enhancing the RQT.

Relevant Quality Factors. The set of quality factors perceived and reported to be relevant in the case company is limited. Among the perceived quality factors, only solution-orientation, lack of atomicity, lack of conciseness, and density received support in at least two statements. Among the reported quality factors, lack of completeness (i.e., missing requirements) stands out as the primary cause of issues in the down-stream development process. Note that the presented quality factors were only ordered by prevalence, not by importance. I.e., a rarely reported quality factor may have a higher impact than a commonly reported quality factor. Follow-up studies measuring the impact of the quality factors are necessary to determine their relative importance.

Mixed Impact. The analysis of the interview data shows that quality factors can have mixed impact on different activities. For example, while a solution-oriented requirement might negatively impact the activities of understanding and verifying it, planning and translating activities become more stable.

Interactions Matter. The analysis of the interview data shows that interactions between quality factors but also between quality and context factors have a significant effect. In particular, quality factors like the novelty of a feature, the experience of the involved engineers, and supplementary communication mediate the effect of quality factors as shown in Fig. 4.

5   Discussion

5.1   Implications

Contribution to Research. The empirical evidence both strengthens existing requirements engineering theories and guides further advances in requirements quality research. The results strengthen the RQT [9,15] in that artifact properties impacting activity properties constitutes requirements quality. Moreover, the results confirm that one quality factor may have different impacts on different activities [9]. Incautiously removing a quality factor from an entity due to the negative impact on one activity may, therefore, also mitigate its positive impact on other activities. The results further strengthen the *Naming the Pain in Requirements Engineering* (NaPiRE) initiative [18] by contributing more granular evidence to the problems of RE relevant to practice. The results of the issue data agree with the conclusion of NaPiRE that missing requirements are among the most impactful quality defects. Similarly, the results of both data sources of the study agree with previous studies that the effect of ambiguity is less relevant in practice than often assumed [5,22]. Also, the results highlight potential gaps in literature. Commonly researched quality factors like passive voice [14,23] did not show up in our data, while often discussed yet seldomly empirically investigated factors like solution-orientation [2] show high prevalence in practice. Finally, the results support the advocacy for context-sensitive research in empirical software engineering [4] by emphasizing context factors as mediators of the impact of artifacts on activities. The results of this study guide further research advances through the approach of identifying relevant factors of requirements quality. The study reduces the vast space of several hundred potential quality factors [16] down to about 30 that are relevant to the specific context of an organization.

Impact on Practice. The operationalization of the RQT in practice becomes feasible due to the reduction of variables to measure. The results steer the next step of research with the case company toward detecting solution-oriented, non-atomic, non-concise, dense, and incomplete requirements. Additionally, effort will be focused on measuring the relevant context factors of involvement, experience, novelty, and supplementary communication. These measurements enable the impact estimation of requirements quality on dependent activities and advance requirements quality research roadmaps [7,15].

Limitations. One remaining gap to overcome is the lack of an overview of requirements-dependent activities and their measurable attributes, as also outlined in previous research roadmaps [7,15]. These attributes constitute the dependent variable in the impact estimation implied by the RQT [15]. The interview data contains 44 statements with either an unclear activity or attribute impacted by a quality factor. Developing a model of requirements-dependent activities and their attributes is a necessary next step to achieve operationalization of the RQT [15].

Another gap is highlighted by the discrepancy between the quality factors identified through the two different data sources. One would expect convergence of the quality factors that were perceived as relevant by involved practitioners and quality factors that were reported to have an impact. This discrepancy might be caused by a lack of saturation in the study or by the lack of detail with which responsible quality factors were reported in the issues. An alternative hypothesis is that the quality factors exist on different scopes: The quality factors that were perceived by software engineers might also be addressed by them, which leaves those that were not perceived to produce an impact on later SE activities and result in an issue. This study is unable to clearly attribute the discrepancy of relevant quality factors to any root cause and requires follow-up studies.

5.2   Threats to Validity

This section discusses threats to validity according to Runesson et al. [24] and Wohlin et al. [29] and, additionally, addresses concerns of trustworthiness of the thematic synthesis [6].

Regarding conclusion validity, the *reliability of measures* is a prevalent threat given the subjective coding process. It has been minimized by involving independent raters and calculating inter-rater agreement where applicable. Still, the complexity of requirements quality may render one hour interviews incapable of properly eliciting all relevant quality factors and their impact, as indicated by the lack of saturation in the data. Follow-up studies focusing on specific aspects of requirements quality may be necessary to complement this general case study. Similarly, the lack of control over the issue data questions their reliability. The discrepancy of the quality factors and their granularity reported in issues when compared to those elicited by interviewees hints at the fact that the root cause of issues were not documented in sufficient detail. Involving a senior engineer from the case company addressed this issue and provided an adequate *confirmability* of the data, but other sources of software telemetry would be necessary to answer RQ2 on a more granular level. Finally, the independent coding of the interview data by the senior researcher posed one more threat to validity. For coding the activities and attributes, they were instructed to use the codes that emerged from the first round of coding performed by the author. Consequently, the second round of coding was deductive, while the first one was inductive. This, potentially, introduced bias, as the second coding phase was constrained by the codes from the first phase. The two researchers discussed this circumstance and the second round of coding did not raise any issues regarding missing codes. Hence, this threat is assumed to be minimal.

Regarding internal validity, the interview data suffers from *selection* bias. The interview participants were sampled by the industry contact and were all members of the same development team in the organization. However, the participants show a wide variation of background and experience, which leads to assume that this threat is minimal.

Regarding construct validity, the study suffers from *inadequate preoperational explication of constructs*, i.e., immaturity of some of the concepts of the theoretical framework (the RQT). In particular, the activity group within the RQT (activities and attributes) is insufficiently explored in requirements quality research [7,15]. As a consequence, several interview statements failed to specify the activity and attribute impacted by requirements quality. The threat was minimized by excluding this data from the analysis.

Regarding external validity, the inference of this study is not generalizable or transferable [6] by design of the case study method. Additional research replicating this study in other companies are necessary to generalize the results.

5.3 Future Work

While the results of a case study are not generalizable per se [6], they can contribute to more general knowledge by being aligned with theory [25] while remaining explicit about the study context [4]. Results of future studies replicating this case study in different contexts can be assembled in the common frame of the RQT [15] to compose a universal, context-sensitive understanding of requirements quality [7]. Follow-up studies quantifying the impact of identified relevant quality factors [14] will enable the theory to additionally compare quality factors in terms of impact, not only prevalence. Allowing this assembly requires terminological consistency that has not quite been achieved yet. The different granularity and texture of the quality factors reported in this study testify that properties of requirements artifacts are yet underresearched. Roadmaps of requirements quality research call for community-driven initiatives at harmonizing the concept of requirements quality factors [7,15,16].

6   Conclusion

This case study demonstrates the application of the requirements quality theory (RQT) to identify relevant factors of requirements quality. By analyzing both interview and issue report data, we identified 17 relevant quality as well as 11 interaction effects among them and with context factors. The study contributes empirical evidence to the relevance of these factors and their effects in the case company. The study emphasizes that (1) some requirements quality factors are more prevalent in practice than others, (2) they may have a simultaneous negative and positive impact on different activities, and (3) context factors mediate their impact. This research promotes requirements quality research by advancing existing research roadmaps [7] toward a quantified impact estimation of requirements quality in practice [15].

Acknowledgements. This work was supported by the KKS foundation through the S.E.R.T. Research Profile project at Blekinge Institute of Technology. I thank Michael Unterkalmsteiner for independently performing the coding task. Furthermore, I owe great thanks to Parisa Yousefi, Charlotte Ljungman, and Fabiano Sato from Ericsson for their continuous support that made this work possible in the first place.

References

1. Bennett, E.M., Alpert, R., Goldstein, A.: Communications through limited-response questioning. Public Opin. Q. **18**(3), 303–308 (1954)
2. Board, I., Committee, I., et al.: IEEE recommended practice for software requirements specifications. In: Institute of Electrical and Electronics Engineers (1998)
3. Boehm, B.W., Papaccio, P.N.: Understanding and controlling software costs. IEEE Trans. Software Eng. **14**(10), 1462–1477 (1988)
4. Briand, L., Bianculli, D., Nejati, S., Pastore, F., Sabetzadeh, M.: The case for context-driven software engineering research: generalizability is overrated. IEEE Softw. **34**(5), 72–75 (2017)
5. de Bruijn, F., Dekkers, H.L.: Ambiguity in natural language software requirements: a case study. In: Wieringa, R., Persson, A. (eds.) REFSQ 2010. LNCS, vol. 6182, pp. 233–247. Springer, Heidelberg (2010). https://doi.org/10.1007/978-3-642-14192-8_21
6. Cruzes, D.S., Dyba, T.: Recommended steps for thematic synthesis in software engineering. In: 2011 International Symposium on Empirical Software Engineering and Measurement. pp. 275–284. IEEE (2011)
7. Femmer, H.: Requirements quality defect detection with the Qualicen requirements scout. In: REFSQ Workshops (2018)
8. Femmer, H., Fernández, D.M., Wagner, S., Eder, S.: Rapid quality assurance with requirements smells. J. Syst. Softw. **123**, 190–213 (2017)
9. Femmer, H., Mund, J., Fernández, D.M.: It's the activities, stupid! A new perspective on re quality. In: 2015 IEEE/ACM 2nd International Workshop on Requirements Engineering and Testing, pp. 13–19. IEEE (2015)
10. Feng, G.C.: Mistakes and how to avoid mistakes in using intercoder reliability indices. Methodol. Eur. J. Res. Methods Behav. Soc. Sci. **11**(1), 13 (2015)
11. Fernandez, D.M., Wagner, S., Lochmann, K., Baumann, A., de Carne, H.: Field study on requirements engineering: investigation of artefacts, project parameters, and execution strategies. Inf. Softw. Technol. **54**(2), 162–178 (2012)
12. Franch, X., et al.: How do practitioners perceive the relevance of requirements engineering research? IEEE Trans. Softw. Eng. (2020)
13. Franch, X., Palomares, C., Quer, C., Chatzipetrou, P., Gorschek, T.: The state-of-practice in requirements specification: an extended interview study at 12 companies. Requirements Eng., 1–33 (2023)
14. Frattini, J., Fucci, D., Torkar, R., Mendez, D.: A second look at the impact of passive voice requirements on domain modeling: Bayesian reanalysis of an experiment. In: 1st International Workshop on Methodological Issues with Empirical Studies in Software Engineering (WSESE2024). ACM (2024)
15. Frattini, J., Montgomery, L., Fischbach, J., Mendez, D., Fucci, D., Unterkalmsteiner, M.: Requirements quality research: a harmonized theory, evaluation, and roadmap. Requirements Eng. (2023)
16. Frattini, J., Montgomery, L., Fischbach, J., Unterkalmsteiner, M., Mendez, D., Fucci, D.: A live extensible ontology of quality factors for textual requirements. In: 2022 IEEE 30th International Requirements Engineering Conference (RE), pp. 274–280. IEEE (2022)
17. Host, M., Runeson, P.: Checklists for software engineering case study research. In: First International Symposium on Empirical Software Engineering and Measurement (ESEM 2007), pp. 479–481. IEEE (2007)

18. Méndez, D., et al.: Naming the pain in requirements engineering: contemporary problems, causes, and effects in practice. Empir. Softw. Eng. **22**, 2298–2338 (2017)
19. Miles, M.B., Huberman, A.M.: Qualitative Data Analysis: An Expanded Sourcebook. Sage (1994)
20. Montgomery, L., Fucci, D., Bouraffa, A., Scholz, L., Maalej, W.: Empirical research on requirements quality: a systematic mapping study. Requirements Eng. **27**(2), 183–209 (2022)
21. Petersen, K., Wohlin, C.: Context in industrial software engineering research. In: 2009 3rd International Symposium on Empirical Software Engineering and Measurement, pp. 401–404. IEEE (2009)
22. Philippo, E.J., Heijstek, W., Kruiswijk, B., Chaudron, M.R.V., Berry, D.M.: Requirement ambiguity not as important as expected—results of an empirical evaluation. In: Doerr, J., Opdahl, A.L. (eds.) REFSQ 2013. LNCS, vol. 7830, pp. 65–79. Springer, Heidelberg (2013). https://doi.org/10.1007/978-3-642-37422-7_5
23. Rosadini, B., et al.: Using NLP to detect requirements defects: an industrial experience in the railway domain. In: Grünbacher, P., Perini, A. (eds.) REFSQ 2017. LNCS, vol. 10153, pp. 344–360. Springer, Cham (2017). https://doi.org/10.1007/978-3-319-54045-0_24
24. Runeson, P., Host, M., Rainer, A., Regnell, B.: Case Study Research in Software Engineering: Guidelines and Examples. Wiley (2012)
25. Sjøberg, D.I., Dybå, T., Anda, B.C., Hannay, J.E.: Building theories in software engineering. In: Shull, F., Singer, J., Sjøberg, D.I.K. (eds.) Guide to Advanced Empirical Software Engineering, pp. 312–336. Springer, London (2008). https://doi.org/10.1007/978-1-84800-044-5_12
26. Strauss, A., Corbin, J.: Basics of Qualitative Research. Sage Publications (1990)
27. Wagner, S., et al.: Status quo in requirements engineering: a theory and a global family of surveys. ACM Trans. Softw. Eng. Methodol. (TOSEM) **28**(2), 1–48 (2019)
28. Wohlin, C.: Case study research in software engineering-it is a case, and it is a study, but is it a case study? Inf. Softw. Technol. **133**, 106514 (2021)
29. Wohlin, C., Runeson, P., Höst, M., Ohlsson, M.C., Regnell, B., Wesslén, A.: Experimentation in Software Engineering. Springer, Heidelberg (2012). https://doi.org/10.1007/978-3-642-29044-2

Quality Requirements

Assessing the Understandability and Acceptance of Attack-Defense Trees for Modelling Security Requirements

Giovanna Broccia[1]([✉]) [ID], Maurice H. ter Beek[1] [ID], Alberto Lluch Lafuente[2] [ID], Paola Spoletini[3] [ID], and Alessio Ferrari[1] [ID]

[1] ISTI-CNR, Pisa, Italy
{giovanna.broccia,maurice.terbeek,alessio.ferrari}@isti.cnr.it
[2] DTU, Lyngby, Denmark
albl@dtu.dk
[3] Kennesaw State University, GA, USA
pspoleti@kennesaw.edu

Abstract. *Context and Motivation* Attack-Defense Trees (ADTs) are a graphical notation used to model and assess security requirements. ADTs are widely popular, as they can facilitate communication between different stakeholders involved in system security evaluation, and they are formal enough to be verified, e.g., with model checkers. *Question/Problem* While the quality of this notation has been primarily assessed quantitatively, its understandability has never been evaluated despite being mentioned as a key factor for its success. *Principal idea/Results* In this paper, we conduct an experiment with 25 human subjects to assess the understandability and user acceptance of the ADT notation. The study focuses on performance-based variables and perception-based variables, with the aim of evaluating the relationship between these measures and how they might impact the practical use of the notation. The results confirm a good level of understandability of ADTs. Participants consider them useful, and they show intention to use them. *Contribution* This is the first study empirically supporting the understandability of ADTs, thereby contributing to the theory of security requirements engineering.

Keywords: security requirements · Attack-Defense Trees · understandability evaluation · empirical user study · Method Evaluation Model

1 Introduction

The definition of security requirements entails the representation and analysis of envisioned threats and mitigation solutions, oriented to eventually define a security policy [10]. Several notations have been proposed in requirements engineering (RE) to model and analyse security requirements, such as extensions of well-known notations (e.g., Secure I* [21] and Secure UML [22]) and other comprehensive notations with analysis capabilities (e.g., the Socio-Technical Security

D. Mendez and A. Moreira (Eds.): REFSQ 2024, LNCS 14588, pp. 39–56, 2024.
https://doi.org/10.1007/978-3-031-57327-9_3

Modelling Language (STS-ML) [29] and the Restricted Misuse Case Modeling (RMCM) approach [23]).

Among this variety of proposals, Attack-Defense Trees (ADTs) offer a graphical notation used to model and assess the security requirements of systems or assets. They provide a representation of possible actions an attacker might take to attack a system and the measures that a defender can employ to protect the system [15]. The purposes of ADTs are multiple. In addition to providing a threat modelling methodology, they can be used for quantitatively assessing the security of a system (e.g., with model checking). Moreover, ADTs are useful for facilitating communication between stakeholders from different fields and with different backgrounds (e.g., domain experts, security experts).

Several studies have shown how graphical notations are more comprehensible by humans than textual notations [32,35]. However, although ADTs have been claimed as one of the most popular graphical models for system security analysis [11], extremely easy to use also for novice users [38], and as an easily understandable human-readable notation [8], no user study has been proposed to verify these hypotheses. Albeit this research direction holds promise and would be helpful in evaluating their effectiveness [11,19,20]. Indeed, beyond the realm of attack trees, there exists a substantial body of empirical research literature focused on security modelling and assessment [6,17,18]. These kinds of studies are particularly beneficial given the centrality of humans in system security—both for possible insider attacks and for human errors that make the system vulnerable [8].

In this paper, we present the first experiment that aims at investigating the quality of the ADT notation, both in terms of understandability and in terms of user acceptance. We designed the study based on the Method Evaluation Model (MEM) [27], a model used to evaluate information technologies, which extends the Technology Acceptance Model (TAM) [7]. We adapt MEM following the approach by Abrahão [1] and identify two classes of variables: performance-based and perception-based. The performance-based variables aim at assessing the understandability of ADTs, while perception-based variables seek to evaluate the users' acceptance of ADTs.

Our results show that: (1) ADTs are sufficiently understandable; (2) ADTs are perceived as easy to use and useful, and participants express the intention to use them; (3) there is a relationship between perceived usefulness and intention to use; (4) there are no significant relationships between various performance-based measures of understandability (effectiveness and efficiency) and perception-based variables (ease of use, usefulness, intention to use), except in the following cases: (a) perceived ease of use has a positive relationship with effectiveness, i.e., those who make fewer mistakes in different ADT understandability tasks generally consider the notation easier; (b) those who *apply* the method better in practice also consider it more useful; (c) those who make fewer mistakes when *observing* the notation used in realistic contexts, consider the method easier. Our replication package is publicly available [4].

Related Work. Several notations have been proposed in RE to model and analyse security requirements [13,26,34,37]. Some of these notations are extensions of existing notations, like Secure I* [21], KAOS [30]), Secure UML [22], Misuse cases [33], and Secure Tropos [12].

Other attempts, some based on the languages above, also offer analysis capability. In particular, the ones mentioned in the Introduction. STS-ML [29] is an actor- and goal-oriented security requirements modelling language based on Tropos, able to capture system security needs and requirements at the organisational level and reason about corporate assets, social dependencies, and trust properties. RMCM [23] is a use case-driven modelling method that uses misuse case diagrams [33] to support the specification of security and privacy requirements of multi-device software ecosystems in a structured and analysable form. The Risk-based Security Requirements (RBSR) model [9] associates security requirements with specific weaknesses and risk profiles that can vary over time and provides mitigation accordingly to these variations. Finally, [39] introduces a threat-based security framework and its Business Process Model and Notation (BPMN) extension to model the security threat and support risk analysis.

Labunets et al. observed a difference in the representation of security risk assessment between academic proposals and industry standards. Academic approaches favour graphical notation, while the industry leans towards tabular models. Several studies were conducted to compare the effectiveness of graphical and tabular models. [17] proved that both methods are equally effective. In [18], a comparative analysis of visual and textual risk-based approaches revealed that the visual method is more effective for identifying threats, while the textual method is slightly better for eliciting security requirements.

In [19], the results of an empirical evaluation conducted to determine the effectiveness of two attack modelling techniques, an adapted attack graph method and the fault tree standard, are reported. The results indicate that the attack graph method is more effective than the fault tree method.

2   Attack-Defense Trees

The assessment of system security through graphical tree structures originated in 1960 with fault tree analysis [36], and gradually spread with the usage of similar structures such as attack trees [24,31]. To manage the dynamic nature of system security, Attack-Defense Trees (ADTs) [15] were introduced, extending attack trees with defense strategies and quantitative risk assessment [3,14]. ADTs model attack-defense scenarios, namely 2-player games between a proponent and an opponent.

Formally, ADTs are rooted trees with labelled nodes of two opposite types: attack nodes and defense nodes, representing the goals of the attacker and the defender, respectively. The root can be either type: if the root is an attack node, the proponent is an attacker; conversely, if the root is a defense node, the proponent is a defender. The main goal can be refined into sub-goals, described by its child nodes of the same type. The refinement can be either conjunctive (i.e., all sub-goals must be achieved to achieve the parent goal) or disjunctive (i.e., at least one of the sub-goals must be achieved to reach the parent goal). A node with no children of the same type is called a non-refined node, and it represents a basic/atomic action. Each node may have one child of the opposite type, representing a countermeasure to its (sub-)goal. Essentially, an attack node may have a number of children that refines the attack and a single defense node that fends it off. Conversely, a defense node may have a number of children which refines the defense, and a single attack node that counterattacks it.

To demonstrate the features of ADTs, we present a simple fictitious scenario describing the theft of the Mona Lisa painting (cf. Fig. 1). To steal the painting, two kinds of attacks can be carried out: enter the Louvre museum by the door or by the window. Figure 1 shows in detail only the door branch (further attacks and defenses could easily be added). To secure the door, the museum can use an alarm; however, the attacker can perform a counterattack by forcing the alarm system. To do so, the attacker needs to get both the username and the password.

Evaluation of ADTs has so far considered issues like the consistency

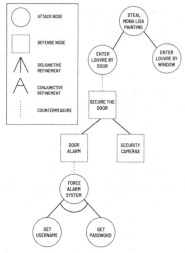

Fig. 1. ADT for theft of Mona Lisa.

between an ADT and the system and the impact of repeated labels on results [2,16]. As far as we know, there is no work in the literature that has focused on the assessment of the comprehensibility of ADTs (neither of attack trees). Albeit their comprehensibility is usually assessed as a factor of success [8,11,38].

3   Method Evaluation and Technology Acceptance Model

The Method Evaluation Model (MEM) [27] is a model used to evaluate new information technologies. According to MEM, the usage of new technologies is influenced by a set of *perception-based* variables and *performance-based* variables.

The perception-based variables are used to gauge the level of *acceptance* of the technology and include the perceived ease of use (PEOU), which measures how easy the technology is perceived to be, the perceived usefulness (PU), which measures how useful the technology is perceived to be, and the intention to use (ITU), which measures the extent to which users intend to use the technology in the future. The performance-based variables consist of efficiency and effectiveness, which measure the effort required to use the technology and how well the technology has been used to reach the goals, respectively. Essentially, the adoption of a new technology depends not only on whether it is actually effective but also on whether the users perceive it to be effective.

MEM has been applied in the fields of RE [1] and language comprehension [5]. In both studies, the performance-based variables (efficiency and effectiveness) have been adapted to measure the *understandability* of requirement models and language constructs, respectively. In practice, the performance-based variables are understandability effectiveness and understandability efficiency, computed based on the results obtained by sample subjects in problem-solving tasks. This paper adopts this approach and further decomposes the variables into fine-grained dimensions (cf. Sect. 4). In line with MEM, we evaluate if these variables are related to perception-based variables.

4 Study Design

Our experiment aims to study the degree of ADTs understandability and users' acceptance. We also study if there is a relationship between the degree of acceptance of the notation and its understandability.

4.1 Variables, Research Questions, and Tests

Acceptance and Understability Dimensions. Users' acceptance is based on the MEM model presented in Sect. 3. In particular, we evaluate *acceptance* using the three perception-based variables from the MEM (PEOU, PU, and ITU).

Understandability is evaluated in terms of effectiveness and efficiency based on the results of sample subjects in some problem-solving tasks (as suggested by the literature, e.g., [28]). For both effectiveness and efficiency, we further distinguish between fine-grained understandability, which considers three different dimensions of understandability separately, and coarse-grained understandability, which measures the average across the dimensions. The dimensions are:

UNC Understandability not in context measures the comprehensibility of ADTs *syntax*. It assesses users' ability, after ADT training, to identify correct ADT construction, recognise nodes (for attack and defense), refinements (conjunctive and disjunctive), countermeasures, and understand sequential actions and their temporal order in ADTs.

UIC Understandability in context measures the comprehensibility of ADTs *semantics*. It assesses users' ability, after training, to answer questions about both existing and instantiated ADTs and to recognise if an ADT accurately models a specific behaviour in a given scenario.

TRF Transferability measures the practical use of the notation, evaluating users' ability, after training, to create or modify ADTs. This includes recognising the appropriate elements to add to the tree for modelling specific behaviour and knowing where to place these elements.

Research Questions. We aim to answer the following research questions:

RQ1. *How well users understand ADTs?* This RQ aims to understand the level of effectiveness and efficiency with which users comprehend ADTs.

RQ2. *What is the degree of acceptance of ADTs by users?* This RQ aims to understand how much users perceive the notation as easy to use and useful and to what extent they intend to use ADTs in the future.

RQ3. *What is the relationship between ease of use/usefulness of the notation and intention to use it in the future?* Differently from RQ1, which focuses on each perception-based variable independently, this RQ aims at checking whether there is a relationhip among the variables, and in particular, if ease of use and usefulness are related to intention to use.

RQ4. *What is the relationship between the overall ADT understandability and the users' perception of ADTs' ease of use and usefulness?* With this RQ, we check whether users who perform best in understanding the notation also tend to evaluate the ADTs as easier and more useful.

RQ5. *What is the relationship between the different dimensions of understandability and the users' perception of ADTs' ease of use and usefulness?* Here, we want to check if there is an understandability dimension that is related to the perception of users in terms of ease of use and usefulness.

Variables for Acceptance and Understandability. We measure the three perception-based variables (PEOU, PU, and ITU) through an instrument adapted from MEM [27], namely a questionnaire composed of a set of statements for each variable. We shuffle the statements and add their negated version to avoid systematic response bias (i.e., both the statements "ADTs are easy to learn" and "ADTs are not easy to learn" are present) [1]. Users need to evaluate each statement on a Likert scale from 1 (*strongly agree*) to 5 (*strongly disagree*). Table 1 shows the list of positive statements for PEOU, PU, and ITU. Each variable is computed as the mean of its statements points (the points for negative statements are counted as 6 minus the points given as the answer).

Understandability dimensions are measured through specific tasks:

1. UNC is measured through a set of true/false questions on domain-agnostic ADT fragments (A, B, C instead of names), to ensure that users' responses are not influenced by knowledge of the domain.
2. UIC is evaluated through a set of yes/no questions on instantiated ADTs fragments.
3. TRF is measured through a number of instantiated ADTs fragments to extend with a set of requests.

For each of these dimensions, we compute effectiveness as the number of correct answers over the number of questions and efficiency as effectiveness over time [1]. Therefore, we have six different variables: *UNC effectiveness, UNC efficiency, UIC effectiveness, UIC efficiency, TRF effectiveness,* and *TRF efficiency.* For what concerns total understandability, we compute *understandability effectiveness* as the mean of the effectiveness of the three dimensions and *understandability efficiency* as the mean of the efficiency of the three dimensions.

Table 1. Perception-based statements (positive statements).

		Statements
PEOU	1	It was easy for me to understand what the ADTs represented
	2	ADTs are simple and easy to understand
	3	ADTs are easy to learn
	4	Overall, the ADTs were easy to use
PU	1	Overall, I think that ADTs provide an effective means for describing security threats and countermeasures
	2	I believe that ADTs have enough expressiveness to represent security threats and countermeasures
	3	Overall, I find ADTs to be useful
	4	I believe that ADTs are useful for representing security threats and countermeasures
	5	Using ADTs would improve my performance in describing security threats and countermeasures
	6	I believe that ADTs are organised, clear, concise, and unambiguous
	7	I believe the use of ADTs would reduce the time required to represent security threats and countermeasures.
ITU	1	If I were to work for a company in the future, I would use ADTs to specify security threats and countermeasures
	2	I intend to use ADTs in the future if given the opportunity
	3	I would recommend the use of ADTs to security practitioners
	4	It would be easy for me to become skilled in using ADTs

Hypothesis Testing. To answer the research questions, we test a number of NULL hypotheses (cf. Table 2). Not all the combinations of variables are considered, following the approach by Abraõ [1], who relates ITU to PU and PEOU only, and not to performance-based variables.

4.2 Study Phases

The study is conducted online (material in [4]) and structured in 6 phases.

Phase 1 – Recruitment. Participants are contacted through a recruitment e-mail with all the information needed to perform the study. Specifically, links to a video training, a spreadsheet file where to get their identifier and the link to the test, the pre- and post-test questionnaires, the consent form, and study instructions.

Phase 2 – Binding. To ensure anonymity, participants are provided with a unique alphanumeric identifier via a spreadsheet file with a link to their test document (there is a different document for each participant). They are instructed to keep the identifier for the entire test, preserve the link to the test document to be used in a subsequent phase, and use incognito mode to protect their identity.

Phase 3 – Training. Before starting the test, we ask participants to watch a video that presents the ADT notation. The video is available online (https://youtu.be/KLIH-yultgI) and it contains all the information needed to complete the test. Participants are asked to use this support only once before they begin the test.

Phase 4 – Pre-test Questionnaire. We ask participants to fill out an online questionnaire whose link has been sent by e-mail during the recruiting phase. The questionnaire collects information about gender, age, education, employment, work area, level of knowledge of ADTs, and education on ADTs. Participants have to mark the questionnaire with the identifier received during the binding phase (Phase 2).

Table 2. Hypotheses for each research question.

RQ1	$H1_0$	Users are not effective in understanding ADTs
	$H2_0$	Users are not efficient in understanding ADTs
RQ2	$H3_0$	ADTs are perceived as difficult to use
	$H4_0$	ADTs are perceived as not useful
	$H5_0$	There is no intention to use the ADT in the future
RQ3	$H6_0$	There is no relationship between perceived ease of use and perceived usefulness
	$H7_0$	There is no relationship between perceived usefulness and intention to use
	$H8_0$	There is no relationship between perceived ease of use and intention to use
RQ4	$H9_0$	There is no relationship between understandability effectiveness and perceived ease of use
	$H10_0$	There is no relationship between understandability effectiveness and perceived usefulness
	$H11_0$	There is no relationship between understandability efficiency and perceived ease of use
	$H12_0$	There is no relationship between understandability efficiency and perceived usefulness
RQ5	$H13_0$	There is no relationship between understandability not in context effectiveness and perceived ease of use
	$H14_0$	There is no relationship between understandability not in context effectiveness and perceived usefulness
	$H15_0$	There is no relationship between understandability not in context efficiency and perceived ease of use
	$H16_0$	There is no relationship between understandability not in context efficiency and perceived usefulness
	$H17_0$	There is no relationship between understandability in context effectiveness and perceived ease of use
	$H18_0$	There is no relationship between understandability in context effectiveness and perceived usefulness
	$H19_0$	There is no relationship between understandability in context efficiency and perceived ease of use
	$H20_0$	There is no relationship between understandability in context efficiency and perceived usefulness
	$H21_0$	There is no relationship between transferability effectiveness and perceived ease of use
	$H22_0$	There is no relationship between transferability effectiveness and perceived usefulness
	$H23_0$	There is no relationship between transferability efficiency and perceived ease of use
	$H24_0$	There is no relationship between transferability efficiency and perceived usefulness

Phase 5 – Test. We ask participants to fill out the test in all its phases. The test is accessible through the link received during the binding phase (Phase 2); such a link leads to an editable online document (a different document for each participant). The spreadsheet accessed in Phase 2 enables us to bind each document to the ID of the corresponding user. The test is composed of 4 steps:

i **Retention.** Retention measures the comprehension of the training material and the ability to retain knowledge from it. We use this step to keep in the participants' memory the concepts presented in the training video that they will need during the test. The outcome of this step is not utilised in the calculation of understandability. In this step, a list of figures (i.e., all figures in the legend of Fig. 1) is presented and, for each figure, a table with two definition options. Participants are asked to mark the right definition for each figure.

ii **Understandability not in context.** With this step, we want to get how understandable is the syntax of the notation for the participants. In this step, 6 items are presented, and for each of them, we show one or more attack-defense tree fragments and 4 statements. Participants have to check for each of the statements whether it is true or false. Participants are asked to write down the starting (when starting step ii) and finishing time (when completing all the steps).

iii **Transfer.** Transfer measures how much is transferable the knowledge acquired through the training material. In this step, three attack-defense tree fragments are presented and, for each of them, a list of three requests. Participants are asked to modify the tree fragments according to the requests using an editable diagram embedded in the document (the instructions to modify the diagram are written inside the diagram itself). The three ADT fragments used represent common and familiar types of attacks, namely an attack on a bank account, an attack to open a safe lock, and an attack to burgle a house. For each fragment, three requests were made, each with increasing levels of difficulty: (i) participants are asked to add a node to the tree and specify the type of node and its position; (ii) participants are asked to add all the nodes necessary to model a given situation; (iii) participants are asked to modify the tree according to given syntactic and/or semantic constraints. For each of the three items, participants are asked to write down starting and finishing times in the appropriate lines.

iv **Understandability in context.** With this step, we want to perceive to what extent users, after a training phase on ADTs, are able to answer questions about given ADTs. In this step, three attack-defense tree fragments are presented, and, for each of them, a list of three yes/no questions. Participants are asked to answer the questions by typing in the document "yes" or "no". The three ADT fragments used are extended versions of the fragments used in the Transfer step (cf. step iii). For each of the three items, participants are asked to write down starting and finishing times in the appropriate lines.

Users are not bound by a specific time frame for the test phase, but allocating 40 min is deemed sufficient for completing phases ii, iii, and iv (according to the authors ter Beek and Lluch Lafuente, who are ADT experts [3]). This duration considers the time required for reading and analysing questions, processing ADT fragments, providing accurate answers, and adapting to the platform used.

Phase 6 – Post-test Questionnaire. We ask participants to fill out an online questionnaire whose link has been sent by e-mail during the recruiting phase. We use this phase to measure the perception-based variables (namely, PEOU, PU, and ITU) through a set of statements users need to rate from 1 to 5. The questionnaire contains 8 statements concerning PEOU, 14 statements on PU, and 8 statements concerning ITU (see Table 1). Participants have to mark the questionnaire with the identifier received during the binding phase (Phase 2).

5   Study Execution

The experimental study protocol containing the definition of the study phases, its rationale, as well as the data analysis process has been submitted to the ethical committee of the Italian National Research Council (CNR), which authorised the administration of the test. To take part in the study, participants are asked to sign an informed consent for the processing of personal data.

Participants. In total, 25 participants took part in the study: computer science students, Ph.D. students, and professors; researchers in the field of software engineering, formal methods, and security; participants belong to Kennesaw State

Table 3. Descriptive statistics.

Variables	Median	Mean	Std. dev.	Min.	Max.
PEOU	4.25	4.18	0.563	2.875	5
PU	4	3.92	0.37	2.929	4.571
ITU	3.875	3.88	0.403	3	5
UNC effectiveness	0.750	0.783	0.083	0.625	0.958
UNC efficiency	0.094	0.103	0.046	0.024	0.188
UIC effectiveness	0.889	0.907	0.175	0.111	1
UIC efficiency	0.250	0.264	0.135	0.009	0.500
TRF effectiveness	0.667	0.613	0.267	0	1
TRF efficiency	0.023	0.026	0.015	0	0.049
understandability effectiveness	0.792	0.768	0.134	0.287	0.986
understandability efficiency	0.118	0.131	0.059	0.011	0.241

University, CNR, University of Pisa, and the Technical University of Denmark. Participants in the study were selected opportunistically based on their availability. They were of both genders (56% men, 40% women, 4% prefers not to answer), aged between 21 and 56 years old. We asked them to self-evaluate their knowledge of ADTs before the test on a 5-point scale from 1 (*no knowledge*) to 5 (*advanced*) and whether they knew similar notations. The results are reported in Figs. 2 and 3, respectively. A total of 80% of the participants did not receive any education on ADTs before the test; the remaining participants attended a university course, a seminar, or self-educated.

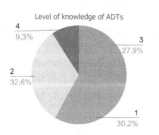

Fig. 2. Level of knowledge of ADTs.   **Fig. 3.** Similar notations known.

Results. Table 3 shows descriptive statistics for all the variables gathered with the test, i.e., the perception-based variables (PEOU, PU, and ITU) and the performance-based variables: (1) understandability not in context effectiveness and (2) efficiency; (3) understandability in context effectiveness and (4) efficiency; (5) transferability effectiveness and (6) efficiency; and (7) understandability effectiveness and (8) understandability efficiency.

The perception-based variables are all above the average value of the Likert scale (i.e., 3), which thus suggests a general degree of acceptance of the notation. The results indicate that users perceive ADTs as easy to use (mean score of 4.18) and as useful (mean value 3.92). Results also suggest that users intend to use the notation in the future (ITU has an average score of 3.88).

The results indicate a generally good level of understandability of ADT notation with an average total understandability effectiveness above 0.76, meaning that $\sim 77\%$ of the questions of the test are correctly answered.

Regarding the different dimensions composing understandability, the results show that understandability in context is the measure that provides the highest contribution (average effectiveness of 0.907), followed by understandability not in context (effectiveness = 0.783) and transferability (effectiveness = 0.613). This suggests that while participants understand the syntax and semantics of ADT fragments, they have more difficulty applying them in practice. For what concerns efficiency, we observe a similar trend, thereby confirming that ADTs "in action" are perceived as more difficult.

Table 4 summarises the relation between variables expressed in the hypotheses addressing each RQs presented in Sect. 4. For each hypothesis, the column "Reject" reports a "T" if the hypothesis has been rejected and an "F" otherwise. Below we discuss in detail only the rejected NULL hypotheses because no conclusions can be made for the others.

RQ1. To answer RQ1, we applied a Wilcoxon signed rank test to check whether effectiveness and efficiency are significantly above the target values of 0.6 (indicating a sufficient performance according to ter Beek and Lluch Lafuente, two ADT experts [3]) and of 0.015 (i.e., 60% of 1 (maximum effectiveness)/40 min (expected completion time)), respectively. We apply a non-parametric test (i.e., Wilcoxon signed rank) because the normality check, performed with the Kolmogorov-Smirnov test, fails for all the variables with p-value well below the 0.05 significance level. The test results show that both variables are significantly higher than the target values for $\alpha = 0.05$, with p-values of 0.000139 and 7.381e-06, respectively, with large effect-size (cf. Table 4). Therefore rejecting H1$_0$ and H2$_0$, and attesting a **sufficient overall understandability of the ADT notation**. Fine-grained effectiveness and efficiency measures are all significantly greater than the respective reference values for both effectiveness and efficiency, with the exception of transferability; we refer to [4] for detailed information.

RQ2. To answer RQ2, we applied a Wilcoxon signed rank test to check whether PEOU, PU, and ITU are significantly above the average value of the Likert scale (i.e., 3). The test results show that all the variables attesting the acceptance are significantly higher than 3 for $\alpha = 0.05$, with p-values of 1.077e-05, 7.109e-06, and 9.282e-06, respectively, with large effect-size (cf. Table 4). Therefore rejecting H3$_0$, H4$_0$, and H5$_0$ and confirming the overall degree of acceptance of the ADT notation as high. As the boxplot in Fig. 4 shows, while ITU and PU have comparable values, PEOU receives the highest score. This suggests that **ease of use is the main characterising quality of ADTs**.

Table 4. Statistics summary. Blue rows indicate NULL hypotheses that have been rejected (p-value < 0.05). The term "effv" indicates effectiveness and the term "effc" indicates efficiency.

RQs	Hyp.	Variables	Reject	p-value	Effect-size
RQ1	$H1_0$	Effectiveness	T	0.000139	1.255714
	$H2_0$	Efficiency	T	$7.381E - 06$	1.983621
RQ2	$H3_0$	PEOU	T	$1.08E - 05$	2.097433
	$H4_0$	PU	T	$7.11E - 06$	2.485847
	$H5_0$	ITU	T	$9.28E - 06$	2.185815

RQs	Hyp.	Relation between variables	Reject	Eq.	p-value
RQ3	$H6_0$	PEOU → PU	F	PU = 3.6 + 0.073 * PEOU	0.5962
	$H7_0$	PU → ITU	T	ITU = 0.5 + 0.86 * PU	2.44E-06
	$H8_0$	PEOU → ITU	F	ITU = 2.9 + 0.24 * PEOU	0.108
RQ4	$H9_0$	und. effv → PEOU	T	PEOU = 2.8 + 1.8 * und. effv	0.03677
	$H10_0$	und. effv → PU	F	PU = 3.2 + 0.97 * und. effv	0.08483
	$H11_0$	und. effc → PEOU	F	PEOU = 3.8 + 2.7 * und. effc	0.1752
	$H12_0$	und. effc → PU	F	PU = 4 - 0.5 * und. effc	0.8492
RQ5	$H13_0$	UNC effv → PEOU	F	PEOU = 4.5 - 0.74* UNC effv	0.7578
	$H14_0$	UNC effv → PU	F	PU = 4.5 - 0.44 * UNC effv	0.4241
	$H15_0$	UNC effc → PEOU	F	PEOU = 3.9 + 2.9 * UNC effc	0.2606
	$H16_0$	UNC effc → PU	F	PU = 3.9 - 0.22 * UNC effc	0.8952
	$H17_0$	UIC effv → PEOU	T	PEOU = 2.8 + 1.5 * UIC effv	0.02051
	$H18_0$	UIC effv → PU	F	PU = 3.5 + 0.43 * UIC effv	0.3332
	$H19_0$	UIC effc → PEOU	F	PEOU = 3. 9 + 1.1 * UIC effc	0.2168
	$H20_0$	UIC effc → PU	F	PU = 4 - 0.16 * UIC effc	0.7812
	$H21_0$	TRF effv → PEOU	F	PEOU = 3.7 + 0.73 * TRF effv	0.08802
	$H22_0$	TRF effv → PU	T	PU = 3.5 + 0.62 * TRF effv	0.02494
	$H23_0$	TRF effc → PEOU	F	PEOU = 3.9 + 10 * TRF effc	0.2105
	$H24_0$	TRF effc → PU	F	PU = 3.8 + 3.9 * TRF effc	0.4685

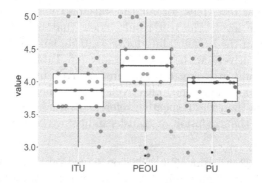

Fig. 4. Result for RQ2: boxplot of users' acceptance variables.

RQ3. To answer RQ3, we fit a regression linear model between PEOU and PU, and between both PEOU and PU and ITU. As shown in Fig. 5a, the test results attest that there is a significant positive relationship between PU and ITU (p-value = 2.44e−06). We can thus reject $H7_0$ and suggest that **users intend to use the notation in the future more for its usefulness than for its easiness.**

RQ4. To check if there is a relationship between the understandability of the notation and the users' perceptions about its easiness and usefulness, we test $H9_0$–$H12_0$ by fitting a linear model between PEOU and PU and understandability effectiveness, and between PEOU and PU and understandability efficiency. Our results show a significant positive relationship between effectiveness and perceived ease of use (cf. Fig. 5b), suggesting that **users who perform best in the test tend to evaluate better the notation in terms of easiness.**

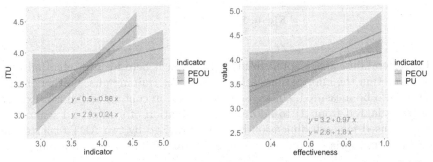

(a) Relationship between PEOU and PU and ITU.

(b) Relationship between understandability effectiveness and PEOU and PU.

Fig. 5. Results for RQ3 and RQ4.

RQ5. Finally, to understand whether one of the understandability dimensions affects most the perceived easiness and usefulness, we fit a regression linear model between the perception-based variables (PEOU and PU), and the effectiveness and efficiency of the three understandability dimensions (understandability not in context, understandability in context, and transferability). Our results show that the effectiveness of understandability in context and of transferability both have a significant positive relationship with PEOU and PU, respectively (cf. Fig. 6a and 6b). We can thus reject $H17_0$ and $H22_0$, and confirm that users who observed instantiated trees and understand their meaning tend to evaluate the notation as easier, while users who apply the method better by extending the tree correctly tend to evaluate it as more useful. This suggests that **users who successfully use the notation in practice, tend to appreciate it more.**

5.1   Threats to Validity

Construct Validity. Users' acceptance was assessed through existing models [27] and adapted to the ADT notation according to [1]. The usage of effectiveness and efficiency for understandability performance is widely used in the literature (cf., e.g., [1,5,27]). For what concerns the understandability dimensions, retention

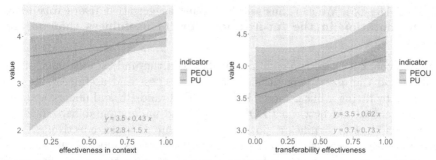

(a) Relationship between understandability in context effectiveness and PEOU and PU. (b) Relationship between transferability effectiveness and PEOU and PU.

Fig. 6. Results for RQ5.

and transferability are adapted from [1,25], even if here we use retention as a means to retain the information gathered from the training phase rather than a dimension to be measured. Understandability not in context and in context are measures adapted from [1] to address the evaluation of syntax and semantics. The tasks used for each dimension have been revised by two ADT experts and considered appropriate to evaluate the understandability of the notation.

Internal Validity. To prevent systematic response bias in user acceptance questionnaires, we mixed positive and negative statements. Moreover, to minimise participant response bias and limit the possible tendency of users to provide positive answers to please the researchers, the experiment was conducted completely online; thus, none of the users met the experimenter. This approach not only preserved participant anonymity but also created a more naturalistic setting, minimising biases introduced by participants' awareness of being observed and diminishing the Hawthorne Effect. The support used during the test (e.g., the editable online document and diagram) may have influenced users' performance. To study this hypothesis, further investigation with users must be carried out to grasp their difficulties with the support.

External Validity. The selected participants encompass diverse genders and experience levels, enhancing generalisability. Participants were opportunistically chosen from the academic field, varying in seniority. However, their representation may not fully encompass all ADT user classes, influencing study results. Further research involving users from different fields is needed to confirm the applicability of conclusions across all user classes. It should also be noted that this study is a controlled experiment, which aims to maximise internal validity and does not evaluate ADT users in a realistic setting, where contextual factors play a relevant role. Therefore, case studies are needed to confirm that our conclusions apply in a real-life security analysis environment.

6   Conclusion and Future Work

In this paper, we presented the first empirical study to assess the quality of ADTs in terms of users' acceptance and understandability. Our evaluation measures how well the notation can be used in practice. In particular, our study focused on assessing users' perceptions variables that attest the notation appreciation in terms of ease of use, usefulness, and intention to use, and of performance variables that attest the degree of understandability of the notation in terms of effectiveness and efficiency. Understandability has also been studied according to three different fine-grained dimensions, and the relation between all these variables has been evaluated through multiple statistical tests.

Our results suggest that the ADT notation is sufficiently understood and greatly appreciated by users, specifically, the main aspect characterising its quality is its ease of use. Overall, the notation has a good level of understandability with a total average effectiveness above 0.76. Among its dimensions, we note better performance in more practical tasks (i.e., those related to observing and extending instantiated trees). Concerning relationships among the variables, we note that general understandability and understandability in context have a relationship with the perceived ease of use and that the ability to apply ADT in practice has a relationship with the perceived usefulness.

In future research, we plan to address user challenges in the test by conducting interviews to assess the impact of the platform on performance. To enhance result accuracy, we will broaden our subject pool, including users from diverse classes, such as those in the security field. We also intend to compare user performance and perceptions across ADTs and other security requirements modelling techniques, preferably textual methods. Additionally, our analysis will encompass various commercial and academic ADT tools.

Acknowledgements. Research supported by the Italian MUR–PRIN 2020TL3X8X project T-LADIES (Typeful Language Adaptation for Dynamic, Interacting and Evolving Systems); by Innovation Fund Denmark and the Digital Research Centre Denmark, through the bridge project "SIOT - Secure Internet of Things - Risk analysis in design and operation"; by Industriens Fond through the project "Sb3D: Security-by-Design in Digital Denmark"; and by the EU Project CODECS GA 101060179. The authors would like to thank all the participants of the study.

References

1. Abrahão, S., Insfrán, E., Carsí, J.A., Genero, M.: Evaluating requirements modeling methods based on user perceptions: a family of experiments. Inf. Sci. **181**(16), 3356–3378 (2011)
2. Audinot, M., Pinchinat, S., Kordy, B.: Is my attack tree correct? In: Foley, S.N., Gollmann, D., Snekkenes, E. (eds.) ESORICS 2017. LNCS, vol. 10492, pp. 83–102. Springer, Cham (2017). https://doi.org/10.1007/978-3-319-66402-6_7
3. ter Beek, M.H., Legay, A., Lluch Lafuente, A., Vandin, A.: Quantitative security risk modeling and analysis with RisQFLan. Comput. Secur. **109**, 102381 (2021)

4. Broccia, G., ter Beek, M.H., Lluch Lafuente, A., Spoletini, P., Ferrari, A.: Assessing the Understandability of Attack-Defense Trees for Modelling Security Requirements: an Experimental Investigation - Supplementary Material. https://doi.org/10.5281/zenodo.10136730

5. Broccia, G., Ferrari, A., ter Beek, M., Cazzola, W., Favalli, L., Bertolotti, F.: Evaluating a language workbench: from working memory capacity to comprehension to acceptance. In: Proceedings 31st International Conference on Program Comprehension (ICPC), pp. 54–58. IEEE (2023)

6. Buyens, K., De Win, B., Joosen, W.: Empirical and statistical analysis of risk analysis-driven techniques for threat management. In: Proceedings 2nd International Conference on Availability, Reliability and Security (ARES), pp. 1034–1041. IEEE (2007)

7. Davis, F.D.: Perceived usefulness, perceived ease of use, and user acceptance of information technology. MIS Q. **13**, 319–340 (1989)

8. Eisentraut, J., Holzer, S., Klioba, K., Křetínský, J., Pin, L., Wagner, A.: Assessing security of cryptocurrencies with attack-defense trees: proof of concept and future directions. In: Cerone, A., Ölveczky, P.C. (eds.) ICTAC 2021. LNCS, vol. 12819, pp. 214–234. Springer, Cham (2021). https://doi.org/10.1007/978-3-030-85315-0_13

9. Ezenwoye, O., Liu, Y.: Risk-based security requirements model for web software. In: Proceedings 30th International Requirements Engineering Conference Workshops (REW), pp. 232–237. IEEE (2022)

10. Fabian, B., Gürses, S., Heisel, M., Santen, T., Schmidt, H.: A comparison of security requirements engineering methods. Requir. Eng. **15**, 7–40 (2010)

11. Gadyatskaya, O., Trujillo-Rasua, R.: New directions in attack tree research: catching up with industrial needs. In: Liu, P., Mauw, S., Stølen, K. (eds.) GraMSec 2017. LNCS, vol. 10744, pp. 115–126. Springer, Cham (2018). https://doi.org/10.1007/978-3-319-74860-3_9

12. Giorgini, P., Mouratidis, H., Zannone, N.: Modelling Security and Trust with Secure Tropos. In: Integrating Security and Software Engineering: Advances and Future Visions, chap. 8, pp. 160–189. IGI Global (2007)

13. Iankoulova, I., Daneva, M.: Cloud computing security requirements: A systematic review. In: Proceedings 6th International Conference on Research Challenges in Information Science (RCIS), pp. 1–7. IEEE (2012)

14. Kordy, B., Kordy, P., Mauw, S., Schweitzer, P.: ADTool: security analysis with attack–defense trees. In: Joshi, K., Siegle, M., Stoelinga, M., D'Argenio, P.R. (eds.) QEST 2013. LNCS, vol. 8054, pp. 173–176. Springer, Heidelberg (2013). https://doi.org/10.1007/978-3-642-40196-1_15

15. Kordy, B., Mauw, S., Radomirović, S., Schweitzer, P.: Foundations of attack–defense trees. In: Degano, P., Etalle, S., Guttman, J. (eds.) FAST 2010. LNCS, vol. 6561, pp. 80–95. Springer, Heidelberg (2011). https://doi.org/10.1007/978-3-642-19751-2_6

16. Kordy, B., Wideł, W.: On quantitative analysis of attack–defense trees with repeated labels. In: Bauer, L., Küsters, R. (eds.) POST 2018. LNCS, vol. 10804, pp. 325–346. Springer, Cham (2018). https://doi.org/10.1007/978-3-319-89722-6_14

17. Labunets, K., Massacci, F., Paci, F.: On the equivalence between graphical and tabular representations for security risk assessment. In: Grünbacher, P., Perini, A. (eds.) REFSQ 2017. LNCS, vol. 10153, pp. 191–208. Springer, Cham (2017). https://doi.org/10.1007/978-3-319-54045-0_15

18. Labunets, K., Massacci, F., Paci, F., Tran, L.M.S.: An experimental comparison of two risk-based security methods. In: Proceedings 7th International Symposium

on Empirical Software Engineering and Measurement (ESEM), pp. 163–172. IEEE (2013)
19. Lallie, H.S., Debattista, K., Bal, J.: An empirical evaluation of the effectiveness of attack graphs and fault trees in cyber-attack perception. IEEE Trans. Inf. Forensics Secur. **13**(5), 1110–1122 (2018)
20. Lallie, H.S., Debattista, K., Bal, J.: A review of attack graph and attack tree visual syntax in cyber security. Comput. Sci. Rev. **35**, 100219 (2020)
21. Liu, L., Yu, E.S.K., Mylopoulos, J.: Secure-I*: engineering secure software systems through social analysis. Int. J. Softw. Inform. **3**(1), 89–120 (2009)
22. Lodderstedt, T., Basin, D., Doser, J.: SecureUML: a UML-based modeling language for model-driven security. In: Jézéquel, J.-M., Hussmann, H., Cook, S. (eds.) UML 2002. LNCS, vol. 2460, pp. 426–441. Springer, Heidelberg (2002). https://doi.org/10.1007/3-540-45800-X_33
23. Mai, P.X., Goknil, A., Shar, L.K., Pastore, F., Briand, L.C., Shaame, S.: Modeling security and privacy requirements: a use case-driven approach. Inf. Softw. Technol. **100**, 165–182 (2018)
24. Mauw, S., Oostdijk, M.: Foundations of attack trees. In: Won, D.H., Kim, S. (eds.) ICISC 2005. LNCS, vol. 3935, pp. 186–198. Springer, Heidelberg (2006). https://doi.org/10.1007/11734727_17
25. Mayer, R.E.: Models for understanding. Rev. Educ. Res. **59**(1), 43–64 (1989)
26. Mellado, D., Blanco, C., Sanchez, L.E., Fernández-Medina, E.: A systematic review of security requirements engineering. Comput. Stand. Interfaces **32**(4), 153–165 (2010)
27. Moody, D.L.: Dealing with Complexity: A Practical Method for Representing Large Entity Relationship Models. Ph.D. thesis, University of Melbourne (2001)
28. Oliveira, D., Bruno, R., Madeiral, F., Castor, F.: Evaluating code readability and legibility: an examination of human-centric studies. In: Proceedings 36th International Conference on Software Maintenance and Evolution (ICSME), pp. 348–359. IEEE (2020)
29. Paja, E., Dalpiaz, F., Giorgini, P.: Modelling and reasoning about security requirements in socio-technical systems. Data Knowl. Eng. **98**, 123–143 (2015)
30. Salehie, M., Pasquale, L., Omoronyia, I., Ali, R., Nuseibeh, B.: Requirements-driven adaptive security: protecting variable assets at runtime. In: Proceedings 20th International Requirements Engineering Conference (RE), pp. 111–120. IEEE (2012)
31. Schneier, B.: Attack Trees. Dr. Dobb's J. (1999)
32. Sharafi, Z., Marchetto, A., Susi, A., Antoniol, G., Guéhéneuc, Y.G.: An empirical study on the efficiency of graphical vs. textual representations in requirements comprehension. In: Proceedings 21st International Conference on Program Comprehension (ICPC), pp. 33–42. IEEE (2013)
33. Sindre, G., Opdahl, A.L.: Eliciting security requirements with misuse cases. Requir. Eng. **10**, 34–44 (2005)
34. Souag, A., Mazo, R., Salinesi, C., Comyn-Wattiau, I.: Reusable knowledge in security requirements engineering: a systematic mapping study. Requir. Eng. **21**, 251–283 (2016)
35. Stein, D., Hanenberg, S., Unland, R.: A graphical notation to specify model queries for MDA transformations on UML models. In: Aßmann, U., Aksit, M., Rensink, A. (eds.) MDAFA 2003-2004. LNCS, vol. 3599, pp. 77–92. Springer, Heidelberg (2005). https://doi.org/10.1007/11538097_6
36. Vesely, W.E., Goldberg, F.F., Roberts, N.H., Haasl, D.F.: Fault Tree Handbook. Technical Report NUREG-0492, Nuclear Regulatory Commission, USA (1981)

37. Villamizar, H., Kalinowski, M., Viana, M., Fernández, D.M.: A systematic mapping study on security in agile requirements engineering. In: Proceedings 44th Euromicro Conference on Software Engineering and Advanced Applications (SEAA), pp. 454–461. IEEE (2018)

38. Wideł, W., Audinot, M., Fila, B., Pinchinat, S.: Beyond 2014: formal methods for attack tree-based security modeling. ACM Comput. Surv. **52**(4), 75:1-75:36 (2019)

39. Zareen, S., Akram, A., Khan, S.A.: Security requirements engineering framework with BPMN 2.0.2 extension model for development of information systems. Appl. Sci. **10**(14), 4981 (2020)

Learning to Rank Privacy Design Patterns: A Semantic Approach to Meeting Privacy Requirements

Guntur Budi Herwanto[1,2]($\boxtimes$) (ID), Gerald Quirchmayr[2] (ID), and A. Min Tjoa[2,3] (ID)

[1] Universitas Gadjah Mada, Yogyakarta, Indonesia
gunturbudi@ugm.ac.id
[2] University of Vienna, Vienna, Austria
[3] Vienna University of Technology, Vienna, Austria

Abstract. **[Context and Motivation]** Privacy requirements engineering is a critical aspect of software design to ensure that user data is protected in accordance with both regulatory and privacy objectives. The privacy requirements identified through this process can be addressed using various privacy design patterns. **[Question/Problem]** Identifying and implementing the most suitable privacy design patterns poses a major challenge for developers. They need to meticulously examine a wide range of options, which makes it challenging to quickly and effectively choose and justify the best solutions. **[Key Ideas/Results]** To address this gap, we developed a machine learning model that focuses on semantic text features and learning-to-rank algorithms to recommend privacy design patterns that meet specified privacy requirements. **[Contribution]** The main contribution of this paper is the development of a recommendation system for privacy design patterns based on privacy requirements using only text-based attributes. Our system's reliance on text as the sole input guarantees its broad applicability, avoiding the constraints of fixed mappings prevalent in previous methodologies. The performance of the model has shown encouraging results in understanding the semantic meaning of privacy requirements and mapping them to privacy design patterns, indicating its suitability for inclusion in the privacy engineering process.

Keywords: privacy requirements engineering · privacy design pattern · natural language processing · learning to rank

1 Introduction

The enforcement of the General Data Protection Regulation (GDPR) has increased the focus on principles such as privacy by design and privacy by default, requiring developers to embed privacy considerations throughout the development process [6]. However, there's a gap in understanding how to technically implement these principles, leading to the emergence of privacy engineering [32].

During the requirements phase, developers should identify the privacy needs of the system. These needs come from various sources, such as risks associated

D. Mendez and A. Moreira (Eds.): REFSQ 2024, LNCS 14588, pp. 57–73, 2024.
https://doi.org/10.1007/978-3-031-57327-9_4

with the use of personal data [9,22], privacy principles [27], and legal mandates [3]. As legal regulations and cybersecurity complexities escalate, developers are faced with an increasing number of requirements to manage. For example, the ProPAN methodology lists 13 potential categories of privacy requirements [22]. Addressing these manually is tedious, and although there is a mapping table between privacy requirements and privacy design solutions, it can be a problem for developers to choose the exact solution. [2]. In addition, developers may not have the breadth of knowledge necessary to fully understand privacy implications, or may overlook critical privacy requirements amidst a variety of development tasks [29]. These factors underscore the need for an automated approach to help developers navigate privacy requirements and select appropriate privacy design solutions.

Research in Privacy Enhancing Technologies (PET) and Privacy Design Patterns (PrDPs) aims to address privacy requirements (PRs) [8]. PrDPs provide documented answers to recurring problems and are often structured in catalogs [8,10]. However, existing catalogs do not provide an agile way for developers to identify the most relevant privacy patterns for their projects. Moreover, they do not consistently incorporate the latest developments in PETs.

In this paper, we propose a novel recommendation system that assists in the selection of PrDPs by analyzing the textual content of PRs. We recognize that text, while simple in form, contains rich and detailed meanings specifically relevant to PrDPs. No current solutions provide this linkage, and our research aims to fill this gap by leveraging the capabilities of natural language understanding to semantically match PRs with PrDPs. In addition, our method simplifies the integration of current privacy solutions, even without structured patterns, facilitating the adoption of privacy design methodologies. By automating the correlation of PRs with PrDPs using advanced machine learning techniques, we take a significant step toward implementing agile privacy-by-design practices [1].

2   Background and Related Work

Privacy requirements engineering (PRE) is a process of identifying the privacy needs of a system to be built [22]. Prior research has explored the transformation of these principles into specific requirements [22,27]. Meis and Heisel [22], in addition to Pfitzmann and Hansen [27] outlined 13 main requirements, including Data Confidentiality, Undetectability, and Data Unlinkability. Additionally, requirements can arise from system-specific risks, using a risk-based approach. For instance, the LINDDUN method [9] employs threat modeling to define requirements. A combination of risk-based and goal-based methods is optimal for creating PRs [25].

In the field of PRE, various solutions such as PETs, PrDPs, and privacy-aware architectures have been proposed to tackle PRs and associated threats. Deng et al. [9] suggested the use of PETs to counter threats identified by the LINDDUN framework. There are efforts to consolidate these PETs into standardized patterns [13,23]. To aid in the engineering phase, it is recommended to categorize PETs using specific templates [8,23]. The template by Meis and Heisel [23] emphasizes integration aspects, while Colesky's template [8] focuses on both soft and hard privacy goals, leading to what is known as PrDPs. Although there are around 70 cataloged patterns, selecting the most appropriate PrDPs and automating this selection remains a challenge [2,5]. An example of a PrDP is illustrated in Table 1.

Table 1. Privacy Pattern Example: Who's Listening

Pattern Name
Who's Listening
Summary
Inform users of content where other users or unauthenticated persons having accessed the same content are listed, and may access any further disclosures
Context
Users of a service regularly shares its usage with other users. Sometimes these are users they know personally, and sometimes these are anonymous, unauthenticated persons. This occurs particularly in shared or collaborative environments where content is generated. Knowledge of the contributions of other users contributes to additional or refined content in general
Problem
Users do not know if the content they are accessing or have disclosed has been accessed or modified by others, nor if it is someone they know
Forces and Concerns
- Users want to know who can access their disclosures and those of others - Users want to know that specific other users have accessed or modified content - Controllers do not want users to be unaware of who can see their disclosures - Controllers want to log access to prevent abuse
Solution
Provided that users know their access is not private, inform them of other users, even unauthenticated, which are also accessing the content in question

Considering the diverse range of PrDPs available, it becomes crucial to have an effective method for identifying the most suitable options for specific PRs. This selection process depends heavily on understanding the context and aligning it with the criteria of PrDPs. Previous efforts have provided decision support using rule-based systems [26], ontology databases [11,31], and text relevance methods [16].

Rule-based systems require patterns with clearly defined properties [26], whereas ontology-based approaches organize patterns in an ontology format [11,31]. Text-based methods are crucial, especially since pattern information is often stored in text form. For instance, the ArchReco system [31] employs the TF-IDF algorithm to assess the significance of words within a document collection. Hussain [16], on the other hand, uses a supervised machine learning technique called Learning to Rank (LeToR) for text analysis. LeToR specializes in sorting and prioritizing information based on relevance, using labeled training data that includes search queries, potential candidates, and their relevance ratings [18]. Our approach adopts LeToR for Privacy Design Patterns (PrDPs), favoring context-aware recommendations over traditional methods like TF-IDF [31]. This approach aims to streamline the privacy requirements engineering with privacy design engineering.

3   Recommending Privacy Design Patterns

Matching privacy requirements (PRs) with the appropriate privacy design patterns (PrDPs) presents a significant challenge due to the often indirect textual relationship between PRs and PrDPs. Basic approaches that rely solely on textual similarity may not effectively capture the nuanced context necessary for identifying the most suitable design pattern. To address this issue, we employ supervised learning techniques, leveraging their ability to learn from labeled data to more accurately predict the relevance of PrDPs to specific PRs. This method allows for a more nuanced understanding of the complex relationships between privacy requirements and design patterns, beyond mere word structure.

We have developed two key machine learning strategies to enhance the system's recommendation accuracy: a classification model that categorizes privacy requirements into distinct groups, and a Learning-To-Rank (LeToR) algorithm that orders design patterns by their relevance to the given requirements. The distinction between hard and soft privacy requirements is a critical aspect of our classification model. Hard privacy requirements focus on technical aspects of data protection, such as anonymity and encryption, while soft privacy requirements relate to policy and user consent aspects [2]. This differentiation is crucial because it allows our system to tailor recommendations more effectively by understanding the type of privacy protection each requirement seeks. By classifying PRs into these categories, we provide the LeToR model with valuable input features that significantly refine the final pattern recommendations, ensuring that they align more closely with the specific needs of the privacy requirement in question.

An illustrative example in Fig. 1 visually explains the input and output process of our recommendation system, highlighting how these two stages classification and ranking work together to improve the system's ability to suggest context-appropriate PrDPs.

3.1   Datasets

Our approach mainly involves datasets containing Privacy Requirements (PRs) and Privacy Design Patterns (PrDPs). For the classification model, we utilize PRs tagged with their corresponding goals, either 'hard' or 'soft', to enable binary classification, along with a set of 13 distinct objectives for multiclass classification. In the context of the LeToR model, each pairing of a PR and a PrDP is assigned a specific relevance score, which indicates the level of relevance or suitability of the PrDP to the PR.

privacy requirements

*"As a Parent, I want the **Child data** that used in Message Child's Counselors to **be protected from being linked directly or indirectly** to other personal data within or outside of our system, so that an attacker cannot link it to the identity of subject in **Child data"***

Privacy Requirement Classification ➡ Learning to Rank Privacy Design Pattern ➡

privacy design pattern recommendations

Onion Routing
This pattern provides unlinkability between senders and receivers by encapsulating the data in different layers of encryption, limiting the knowledge of each node along the delivery path. ③

Strip Invisible Metadata
Strip potentially sensitive metadata that isn't directly visible to the end user. ②

Psuedonymous Identity
I hide the identity by using a pseudonym and ensure a pseudonymous identity that can not be linked with a real identity during online interactions. ①

Personal Data Store
Subjects keep control on their personal data that are stored on a personal device.

Trust Evaluation of Services Sides
A visual highlight provided by an authority which signals the extent to which given privacy criteria are fulfilled. It should be clearly placed and easily found, with links to additional information.

Aggregation Gateway
Encrypt, aggregate and decrypt at different places.

privacypatterns.org

Fig. 1. Practical example showcasing the recommendation of PrDPs utilizing the learning-to-rank approach.

Privacy Requirements. We gather PRs data from three different PRE methods [15,21,24]. Examples of PRs from each method can be found in Table 2.

1. PrivacyStory [15] offers a tool for generating PRs. In total, we are able to generate 145 PRs based on two projects.
2. ProPAN [21] offers 16 requirements from a case study and additional text templates for each privacy requirement category.
3. Security Compass [24] presents 39 requirements in their paper.

Table 2. Examples of privacy requirements for the three data sets: The first row is an example of a hard goal (unlinkability) and the second row is an example of a soft goal (transparency)

PrivacyStory [15]	ProPAN [22]	Security Compass [24]
As a Parent, I want the Child data that used in Enroll Children to be protected from being linked directly or indirectly to other personal data within or outside of our system, so that an attacker cannot link it to the identity of subject in Child data	For each pair of personal data links of the data Subject, counterstakeholders to whom the personal data personal Data are related to shall at most be able to link instances of the two elements of the pair to each other with linkability single	As a user, the goal is to be able to provide/withdraw consent to/from the usage of data in identifiers that can be associated with individuals when combined with their personal data so that they cannot be identified indirectly by data in identifiers
As a Parent, I want to be informed and consented that the Child data is used in Enroll Children, so that I can exercise my rights when it is used outside of this context.	The person concerned must be informed that his personal data is collected by the system executed by the controller.	As a user, the goal is to be able to easily access information about processing activities that involve individuals' personal data in clear and understandable language so they can exercise their right to view their processed personal data.

Privacy Design Patterns Catalogues. We found the most comprehensive catalog of PrDPs to be the one developed by Colesky et al. [8] which also available online[1]. Although the patterns are not regularly updated, the total number of 71 PrDPs covering the objectives of PRs is sufficient for the proof of concept of our LeToR method.

Relevancy Score Between Privacy Requirements and Design Patterns: LeToR is designed to prioritize the most relevant Privacy Design Patterns (PrDPs) based on the requirements specified in the text. Since this process is supervised, it's essential to define a 'ground truth' that establishes how relevant each privacy requirement is to the PrDPs.

We draw on the initial mapping done by Al-Momani et al. [2] as a basis for evaluating the relevance of specific privacy objectives to privacy patterns. This mapping, found in Al-Momani et al.'s work [2], correlates eight privacy goals-namely anonymity, unlinkability, confidentiality, plausible deniability, undetectability, manageability, intervenability, and transparency-with PrDPs identified in the Colesky collection [8]. Although this provides a useful starting point for matching patterns to privacy goals, it doesn't always reflect the specific context of the requirements. For example, confidentiality-focused patterns such as 'Buddy List' and 'Encryption user-managed keys' may vary in their applicability depending on the scenario, which calls for additional context-specific adjustments. These adjustments are further refined with the help of human annotators.

The refinement of these contextual adjustments and the evaluation of pattern relevance to privacy requirements are carried out by human annotators, notably the first author. Before the annotation process, the first author extensively reviewed all PrDPs to ensure a deep understanding of each pattern's intended application and scope. This comprehensive review is critical for accurately assessing the relevance of PrDPs to the privacy requirements.

During the annotation phase, each privacy requirement's relevance to the various PrDPs is scored on a scale from 1 to 5, where a score of 5 signifies the highest relevance and, consequently, the highest priority for recommendation. This grading system is facilitated through a custom-built annotation tool, designed to streamline the evaluation process and ensure consistent scoring. The outcomes of this detailed annotation exercise form the core of our training dataset, underpinning the supervised learning approach of our recommendation system.

To foster transparency and enable further research, we have made the dataset, annotation tools, and the recommendation system available in a public repository[2]. The methodology for constructing the ground truth, from the initial review of PrDPs to the final annotation of privacy requirements, is depicted in Fig. 2.

---

[1] https://privacypatterns.org/.

[2] https://github.com/gunturbudi/pattern-recommender.

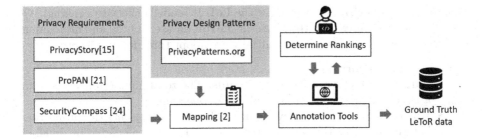

Fig. 2. Establishing LeToR ground truth data.

3.2 Privacy Requirements Classification

Privacy requirements are classified into 13 privacy goals [22,27], and PrDPs cover one or multiple goals. By identifying which privacy goals are linked to each requirement, we can suggest possible PrDPs. LINDDUN's [9] concept of hard privacy comprises unlinkability, anonymity, pseudonymity, plausible deniability, undetectability, unobservability, and confidentiality. The soft privacy of LINDDUN incorporates awareness of user content, policy adherence, and consent compliance. ProPAN's taxonomy, which is similar to LINDDUN's soft privacy, emphasizes transparency and the ability to intervene [22]. Al-Momani maps the correlation between privacy goals and PrDPs, noting that a design pattern can address multiple goals, typically aligning with either hard or soft privacy goals [2]. Using this mapping, we classify PrDPs and PRs as either hard or soft privacy [2].

The classification workflow begins with preprocessing the requirement text, followed by tokenization and representation as a bag of words. Given the specificity of our dataset, we selected the Naive Bayes algorithm [20] for classification which is well-suited for small dataset. To improve the accuracy of our limited dataset, we integrated the EDA method [33].

During the prediction phase, users can input their PRs in textual format and avoid the restrictions of predefined templates [22] or initial mapping [2]. This input is subject to the same preprocessing steps, and the Naive Bayes model trained earlier categorizes the PRs into either hard or soft privacy goals. The result of this classification will serve as one of the feature in identifying the relevance of PrDPs. Furthermore, the system uses the LeToR model to determine their priority.

3.3 Learning-to-Rank Privacy Design Patterns

To construct an LeToR model, two foundational steps must be followed: (1) selecting the appropriate features, also known as feature engineering, and (2) determining the relevance score between the query (PRs) and the documents (PrDPs). We can then apply the LeToR algorithms, which are explained in Sect. 4.

Feature Engineering is pivotal for the LeToR model's effective ranking. We adopt multiple features from the Microsoft LeToR dataset [14,28] and interaction level of semantic features [12]. Specifically, we adopt three feature levels:

1. **Query-level Features:**
   - **Number of Covered Query Terms:** The total number of terms from a PRs that are included in the PrDPs collection.
   - **Ratio of Covered Query Terms:** This ratio represents the proportion of terms in a PRs that are covered by a PrDPs. It is calculated by dividing the number of terms in the PRs that are included by the PrDPs by the total number of terms in the PRs.
   - **Character count** in the privacy requirement.
   - **Query IDF (Inverse Document Frequency):** Measures a term's importance in a PRs, based on its rarity across PrDPs.
   - **Query TF-IDF:** A metric that combines the frequency of a term in a PRs with its IDF.
   - **Category** of PRs: Classifies PRs using binary (two-goals) or multi-class (multiple-goals) methods.
2. **Document-level features:**
   - **Category** of PrDPs: Classifies PrDPs using binary (two-goals) or multi-class (multiple-goals) methods.
3. **Interaction-level features:**
   - **Term frequency (TF):** A measure of the count of each term in a PRs within the respective PrDPs.
   - **Okapi BM-25:** Scores the relevance of a PRs to PrDPs based on TF and IDF.
   - **Semantic similarity:** A cosine similarity of the embedding vectors between PRs and (1) PrDPs title, (2) PrDPs excerpt, (3) PrDPs content. Embedding vectors are generated using sentence transformers [30] that combine a general-purpose model[3] with a domain-specific one designed for legal contexts[4].
   - **Hadamard Product of Embedding:** A vector produced by the element-wise multiplication of the embedding vectors from PRs and PrDPs. The production of these vectors employs the same sentence transformer models as those used for calculating semantic similarity.
   - **Concatenation of Embedding:** A vector formed by the concatenation of the embedding vectors from PRs and PrDPs, maintaining the original vectors' structure. This method also uses the same embedding as the calculation of semantic similarity.

Our feature extraction method exclusively employs text-based attributes to ensure that both PRs and new PrDPs rely solely on textual data. This allows for the straightforward integration of PrDPs into the corpus, eliminating the need for manual curation of additional information, as is required for a PrDPs catalog [8].

---

[3] https://huggingface.co/sentence-transformers/all-mpnet-base-v2.
[4] https://huggingface.co/dean-ai/legal_heBERT_ft.

4 Results and Analysis

In this section, we present the results of our evaluation of the performance of our Learning-To-Rank (LeToR) model. Our analysis focuses on three main areas: evaluating the model's performance using standard metrics, investigating the importance of different features within the LeToR framework, and conducting a pilot example that applies our model to examples from the privacy requirements engineering method.

4.1 Experimental Setup

The goal of our experiment is to identify the performance of the classification model, as well as the optimal Learning-to-Rank (LeToR) algorithm and features for ranking PrDPs.

Classification Model. We are evaluating two classification models: (1) A binary classification model that classifies requirements as either "soft goals" or "hard goals". (2) A multi-class classification model that maps requirements to one of the privacy goals defined in [2]. This class will be used as a feature in LeToR, as mentioned in Sect. 3.3. Multinomial Naive Bayes was utilized as the classification algorithm for both cases. We assessed the classification model using 5-fold cross-validation.

Learning-to-Rank Model. We evaluate our LeToR model using the Normalized Discounted Cumulative Gain (NDCG). NDCG is calculated by first assigning a relevance score to each item in the ranked list, with higher relevance scores indicating more relevant items [7]. This initial score is called DCG. The formula can be seen in (1). Then, the relevance scores are discounted by the element's position in the ranking list, with higher-ranked elements receiving higher discounts. Finally, the discounted relevance scores are summed up and normalized by the maximum possible cumulative discounted score, to give the final NDCG score. The NDCG formula can be seen in (2). NDCG takes into account both relevance and rank position. NDCG aims to ensure that the ranked PrDPs are both relevant to the user and presented in an order that maximizes their usefulness. This can improve the user experience by providing them with PrDPs that are most relevant to the PRs. We use the "@k" value, which indicates the cutting threshold for the recommendation results. This means that only the top k items in the ranking will be considered when calculating the NDCG score. For NDCG, we use $k = \{1, 2, 3, 4, 5, 7, 10\}$.

$$DCG@k = \sum_{i=1}^{k} \frac{rel_i}{log_2(i+1)} \quad (1) \quad NDCG@k = \frac{DCG@k}{iDCG@k} \quad (2)$$

To enhance the model's generalization capability, we implement 5-fold stratified cross-validation on the training dataset, segmenting it into five equal parts of training and testing data. The stratification is conducted according to the binary classification outcomes of 'soft' and 'hard' goals in the context of privacy requirements, ensuring that each fold reflects a proportional distribution of these two categories. This approach guarantees that the evaluation of the model's performance is not skewed by an uneven distribution of the classification targets.

Our experimentation is structured into three distinct data sets. The first set, termed the baseline, includes features excluding the hadamard product and concatenation. This choice stems from the inherent sparsity of both hadamard product and concatenation features, which are derived from sparse sentence vector embeddings. The baseline set offers a clearer interpretation of each feature's individual importance. The second data set incorporates the hadamard product, while the third set combines elements from the baseline with both the hadamard product and concatenation features.

We utilize the LambdaRank algorithm for our LeToR experiment [4]. We have also experimented with other available algorithms, but LambdaRank has proven to be the best. Due to space constraints, we only show LambdaRank. LambdaRank is an algorithm derived from RankNet [4], which is itself a pairwise ranking algorithm using a neural network architecture. The lambda refers to the use of gradients of an approximation to the NDCG metric during training. This means that the model does not just learn to predict the correct ordering of documents, but it optimizes for the highest possible NDCG score, which makes it particularly suited for scenarios where the ranking quality at the top of the list is more important than elsewhere. We utilize LightGBM framework for the implementation of LambdaRank [17].

4.2   Model Performance

Classification Model Performance. The binary classification model achieved notably higher performance, with an accuracy of 96.52% and an F1-measure of 97.31%. The multi-class classification model achieved an accuracy of 83.77% and an F1-measure of 83.40%. Despite the lower performance of the multi-class classifier compared to the binary one, its results are still considered acceptable for incorporation into the LeToR model as a feature, given the potential for other features to compensate for inaccuracies. Moreover, certain misclassifications may be acceptable, as individual PrDPs can simultaneously address multiple goals. For instance, the "Protection against Tracking" pattern covers both unlinkability and anonymity, as shown by Al-Momani et al. [2].

LeToR Model Performance. The baseline model, excluding the Hadamard product and embedding concatenation features, achieves NDCG scores from 72.78% at NDCG@1 to 73.89% at NDCG@10. Adding the Hadamard product ('+ hadamard') enhances NDCG scores significantly, rising from 79.56% to 82.22%. The best results occur when combining the Hadamard product with concatenation ('+ hadamard, concat'), pushing NDCG scores from 80.64% to 83.05%, indicating that integrating the concatenation features refines the model's ranking performance. The detailed NDCG performance can be seen in Table 3.

Table 3. Performance of Learning-to-Rank Algorithm

Data	NDCG@1	NDCG@2	NDCG@3	NDCG@4	NDCG@5	NDCG@7	NDCG@10
baseline	72,78%	70,44%	69,60%	69,93%	70,12%	72,17%	73,89%
+ hadamard	79,56%	79,15%	78,16%	79,48%	79,59%	80,79%	82,22%
+ hadamard, concat	**80,64%**	**80,14%**	**79,54%**	**80,12%**	**80,62%**	**81,30%**	**83,05%**

4.3 Feature Importance of LeToR

In this section, we analyze the significance of features that enhance the performance of our LeTor model discussed in the previous section. Our first focus is on a third model that utilizes both 'hadamard products' and concatenation techniques to expand the baseline model's 36 features to a total of 9,252 features, mainly originating from the embedding vectors of PRs and PrDP. We assess the significance of each feature, categorizing them into three groups: the initial 36 features (baseline), features numbered 37 to 4,644 (hadamard), and features 4,645 to 9,252 (concatenation). We then calculated both the sum and the average importance of these features. The summed importance, depicted in Fig. 3, indicates that the 'hadamard' features are the most significant overall. The importance of the baseline comes last due to its limited number of features. However, when we examined the average importance, we found that the baseline features are individually quite significant to the model's overall effectiveness. This significance is followed by the 'hadamard' and then the 'concatenation' features. The average importance was considered due to the sparse nature of the vector embeddings, which led to many features having zero importance (Fig. 4).

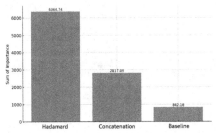

Fig. 3. Sum of Importance   **Fig. 4.** Average of Importance

We examined the importance of individual features using SHAP values [19], which quantify a feature's contribution to the prediction's deviation from the dataset's average. The top features for our best model, as shown in Fig. 5, are mainly semantic similarities between PRs and PrDP 'titles', 'content', or 'excerpts', with a strong emphasis on 'hadamard product' features. Notably, a single cosine similarity value between PRs and PrDP 'titles' was highly influential, suggesting that even simple measures can significantly impact the model's decisions alongside complex features like 'hadamard products'. It is noteworthy that the label "#1" following each feature implies the use of 'general' rather than 'legal' embeddings, suggesting that the latter do not significantly improve the performance of the model, which may warrant further analysis.

In the beeswarm plots, each point represents a SHAP value for a feature concerning a single data instance. The varied SHAP values (in Fig. 6) for 'Hadamard Content #1 562' highlight its significant and diverse influence on model decisions, reflecting the feature's variability. In contrast, the concentration of high SHAP values for 'Title Similarity #1' on the positive end indicates a consistent and strong positive effect on rankings, with higher similarity scores correlating with higher ranks.

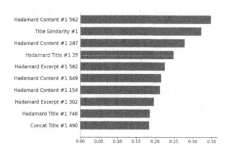

Fig. 5. Global Feature Importance (+hadamard, concat)

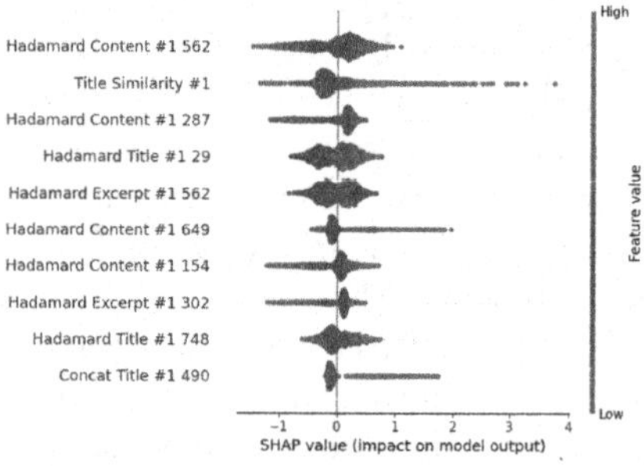

Fig. 6. Local Explanation Summary (+hadamard, concat)

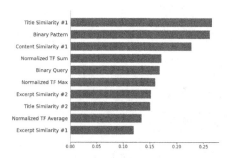
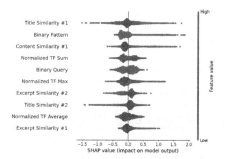

Fig. 7. Global Feature Importance (baseline)

Fig. 8. Local Explanation Summary (baseline)

The baseline model is examined in greater detail as a result of its influence on the individual features. Similar to the previous model, as depicted in Fig. 7, the cosine similarity between semantic embeddings between PRs and PrDPs 'title' is the most significant feature. It is also crucial to emphasize that the binary classification of 'hard' and 'soft' goals on both PRs and PrDPs influences the significance metrics employed in the model. Additionally, the beeswarm plot in Fig. 8 reveals that features like 'Binary Pattern' and 'Content Similarity', with predominantly positive SHAP values, are generally indicative of higher relevance, which in turn, promotes a higher ranking of the item. In contrast, a feature such as 'Binary Query' with mostly negative SHAP values suggests that its occurrence is often perceived by the model as a sign of lower relevance, thereby placing the item lower in the ranking.

4.4   Pilot Example

To evaluate our recommendation system, we applied our model to various test scenarios provided by LINDDUN [9] threat use cases[5]. For each type of threat, we selected a representative example and transformed the descriptions into specific privacy requirements. The outcomes of this process are presented in Table 4.

Table 4. Recommendation based on LINDDUN [9] Use Cases. The number after privacy requirements indicates the 'id' of the use case in the source.

#	Privacy Requirements	Privacy Design Pattern Recommendations
L	Limit the precision of stored location data to ensure individual user activities cannot be uniquely identified (1.5)	(1) Attribute Based Credentials, (2) Pseudonymous Identity, (3) Location Granularity, (4) Decoupling [content] and location information visibility, (5) Onion Routing
I	Implement measures to prevent the combination of different properties from a user's browser to form a unique identifier. (2.7)	(1) Pseudonymous Identity, (2) Attribute-Based Credentials, (3) Obtaining Explicit Consent, (4) Onion Routing, (5) Protection against Tracking
N	Provide users with the option to communicate without revealing their corporate email address to avoid confirming their organizational affiliation. (3.5)	(1) [Support] Selective Disclosure, (2) Pseudonymous Messaging, (3) Pseudonymous Identity, (4) Attribute-Based Credentials, (5) Added-noise measurement obfuscation
D	Ensure that the operational characteristics of health devices do not inadvertently reveal the medical condition or patient status of the user. (4.2)	(1) Attribute-Based Credentials, (2) Active broadcast of presence, (3) Added-noise measurement obfuscation, (4) Awareness Feed, (5) Personal Data Store
DD	Strip any metadata that may contain personal data from documents upon processing. (5.5)	(1) Strip Invisible Metadata, (2) Added-noise measurement obfuscation, (3) Lawful Consent, (4) Obtaining Explicit Consent, (5) Attribute-Based Credentials
U	Generate notices for data subjects using clear and plain language, understandable to a general audience. (6.2)	(1) Active broadcast of presence, (2) Privacy dashboard, (3) Attribute-Based Credentials, (4) Abridged Terms and Conditions, (5) Dynamic Privacy Policy Display
Nc	Ensure System shall ensure that no viewing data is collected until lawful grounds for such collection are established. (7.5)	(1) Lawful Consent, (2) Attribute-Based Credentials, (3) Awareness Feed, (4) Selective access control, (5) Informed Implicit Consent

---

[5] https://linddun.org/cases/.

In this pilot example, we manually evaluated the ranking to determine the ideal order, aiming to calculate the NDCG for values of k ranging from 1 to 5. The results indicated NDCG scores of 90% at k = 1, 88% at k = 2, 81% at k = 3, 87% at k = 4, and 94% at k = 5. Our thorough analysis revealed that the predictions for Data Disclosure (DD) were most accurately ordered, whereas the predictions for Linkability (L) were least accurate. To improve Linkability, we suggest prioritizing "Location Granularity" patterns in the first order as it involves location data directly. The presence of "Attribute-Based Credentials" in all predictions suggests a potential bias, particularly as it should not be included in predictions related to Unawareness (U) and Non-Compliance (Nc). Nevertheless, the recommendations consistently demonstrate desired PrDPs for each PR, indicating promising quality despite the challenges. Future strategies could improve performance by incorporating user feedback into the training process. This would require ongoing refinement of the model to more closely match its recommendations with user expectations.

5   Threats to Validity

Our study is susceptible to different types of validity issues, including construct, internal, external, and conclusion validity [34]. However, we have addressed and accounted for these concerns during the design and analysis stage. To ensure construct validity, we utilized well-established metrics for the LeToR model. We assessed our model using NDCG@k with varying '@k' values. Moreover, we incorporated PRs and PrDPs based on peer-reviewed research in our model. Internal validity, crucial for attributing outcomes to our methods and not to extraneous factors, presented a challenge due to the subjective nature of scoring privacy requirements' relevance to PrDPs. To overcome this, we based initial relevancy on Al-Momani's work [2] and refined it through careful human annotation. External validity pertains to the generalizability of our findings. To address potential sampling bias, we utilized data from three distinct methods of privacy requirements engineering. This approach facilitated the creation of a representative sample and minimized bias. We recognize the challenge of capturing the full spectrum of context-specific factors of design patterns-such as architectural considerations and the nuances of different application types-in our ground truth creation. We see it as an opportunity for refinement. Additionally, to ensure the reliability and soundness of our findings, we conducted multiple experiments and evaluated our model's strengths and limitations. Our datasets, annotations, and codes are publicly available for review and extension by the scientific community. We have taken steps to reduce the risks of internal and external validity, which bolsters the validity of our conclusions.

6   Conclusion

The Privacy by Design (PbD) approach emphasizes the importance of integrating privacy considerations into system designs right from the beginning stages of development. This involves identifying privacy requirements (PRs) and selecting the most appropriate privacy design patterns (PrDPs) to fulfill those requirements. To assist with this process, we have implemented a recommendation system that utilizes Learning-to-Rank (LeToR) alongside semantic text characteristics, to automatically match PRs with suitable privacy PrDPs. Both the classification and LeToR models exhibit satisfactory performance, confirming the potential of our model as a valuable resource in the field of privacy engineering. The model's reliance on text features proves particularly advantageous when other data types are not easily attainable. Moreover, while our study currently utilizes a set of 70 PrDPs as a foundational baseline, this is merely a starting point. As the collection of PrDPs expands, our system is designed to seamlessly integrate additional patterns. This scalability ensures that our solution remains relevant and effective as the number of PrDPs increases over time, addressing both current and future privacy design challenges.

For future enhancements, we aim to integrate our model into the PrivacyStory tool [15] to provide tailored recommendations from PRs. This integration will make the LeToR model part of a holistic tool that will refine its suggestions through feedback, leading to more precise privacy design pattern recommendations. We also plan to use data from real-world privacy requirements, such as personal health care, the Internet of Things, or other scenarios that are considered privacy sensitive.

Acknowledgment. The authors acknowledge the scholarship granted by the Indonesia Endowment Fund for Education (IEFE/LPDP), Ministry of Finance, Republic of Indonesia, and the support received from the University of Vienna, Faculty of Computer Science.

References

1. Aberkane, A.J., Poels, G., Broucke, S.V.: Exploring automated GDPR-compliance in requirements engineering: a systematic mapping study. IEEE Access **9**, 66542–66559 (2021)
2. Al-Momani, A., et al.: Land of the lost: Privacy patterns' forgotten properties: enhancing selection-support for privacy patterns. In: Proceedings of the ACM Symposium on Applied Computing, pp. 1217–1225 (2021). https://doi.org/10.1145/3412841.3441996
3. Bartolini, C., Daoudagh, S., Lenzini, G., Marchetti, E.: GDPR-based user stories in the access control perspective. In: Piattini, M., Rupino da Cunha, P., García Rodríguez de Guzmán, I., Pérez-Castillo, R. (eds.) QUATIC 2019. CCIS, vol. 1010, pp. 3–17. Springer, Cham (2019). https://doi.org/10.1007/978-3-030-29238-6_1
4. Burges, C.J.: From RankNet to LambdaRank to LambdaMART: an overview. Learning **11**(23–581), 81 (2010)
5. Caiza, J.C., Alamo, J.M.D., Guamán, D.S.: A framework and roadmap for enhancing the application of privacy design patterns. In: Proceedings of the 35th Annual ACM Symposium on Applied Computing, pp. 1297–1304 (2020)
6. Cavoukian, A., et al.: Privacy by design: the 7 foundational principles. Inf. Priv. Commissioner Ontario Can. **5**, 12 (2009)

7. Chen, W., Liu, T.Y., Lan, Y., Ma, Z.M., Li, H.: Ranking measures and loss functions in learning to rank. In: Bengio, Y., Schuurmans, D., Lafferty, J., Williams, C., Culotta, A. (eds.) Advances in Neural Information Processing Systems, vol. 22. Curran Associates, Inc. (2009)

8. Colesky, M., Caiza, J.C.: A system of privacy patterns for informing users: Creating a pattern system. In: ACM International Conference Proceeding Series (2018). https://doi.org/10.1145/3282308.3282325

9. Deng, M., Wuyts, K., Scandariato, R., Preneel, B., Joosen, W.: A privacy threat analysis framework: supporting the elicitation and fulfillment of privacy requirements. Requirements Eng. **16**(1), 3–32 (2011)

10. Drozd, O.: Privacy pattern catalogue: a tool for integrating privacy principles of ISO/IEC 29100 into the software development process. In: Aspinall, D., Camenisch, J., Hansen, M., Fischer-Hübner, S., Raab, C. (eds.) Privacy and Identity 2015. IAICT, vol. 476, pp. 129–140. Springer, Cham (2016). https://doi.org/10.1007/978-3-319-41763-9_9

11. Guan, H., Yang, H., Wang, J.: An ontology-based approach to security pattern selection. Int. J. Autom. Comput. **13**(2), 168–182 (2016). https://doi.org/10.1007/s11633-016-0950-1

12. Guo, W., et al.: DeText: a deep text ranking framework with BERT. In: Proceedings of the 29th ACM International Conference on Information & Knowledge Management, pp. 2509–2516 (2020)

13. Hafiz, M.: A pattern language for developing privacy enhancing technologies. Softw.: Pract. Exp. **43**(7), 769–787 (2013)

14. Han, X., Lei, S.: Feature selection and model comparison on Microsoft learning-to-rank data sets. arXiv preprint arXiv:1803.05127 (2018)

15. Herwanto, G.B., Quirchmayr, G., Tjoa, A.M.: PrivacyStory: tool support for extracting privacy requirements from user stories. In: 2022 IEEE 30th International Requirements Engineering Conference (RE), pp. 264–265. IEEE (2022)

16. Hussain, S., et al.: A methodology to rank the design patterns on the base of text relevancy. Soft. Comput. **23**(24), 13433–13448 (2019)

17. Ke, G., et al.: LightGBM: a highly efficient gradient boosting decision tree. In: Advances in Neural Information Processing Systems, vol. 30 (2017)

18. Liu, T.Y., et al.: Learning to rank for information retrieval. Found. Trends Inf. Retrieval **3**(3), 225–331 (2009)

19. Lundberg, S.M., Lee, S.I.: A unified approach to interpreting model predictions. In: Guyon, I., et al. (eds.) Advances in Neural Information Processing Systems, vol. 30, pp. 4765–4774. Curran Associates, Inc. (2017). http://papers.nips.cc/paper/7062-aunified-approach-to-interpreting-model-predictions.pdf

20. McCallum, A., Nigam, K., et al.: A comparison of event models for naive bayes text classification. In: AAAI-98 Workshop on Learning for Text Categorization, Madison, WI, vol. 752, pp. 41–48 (1998)

21. Meis, R.: Problem-based privacy analysis (ProPAn): a computer-aided privacy requirements engineering method. Universitaet Duisburg-Essen (Germany) (2018)

22. Meis, R., Heisel, M.: Computer-aided identification and validation of privacy requirements. Inf. (Switz.) **7**(2), 28 (2016). https://doi.org/10.3390/info7020028

23. Meis, R., Heisel, M.: Pattern-based representation of privacy enhancing technologies as early aspects. In: Lopez, J., Fischer-Hübner, S., Lambrinoudakis, C. (eds.) TrustBus 2017. LNCS, vol. 10442, pp. 49–65. Springer, Cham (2017). https://doi.org/10.1007/978-3-319-64483-7_4

24. Miri, M., Foomany, F.H., Mohammed, N.: Complying with GDPR: an agile case study. ISACA J. **2** (2018)

25. Notario, N., et al.: PRIPARE: integrating privacy best practices into a privacy engineering methodology. In: 2015 IEEE Security and Privacy Workshops, pp. 151–158. IEEE (2015)
26. Pearson, S., Shen, Y.: Context-aware privacy design pattern selection. In: Katsikas, S., Lopez, J., Soriano, M. (eds.) TrustBus 2010. LNCS, vol. 6264, pp. 69–80. Springer, Heidelberg (2010). https://doi.org/10.1007/978-3-642-15152-1_7
27. Pfitzmann, A., Hansen, M.: A terminology for talking about privacy by data minimization: anonymity, unlinkability, undetectability, unobservability, pseudonymity, and identity management (2010)
28. Qin, T., Liu, T.: Introducing LETOR 4.0 datasets. CoRR abs/1306.2597 (2013). http://arxiv.org/1306.2597
29. Rauf, I., et al.: The case for adaptive security interventions. ACM Trans. Softw. Eng. Methodol. (TOSEM) **31**(1), 1–52 (2021)
30. Reimers, N., Gurevych, I.: Sentence-BERT: sentence embeddings using Siamese BERT-networks. In: Proceedings of the 2019 Conference on Empirical Methods in Natural Language Processing. Association for Computational Linguistics (2019). arxiv.org/abs/1908.10084
31. Sielis, G.A., Tzanavari, A., Papadopoulos, G.A.: ArchReco: a software tool to assist software design based on context aware recommendations of design patterns. J. Softw. Eng. Res. Dev. **5**, 1–36 (2017)
32. Spiekermann, S., Cranor, L.F.: Engineering privacy. IEEE Trans. Softw. Eng. **35**(1), 67–82 (2008)
33. Wei, J., Zou, K.: EDA: easy data augmentation techniques for boosting performance on text classification tasks. In: Proceedings of the 2019 Conference on Empirical Methods in Natural Language Processing and the 9th International Joint Conference on Natural Language Processing (EMNLP-IJCNLP), pp. 6382–6388. Association for Computational Linguistics, Hong Kong (2019)
34. Wohlin, C., Runeson, P., Höst, M., Ohlsson, M.C., Regnell, B., Wesslén, A.: Experimentation in software engineering (2012)

A New Usability Inspection Method: Experience-Based Analysis

Anu Piirisild[1]([✉]) [iD], Ana Perandrés Gómez[2] [iD], and Kuldar Taveter[1] [iD]

[1] Institute of Computer Science, University of Tartu, Tartu, Estonia
{anu.piirisild,kuldar.taveter}@ut.ee
[2] Ageing Lab Foundation, Jaén, Spain
ana.perandres@ageinglab.org

Abstract. [**Context and motivation**] For software product development, user-based evaluation is often applied to detect usability issues. User-based testing is not cost-effective at every stage of the project, therefore for the early stage of the development, Usability Inspection Methods (UIM) are widely used where the evaluation is performed by experts. For the later phase, Summative Usability Testing (SUT) methods with end user involvement are used. However, UIMs have some drawbacks that make them complicated to use: finding good experts can be costly, the scope is too limited, and a lack of methods that allow for usability analysis before prototyping. This article introduces the Experience-based Analysis method (EbA) as an UIM that considers those drawbacks. [**Question/problem**] If we can obtain a considerable amount of usability information with EbA, we can use EbA as an effective UIM method. Therefore, the aim of the study is to identify to what extent the EbA can provide the same usability findings that is obtained by testing with users by SUT. [**Results**] An evaluation was done on a case study, where the information obtained by both EbA and SUT was analyzed by Content Analysis method. The results showed that most of the usability findings by SUT report were also identified by EbA. EbA method resulted in quantitatively more findings than SUT. [**Contribution**] The results lead us to recommend using EbA before SUT. As future work, we first plan to find whether the method gives any false positive or negative results. Second, we plan to conduct a survey to find the limitations of the method.

Keywords: Usability Inspection Method · Human behavior · Usability · Requirements engineering

1 Introduction

For software product development there is a acknowledged need for early phase usability analysis [4,12]. Making changes in the early stages of development is less costly than in later stages [12]. Usability is defined in ISO standard as the "extent to which a system, product or service can be used by specified

D. Mendez and A. Moreira (Eds.): REFSQ 2024, LNCS 14588, pp. 74–91, 2024.
https://doi.org/10.1007/978-3-031-57327-9_5

users to achieve specified goals with effectiveness, efficiency and satisfaction in a specified context of use" [12]. User-based testing at every stage of the project is not cost-effective [4,12], therefore for the early stage of development Usability Inspection Methods (UIM) are often used where usability experts or professionals of other types give feedback on the digital or paper-based prototype and likely user interactions [4,8,15,19]. For the later phase of usability analysis, Summative Usability Testing (SUT) methods are used. SUT assumes that the development phase has been completed at a sufficient level to be evaluated [2]. In a SUT, instead of experts, the end users are involved in testing the prototype or finished product. Due to the cost and logistical challenges of recruiting test-users, a SUT can be costly [15]. Using a SUT can be made more cost-effective and bring greater value by applying an UIM beforehand [8,10,12,15]. UIMs can help to plan the SUT by focusing on predicted issues [15]. The aim of an UIM is to detect usability issues before a SUT to make the otherwise costly usability evaluation more accessible for software development projects [8,15,19].

There are several UIMs listed in Sect. 2, each of which identifies usability issues in a different way. However, there is still some dissatisfaction with UIMs that make them complicated to use. The main drawbacks can be generalized into three categories: 1) finding good experts can be challenging and costly [1,4,10]; 2) scope and therefore, the results are too limited and too much focused on design artifact [1,30,31]; 3) a lack of UIM methods that allow for early usability analysis before prototyping [7,26].

In this paper we introduce a new UIM named Experience-based Analysis (EbA) that addresses the drawbacks 1–3 outlined above. We also present results that demonstrate that the EbA method can obtain sufficient usability information as expected from an UIM, EbA can help to plan the SUT by discovering what usability information is missing, and take some workload off from a SUT.

EbA is intended to be used by the people who elicit the requirements. As soon as a new solution is described, requirements engineers can validate whether it would work as such or would it be reasonable to change the requirements before they are handed over for development. EbA is addressing the three drawbacks 1–3 of the UIMs listed above as follows:

1. Knowledge that would otherwise be obtained from experts is collected by the people who drafted the requirements. They are able to perform the inspection by the EbA method where the expected behaviours by the end users are verified to be present in situations that users have experienced in the past.
2. EbA helps to find extensive usability issues, since EbA is not limited by testing the prototype, and the analysis of the expectations enables analysis of the product under development more broadly.
3. EbA does not require creating any prototype for usability inspection since the EbA analysis is intended to be performed by the same person or team who described the requirements. Therefore there is no need to demonstrate the solution to someone else by means of the prototype.

The EbA method was created by the first author of this article in 2021 and was initially evaluated in three real life projects [22]. The results of the evaluation included by [22] showed that using EbA provides valuable usability information, which helps to plan further activities more effectively. In addition, evaluating the EbA in the three projects [22] showed that after getting acquainted with the idea of the product, the EbA analysis would take only 1–2 working days depending on the extent of the planned product and the level of detail of the analysis. In this paper, we introduce the EbA method to the scientific community.

If we want to suggest requirements engineers to use the EbA to learn about usability issues in an early stage of software product development and we presume that the EbA is able to identify usability information that is commonly collected with a SUT, we need to find out to what extent the EbA can obtain the same usability information as a SUT. For this purpose, we formulate the first research question as follows:

RQ1: How does the usability information obtained with the EbA compare to the usability information obtained with a SUT?

Since we know from [22] that one can collect with the EbA a large amount of usability information and we want to understand the quality of such information, we formulate the second research question as follows:

RQ2: To what extent is the usability information obtained by EbA valuable for product owners?

By product owners we mean people who are responsible that the solution under development will work in an efficient way.

To answer RQ1, usability information obtained by both the EbA and a SUT from a complex case study on technologies for older adults was analyzed by the Content Analysis method [17] and the results of the analysis were then compared to each other. As a result, a considerable amount of overlapping usability findings were identified with the two methods. To answer RQ2, two people from the team of the same complex case study evaluated the information obtained by the EbA. They rated 53% of the findings identified by the EbA as valuable. The results suggest to apply the Experience-based Analysis (EbA) method before starting with prototyping and testing with end users to increase the efficiency of development.

In this paper, we did not compare the EbA with the other UIM methods because this falls outside of the scope of this study. This paper focuses on discovering whether the EbA enables detection of a considerable amount of usability information before testing with end users by a SUT method, and if the EbA is able to identify usability information that is commonly collected with a SUT.

The rest of this paper is organized as follows. Section 2 provides an overview of UIMs and results of similar evaluation studies comparing an UIM and a SUT with each other. Section 3 describes the EbA method and its usage. Section 4 provides a short overview of the case study and describes the validation process. Section 5 reports the results of the validation. Section 6 discusses the results. Finally, Sect. 7 concludes the paper and outlines the future work.

2   Background and Related Work

In the relevant research literature, the most frequently used UIM methods are Heuristic Evaluation (HE), Cognitive Walkthrough (CW), and Task Analysis (TA). In HE, three to six usability experts evaluate user interface with a heuristic technique [4,8,15]. In CW, experts walk through the application to detect whether the user interface directs users to act in the expected way [4,15]. The purpose of the TA method is to understand the work done by users and how the emerging product would affect the users [15] and to learn how people would actually perform specific types of tasks [4]. Two other UIMs are Pluralistic Walkthrough (PW) and Feature Inspection (FI). In PW, scenarios are reviewed and discussed with end users [21]. FI is assessing a proposed feature set for understandability, usefulness, and availability for the user [21]. Newer UIM methods, which can be used e.g. for the development of web applications, can be summarized as the methods that perform usability inspection and problem identification for a given task, aspect or question, such as identifying problematic and confusing links [26].

A number of authors have mentioned the following drawbacks of the existing UIMs: 1) involving experienced experts can be challenging and costly [1,4,10]; 2) evaluators need to have a very good knowledge about the intended product and its users, otherwise the inspections can fail [15]; 3) scope and therefore, the results are too limited [1,30,31]; 4) too much focused on design artifact and user interface interaction instead of covering wider context of use [31]; 5) the TA method could be too time-consuming and skill dependent [4]; 6) the HE method has a reliability problem [4]; 7) the CW method has too much documentation and limits the problem domain [1]; 8) interpreting the notion of usability too narrowly [7,31]; 9) most of the methods focus on the design and implementation phase, which is a later phase of the software development process [26].

In addition to UIMs, user-based testing with early-stage prototypes can also be performed. However, user-based testing at every stage of the project is not always cost-effective or practical [12].

In the literature, studies can be found that have compared with each other the UIM and SUT methods and have identified their overlaps. Khajouei and Farahani [16] combined HE as an UIM method and Think Aloud (TA) as a SUT method to see the potential of identifing usability issues by jointly applying both methods. Resulting from the study, by HE they identified more issues related to user satisfaction, learnability, and error prevention, whereas by the SUT TA method they detected more issues concerned with effectiveness and efficiency. Also, the SUT TA method identified problems with user interaction while HE did not detect these problems. Both methods identified 25% of the same usability issues, and HE and the SUT TA method separately identified 39% and 36% of the usability issues, respectively. Tan et al. [29] likewise compared with each other HE and a SUT method. Their conclusion was that both methods discovered 10% of the same issues, while separately HE covered 60% and SUT 30% of the issues.

3   Experience-Based Analysis (EbA) as a Usability Inspection Method: Overview

The EbA method has been created with the purpose of providing requirements engineers with an early indication of whether the solution will function efficiently and identifying potential usability risks across various aspects.

The EbA method was created by applying the Action Design Research method [27] which allows simultaneously to find a solution to a problem and create a tool facilitating the solution of the problem. The preliminary EbA-tool is free to use and can be found online at [23].

The EbA method is based on the premise that people behave according to experiential patterns: if they have behaved in a certain situation and context in a particular way, they are likely to behave in a similar way in other comparable situations [5,6,9,20].

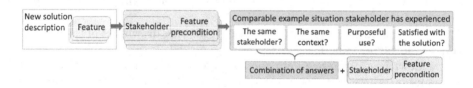

Fig. 1. Experience-based Analysis (EbA) framework [22,23]

The components of the EbA framework are depicted in Fig. 1. All the underlined terms in the following text refer to the terms used in the EbA framework. The process of applying EbA consists of the following steps [22,23]:

Step 1. The description of the planned solution is divided into features, where each feature represents one or more logically related solution capabilities that provide value to a user and are described by a set of functional requirements [32]. Depending on the peculiarity of the technology to be developed or implemented, or elaboration and the structure of the analysis, these solution descriptions can be statements about what the users can do, user stories, use cases, scenarios, functional goals of goal models, process descriptions or other similar models that describe the planned solution and the intended kinds of users. Hereafter and in Fig. 1, these solution descriptions are termed as features.

Example. Let us assume that the platform of an e-service needs to be complemented so that the customers of the e-service could upload files onto the platform instead of sending them separately by e-mail. The product owner of the e-service expresses his opinion as follows: "The platform does not need to limit the format of the files customers upload, because anyway the customers use common software packages. However, the file to be uploaded should be limited to one file and the size of the file can be up to 10 MB." As was stated above, in EbA this kind of requirement is viewed as a feature, which in this case can be formulated as follows: The customers of the e-service can upload one file with no format specified, but a size limit of 10 MB.

Step 2. For each feature it is decided who are the stakeholders involved in the feature or significantly affected by the feature, and what are the essential knowledge, attitudes, skills, awareness, emotions [13], means and activities expected from them within this specific feature, to make this feature efficient. In EbA, such expectations are collected by brainstorming and termed as feature preconditions.

Example. The preconditions for the feature in our running example could be as follows: 1) recipients of the files use the same software package as the customers; 2) the customers are using common software packages; 3) recipients of the files do not need to process the information in the file; 4) the customers can compress files without losing quality of images; 5) the customers know how to include all the information in one file; 6) the customers do not mind including all the information in one file.

Step 3. By using a brainstorming technique, for each feature precondition, the closest comparable example situation should be identified that the stakeholder has experienced in the past and where the same behaviour of the stakeholder appears.

Example. In the running example, each of the six preconditions should be assigned a comparable example situation. For example, precondition 1 is concerned with using software packages, so that the example situation could be: "the recipients of the files have previously opened the attachments sent by the customers via e-mail".

Step 4. It is assessed to what extent the feature precondition and example situation are comparable according to the stakeholders and context, and whether the behavior of the stakeholder was purposeful and what was the satisfaction level of the stakeholder in the comparable example situation. For step 4, the information can be obtained from various sources: published research results and statistics, research articles, opinions left on social media [14], previously collected user behavior and feedback from service providers, information synthesized by the analyst based on public information, and observations in public space. Step 4 is illustrated in Fig. 2 by the frame A and consists of the following substeps:

Step 4.1. The same stakeholder? Is the stakeholder of the feature precondition the same as the stakeholder in the comparable example situation?

Step 4.2. The same context? Is the context of the feature precondition the same as the context in the comparable example situation? The context of usage "comprises a combination of users, goals, tasks, resources, and the technical, physical and social, cultural and organizational environments in which a system, product or service is used" [11] and also, for example, time and purpose [22], and if the usage is by novice or experienced users [33]. It is up to the analyst to decide which elements make up the relevant context for the feature preconditions.

Step 4.3. Purposeful usage? How purposeful was the usage of a solution by the stakeholder in the comparable example situation?

Step 4.4. Satisfied with the solution? Was the stakeholder satisfied with the solution in the comparable example situation?

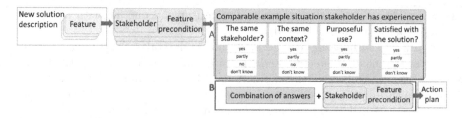

Fig. 2. The EbA framework with the questions and answer options (Frame A) and a combination of the answers (Frame B)

Example. In our example, substeps 4.1–4.4 are applied to the feature precondition number 1 as follows: 4.1.) "Yes", the stakeholder is the same one who previously opened files sent by e-mail. 4.2.) "Yes", the context is the same, because the same kinds of files were created and sent by the customers. 4.3.) It was a "partly" purposeful usage because the recipients of the files claimed they used the Windows operating system, but they occasionally also received files from customers using the IoS operating system that would not open. 4.4.) The answer to the question about satisfaction of the recipients of the files is "no", because the files that would not open caused them additional work.

Step 5. Based on the answers to the questions asked at steps 4.1–4.4, EbA yields a combined answer as a sentence that is composed of designated words and phrases[1]. Further, combining this answer with the <u>Stakeholder</u> and <u>Feature precondition</u> from step 2, the combination presents a prediction about the usability of the feature on the aspect of the precondition under the analysed circumstances. The combination is illustrated in Fig. 2 by the frame B. For example, if the evaluation of the similar situation at steps 4.1–4.4 yields the result "yes-partly-yes-no", the combined components will be "There is dissatisfaction with this *(the core of the feature precondition)*, but in a slightly different context + Stakeholder + Feature precondition". The Stakeholder and the Feature precondition are the same as what we had in step 2. Based on the findings, the analyst can create an action plan. In particular, the analyst can decide whether the stakeholders would have the expected means and would behave in the way expected by the product designers, or if they are not, how could this be improved?

Example. In our example, after performing steps 1–4, our finding by the EbA method can be expressed as follows: "There is dissatisfaction with this *(using the same software)*, but to an extent, + recipients of the files + are using the same software as clients". After the remaining five preconditions of this feature have been analyzed, in context of the feature and the aim of the product, the analyst can decide how this dissatisfaction aspect of the solution could be improved. The example is illustrated by Fig. 3. A larger scale fictional example, where a solution has been analyzed by EbA, can be accessed online [23].

---

[1] The full list of all 256 versions of answers can be accessed online [23].

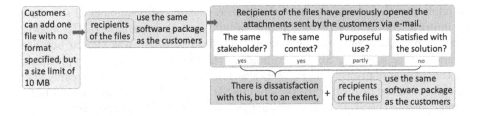

Fig. 3. An example analysed by the EbA framework

The benefits of the EbA method have been described in Sect. 1. The EbA method has the following limitations [22]:

1. You get an answer to what you ask, both in terms of content and scope;
2. Analysing the usability of features and creating an action plan depends on the skills of the person who is analysing the product under development;
3. The analysis of the usability of features depends on the comparable example situation identified and the ability to adequately evaluate it.

4   Validation

Answering RQ1 leads us to know to what extent the usability information that is obtained with a SUT method could have been discovered already with the EbA method, we compared the usability findings yielded by the SUT and EbA methods on the same complex digital service ("case study" hereafter). For the comparison of both methods, we used the SUT and EbA reports of the case study. The SUT was performed by the team of the University of Jaén [25]. The EbA was performed by the first and second author of this paper. The information obtained by either method was analyzed by the Content Analysis (CA) method [17] and then compared to each other. We chose the CA method because CA enables to determine the presence of certain themes within qualitative data. The first author of this paper defined the relevant categories, applied the CA method to identify usability findings, analyzed the discovered information and compared the findings yielded by the SUT and EbA.

Answering RQ2 leads us to know whether the usability findings obtained by EbA are valuable for product owners. The value of the findings was assessed by two stakeholders of the digital service. One of the evaluators, who is also the second author of this paper, is the product owner of the digital service of the case study.

4.1   The Case Study

The case study was chosen from the industry-oriented research and development project Pharaon[2]. The Pharaon project consists of five large-scale case studies

---

[2] https://www.pharaon.eu/.

or pilots the aim of which is to develop an ecosystem of technological solutions to improve the quality of life of older adults and their caregivers. To answer the research questions, the pilot from Andalusia in Spain was chosen because in addition to applying SUT for validation they were also ready to apply EbA. In the Pharaon project, digital services for older adults and their caregivers integrated in the project were evaluated by SUT. The EbA method is meant to be applied before the SUT. Differently, in this validation these methods were applied in the reverse order because the author of the EbA method joined the Pharaon project after the SUT had already been completed. However, we designed the activities of the validation process in such a way that the credibility of the results would be ensured. These activities are described in Sect. 6.1.

4.2   Data Collection

Usability Analysis by SUT. The SUT was performed by the team of the University of Jaén [25]. The aim of the SUT was to highlight bugs, potential bugs, and assess the usability. The team collected feedback about the usage of the digital services during test-sessions in real-life situations according to the requirements represented as use case scenarios (UCS) [28]. Altogether 25 test-scenarios were executed, involving four different technologies provided on one platform. To assess the usability feedback by the end-users, the System Usability Scale (SUS) [3] and After-Scenario Questionnaire (ASQ) [18] interview were applied [25].

Usability Analysis by EbA. The first author of this paper familiarized herself with the Pharaon project by reading the requirements documents of the project. Based on the information acquired from the documents, she considered UCSs as *features* that were used for representing requirements in the Pharaon project. When applying EbA, the first author was not aware of the results of the validation testing by SUT. After she described preconditions and example situations by following the principles of the EbA method as is described in Sect. 3, the second author of this article joined who, being the product owner of the analyzed pilot, knew the local context and services under development. To obtain a sufficient number of usability findings for comparison, under the guidance by the second author, seven features were extracted for comparative analysis that had sufficient amount of usability-related information obtained by SUT. The first author identified 48 feature preconditions for the seven features [24]. These feature preconditions were analyzed together by the first and second author by means of EbA. It is important to emphasize here that neither of the authors is a usability expert by profession. The second author as the project owner was aware of the test report results but she did not use this knowledge when analyzing the feature preconditions. Based on the outcome, the first author of this paper devised an action plan describing which features look fine and what design decisions should be made in the future, which parts of features should be further tested, and which support activities should be planned. In total, each of the 48 preconditions analyzed by EbA led to one usability finding.

4.3   Data Analysis

Before starting the process of data analysis, the relevant data categories for the analysis were defined. The categories and rationale for the selection are explained below for each step of the analysis. The data analysis involved at first extracting the relevant information from the SUT report, and then categorizing the usability findings obtained by SUT and EbA into the appropriate categories according to the CA method [17]. After that, the categorized information obtained by both methods was compared. The process consisted of the following steps:

1. Extracting usability findings from the SUT report representing the seven UCSs considered.

2. Categorizing the usability findings from applying the SUT and EbA into the categories "usability-related issues" or "bugs", depending whether a finding was concerned with an extensive usability issue ("usability" hereafter), or with a technical issue affecting usability ("bug" hereafter). This categorization is necessary for two reasons: (i) to exclude technical issues from the analysis because identifying technical issues is not the purpose of an UIM; (ii) to know how well suited EbA is also for discovering technical issues.

3. Identifying overlaps between the usability findings by SUT and EbA. To find out to what extent EbA can obtain the same information that is otherwise obtained by SUT ("overlap" hereafter), we should first define what do we mean by "the same information". We evaluate this based on two indicators: object and content. The first indicator expresses whether the overlapping findings by both methods address the same object. The object is something that the finding is focused at, e.g. action or feeling. The second indicator expresses whether the content of the conclusions drawn by EbA matches the content of the conclusions drawn by SUT. For example, if a finding by either method identifies that older adults can get stressed because of the fear of not coping with using the technology, the object of both methods is "getting stressed". At the same time, the content of the conclusions by both methods is also the same: "older adults can get stressed". This means that in the given example there was a "full overlap" between the findings by object and content. Identifying the overlaps by object and content allows us to be more specific as to whether the EbA method leads to focusing on the same issues that are otherwise discovered by SUT and does EbA lead to the same conclusions than SUT?

4. Categorizing the usability findings and action plan suggestions from applying SUT and EbA into the categories "technology-oriented" and "non technology-oriented". "Non technology-oriented" findings allow the discovery of usability issues of a product in a broader sense.

5. Categorizing the usability findings by SUT and EbA into the category "need for action" based on what action plan was decided as a conclusion. The motivation for such categorization is preventing misinterpretation of the results, since UIM and SUT reports traditionally focus on discovering usability problems. However, EbA also yields a significant amount of information as to whether the feature precondition is correct, i.e. "it works". If the EbA analysis indicates that the feature precondition is met, it increases confidence to continue with the development of the given feature. The remaining categories enable understanding of the advantages of knowing findings from applying EbA before findings from applying SUT. If a conclusion from applying EbA states that "the technology needs to be complemented" or "the solution does not work this way which means that a different solution should be worked out", early reacting to the situation will be enabled which may avoid re-doing the development process. If "the feature needs a support action" to enables its full value for the product, support actions and resources can be planned at an early stage of the development process. If a finding identifies a critical issue on which we do not have enough information, the issue "needs to be further analyzed or tested" by SUT. This information helps us to plan the test scenarios for the SUT.

6. Categorizing EbA findings into the category "value". The assessment by "value" clarifies whether the usability findings obtained by EbA are valuable for the product owner and other stakeholders. At this step the product owner of the Andalusian pilot, who is also the second author of this paper, along with another member of her team checked all 48 findings identified by EbA and determined their value for the team. They evaluated the usability findings by EbA simultaneously with evaluating the action plan pertaining to these findings proposed by the first author of this paper based on the results of applying EbA. Based on that, all 48 findings by EbA were categorized by "value" into one of the following categories: 1) Important, dealing with it; 2) Important, but cannot deal with it now; 3) Good to know; 4) Irrelevant knowledge; 5) We knew that anyway. The first three categories indicate valuable information, while the remaining two categories indicate the information that was irrelevant or already known.

The original data of the case study can be made public to a limited extent [24], due to the conditions set by the project contract. The analysis of seven features of the case study is available online [24]. The example includes 48 preconditions and comparable example situations. The example also includes categorizations and value evaluations of the findings and suggestions for the action plan.

Based on the information obtained by the data analysis, the research questions RQ1 and RQ2 are answered in Sects. 5 and 6.

5   Results

With respect to the research question RQ1, we found that most of the usability information that is obtained with the SUT method can be discovered also with the EbA method. As is reflected by Fig. 4, from among the 14 findings by SUT, 11 were also identified by EbA, out of which five represented the same object and six partially the same object. The overlap by content between SUT and EbA was similar: five of them drew the same and six partially the same conclusions. There were no situations where the conclusions reached by SUT claimed something totally different from the conclusions reached by EbA.

We also found that in comparison with SUT, the EbA method resulted in quantitatively more and broader usability information, but did not identify any "bugs". In particular, while SUT yielded 14 findings in the "usability" category and 8 findings in the category "bugs", applying the EbA method

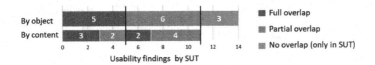

Fig. 4. Overlap of findings by object and content

resulted in 48 findings in the "usability" category and no findings in the category "bug". The "non technology-oriented" aspects were examined as an indicator of broader usability information. The analysis of feature preconditions with EbA provided more information on "non technology-oriented" aspects than "technology-oriented" aspects (67% vs 33%), testing with SUT was the opposite (44% vs 56%). The findings on overlaps belonged to both aspects. The overlap between "technology-oriented" and "non technology-oriented" feature categories is represented in Fig. 5.

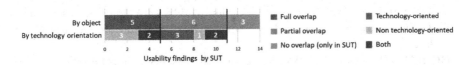

Fig. 5. Overlap of findings in the "technology-oriented" and "non technology-oriented" categories

Both methods provided findings showing what does work in the product, what aspects do not work, and what needs additional support action or change in technology. Figure 6 reflects the balance of the results between the two methods.

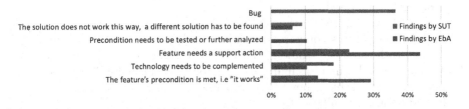

Fig. 6. Percentage distribution of findings in the category "need for action"

With respect to the research question RQ2, we found that more than half of the usability information gathered by EbA – 53% – was evaluated by the product owner and her colleague as valuable, which is also reflected by Fig. 7.

Is valuable 53%			Is not valuable 47%	
Important, dealing with it	Important, but can't deal with it now	Good to know	Irrelevant knowledge	We knew that anyway
14	2	9	6	16

Findings

Fig. 7. Categorization by value of the findings by EbA

6 Discussion

Overall, we can conclude from our study that the EbA method is capable of providing useful usability information in an early phase of development. The information includes findings that are otherwise obtained with a SUT method.

In summary, we answer the research questions (RQs) that were posed in Sect. 1 as follows:

- **RQ1:** The usability information obtained by EbA largely overlaps with the usability information obtained by SUT. In our case study, from among the 14 usability findings by SUT, 11 were also identified by EbA. In comparison with SUT, the EbA method resulted in quantitatively more and broader usability findings.
- **RQ2:** Despite its quantitative abundance, the usability information obtained by EbA can be considered to a large extent as valuable for product owners. More than half of the information about the user behavior gathered by EbA – 53% – was evaluated by the product owner and her colleague as valuable.

According to the research literature on UIMs, UIMs offer a greater value if they are used as complementary methods to SUT [10,12,15,16,29]. This statement also holds in our study, where the analysis by EbA helped discover what usability information was missing, so that the requirements engineering team could select in advance which usability cases had to be tested additionally by SUT or further analyzed.

In Sect. 2, we referred to the two studies where Heuristic Evaluation (HE) as an UIM method was compared to a SUT method [16,29]. When comparing our results with those two studies, it must be pointed out that compared to HE, EbA discovered a lot more usability information that is usually discovered only by HE as an UIM. Table 1 shows the comparison of the validation results by HE with the validation results by EbA.

Table 1. Validation results by HE compared to validation results by EbA

	Only in UIM	Overlap	Only in SUT
UIM HE [16]	39%	25%	36%
UIM HE [29]	60%	10%	30%
UIM **EbA**	72%	22%	6%

According to the research literature, UIMs can detect more general usability problems than SUT [8,16,29]. In our case study, EbA provided quantitatively more information than SUT. This is presumably due to the nature of either method. In particular, SUT involves specific test scenarios to be performed by the end users testing the service, which may not cover the entire scope of the features included by the scenarios. On the other hand, by presenting all the features of the solution chosen by the analyst, EbA directs the analyst to thinking about each feature separately as a whole with respect to its behavioral preconditions. This explains a larger number of usability findings by EbA compared with SUT. Another important aspect is that while SUT can test the features that exist in the prototype at the given moment, EbA provides an opportunity to analyze also those features that have not yet been included in the prototype. In addition, while SUT by definition considers the technology to be used, EbA does not consider it. Therefore, EbA can be used for obtaining usability information also in cases where technological solutions have not yet been fully decided.

The analysis by EbA takes 1–2 working days per project depending on the volume of the planned product and the level of detail of the analysis [22]. Practitioners can decide whether spending 1–2 working days on EbA can avoid some costs that are otherwise spent on prototyping and testing with end-users.

6.1   Threats to Validity

A threat to internal validity is concerned with the scope of the study. Seven use case scenarios (UCS) from one case study were analysed, which is sufficient to draw initial conclusions, but certainly the scope needs to be expanded by subsequent studies.

Another threat to internal validity is concerned with the person who was involved in the validation process. In this study, the same person who invented the EbA method defined the relevant categories, applied the Content Analysis (CA) method to identify usability findings, analyzed the discovered information, and compared the findings yielded by SUT and EbA. The quality of the CA and comparative analysis was ensured by two follow-up data checks after two months' intervals. Next studies need to involve more people for quality control. The same threat to internal validity also comprises how much usability findings this one person was able to discover from the SUT report. Results from SUT were included if the SUT report contained comments explaining the strengths and shortcomings of the tested UCSs. The SUT report of the case study also includes the results from conducting the SUS and ASQ questionnaires. However, these results were difficult to incorporate in analysing usability aspects because they yielded numerical values which are hard to employ for concluding which usability aspects should be improved. For excample, if the SUT report includes a test scenario "Make a videocall" and according to the results of conducting the ASQ questionnaire, the evaluation of this functionality is 5.8 without any further comments on what was missing or what was positive, that information was too general to apply for identifying any usability issues.

The third threat to internal validity was that the SUT results of the case study were presented before the EbA analysis was made. To prevent the risk of using the SUT test report results for evaluating the case study features with EbA, we used the interview format where the first author of this paper, who was not aware of the SUT results, asked questions in a directed manner so that finding answers for validation would be outside of the scope of the test-report. She also made sure to re-rail the discussion when the thought of the respondents seemed to drift to the test-report.

Knowing the SUT results can cause one more threat to internal validity as the value of the findings from applying EbA were evaluated by two people from the Andalusian pilot project of the case study, who were aware of the results of the previously accomplished validation by SUT. However, this did not decrease an advantage of applying EbA, because according to the evaluators, if the EbA analysis and evaluation of the findings had been completed before rather than after SUT, several "we knew that anyway" findings would rather have appeared as "important, dealing with it" findings. This would have increased rather than decreased the share of the "valuable" findings by EbA, which are the findings belonging to the first three categories represented in Fig. 7.

A threat to external validity is the generalizability of our results, considering the particular case study from the Pharaon project that we chose for the validation. We will mitigate this threat by applying the EbA method and conducting the same kind of comparative analysis for other digital products and services.

7   Conclusions

This paper introduced a new UIM that addresses the following drawbacks of the existing UIMs: (i) finding experts can be complicated; (ii) the scope is too limited; (iii) there is the lack of UIM techniques that would allow for early usability analysis before prototyping. This paper also presented the results of the evaluation where the EbA method was compared with a classic SUT method with the aim to find out to what extent EbA yields the same findings that is obtained by SUT and whether this information is valuable for the product owner.

Based on the results of the validation, we can conclude that EbA is providing extensive usability information already in an early phase of requirements engineering when prototyping and SUT have not yet been planned. Despite a large number of findings, slightly more than a half of the usability findings obtained by the EbA method were considered as valuable by the two evaluators from the Andalusian pilot.

With EbA, a considerable number of the same usability findings were identified as with SUT and the usability information obtained by both methods partially or fully overlapped. Therefore, we can conclude that EbA can reduce the workload of the usability testing by SUT by detecting usability issues before SUT, and thus EbA can make the development process more cost-effective. Consequently, EbA should be considered as a complementary method to SUT, because EbA cannot detect all of the usability issues.

Important areas for our future work are the following ones: 1) To investigate if the EbA method yields false positive or false negative results. 2) To perform a larger scale validation of EbA by applying it to different types of digital products and services and by various stakeholders.

Acknowledgements. The research work reported in this paper has received funding from the Pilots for Healthy and Active Ageing (Pharaon) project of the European Union's Horizon 2020 research and innovation programme under the grant agreement no. 857188. The authors are expressing their gratitude to María Parraga Vico from the Ageing Lab Foundation and the University of Jaen from Spain.

References

1. Aamir, M.J., Mansoor, A.: Testing web application from usability perspective. In: 2013 3rd IEEE International Conference on Computer, Control and Communication (IC4), pp. 1–7. IEEE (2013). https://doi.org/10.1109/IC4.2013.6653765
2. Bill, A., Tom, T.: Measuring the User Experience: Collecting, Analyzing, and Presenting Usability Metrics. Interactive Technologies, Morgan Kaufmann, Massachusetts (2013)
3. Brooke, J.: SUS: a retrospective. J. Usability Stud. **8**(2), 29–40 (2013)
4. Cheng, L.C., Mustafa, M.: A reference to usability inspection methods. In: Hassan, O., Abidin, S., Legino, R., Anwar, R., Kamaruzaman, M. (eds.) International Colloquium of Art and Design Education Research (i-CADER 2014), pp. 407–419. Springer, Singapore (2015). https://doi.org/10.1007/978-981-287-332-3_43
5. Clark, A.: Whatever next? Predictive brains, situated agents, and the future of cognitive science. Behav. Brain Sci. **36**(3), 181–204 (2013)
6. Fazio, R.H., Zanna, M.P.: Direct experience and attitude-behavior consistency. In: Advances in Experimental Social Psychology, vol. 14, pp. 161–202. Elsevier (1981). https://doi.org/10.1016/S0065-2601(08)60372-X
7. Fernandez, A., Insfrán, E., Abrahão, S.: Usability evaluation methods for the web: a systematic mapping study. Inf. Softw. Technol. **53**(8), 789–817 (2011). https://doi.org/10.1016/j.infsof.2011.02.007
8. Helander, M.G., Landauer, T.K., Prabhu, P.V. (eds.): Handbook of Human-computer Interaction. Elsevier, Amsterdam (1997)
9. Hohwy, J.: The self-evidencing brain. Noûs **50**(2), 259–285 (2016)
10. Hollingsed, T., Novick, D.G.: Usability inspection methods after 15 years of research and practice. In: Proceedings of the 25th Annual ACM International Conference, pp. 249–255 (2007). https://doi.org/10.1145/1297144.1297200
11. International Organization for Standardization: Ergonomics of human system interaction: Usability, definitions and concepts. Standard, International Organization for Standardization, Geneva, CH (2018)
12. International Organization for Standardization: Ergonomics of human-system interaction-part 210: Human-centred design for interactive systems. Standard, International Organization for Standardization, Geneva, CH (2019)

13. Iqbal, T., Anwar, H., Filzah, S., Gharib, M., Mooses, K., Taveter, K.: Emotions in requirements engineering: a systematic mapping study. In: 2023 IEEE/ACM 16th International Conference on CHASE, pp. 111–120. IEEE (2023). https://doi.org/10.1109/CHASE58964.2023.00020

14. Iqbal, T., Khan, M., Taveter, K., Seyff, N.: Mining reddit as a new source for software requirements. In: 2021 IEEE 29th International Requirements Engineering Conference (RE), pp. 128–138. IEEE (2021)

15. Jacko, J.A.: Human-Computer Interaction Handbook: Fundamentals, Evolving Technologies, and Emerging Applications, 3rd edn. CRC Press Inc., Boca Raton (2012)

16. Khajouei, R., Farahani, F.: A combination of two methods for evaluating the usability of a hospital information system. BMC Med. Inform. Decis. Mak. **20**(1), 1–10 (2020). https://doi.org/10.1186/s12911-020-1083-6

17. Krippendorff, K.: Content Analysis: An Introduction to Its Methodology. Sage Publications, Thousand Oaks (2018)

18. Lewis, J.R.: Psychometric evaluation of an after-scenario questionnaire for computer usability studies: the ASQ. SIGCHI Bull. **23**(1), 78–81 (1991). https://doi.org/10.1145/122672.122692

19. Mack, R., Nielsen, J.: Usability inspection methods: report on a workshop held at CHI'92, Monterey, CA, may 3–4, 1992. ACM SIGCHI Bull. **25**(1), 28–33 (1993)

20. Murray, K.B., Häubl, G.: Explaining cognitive lock-in: the role of skill-based habits of use in consumer choice. J. Consum. Res. **34**(1), 77–88 (2007). https://doi.org/10.1086/513048

21. Nielsen, J.: Summary of usability inspection methods (1994). https://www.nngroup.com/articles/summary-of-usability-inspection-methods. Accessed 1 Oct 2023

22. Piirisild, A.: Analogy-based technology effectiveness prediction model and tool. Master's thesis, University of Tartu (2021). https://hdl.handle.net/10062/72631

23. Piirisild, A.: Tool for the experience-based analysis (EBA) method (2024). https://doi.org/10.23673/re-453

24. Piirisild, A., Gómez, A.P.: Experience-based analysis method - the evaluation data (2024). https://doi.org/10.23673/re-454

25. Polo-Rodríguez, A., et al.: Challenges of ubiquitous and wearable solutions to address active ageing in the Andalusian community. J. Univ. Comput. Sci. **28**(11), 1221 (2022)

26. Rivero, L., Barreto, R., Conte, T.: Characterizing usability inspection methods through the analysis of a systematic mapping study extension. CLEI Electron. J. **16**(1), 12–12 (2013)

27. Sein, M.K., Henfridsson, O., Purao, S., Rossi, M., Lindgren, R.: Action design research. MIS Q., 37–56 (2011)

28. Sterling, L., Taveter, K.: The Art of Agent-Oriented Modeling. MIT Press, Cambridge (2009)

29. Tan, W., Liu, D., Bishu, R.: Web evaluation: heuristic evaluation vs. user testing. Int. J. Ind. Ergon. **39**(4), 621–627 (2009). https://doi.org/10.1016/j.ergon.2008.02.012

30. Tarkkanen, K., Harkke, V.: Scope of usability tests in is development. AIS Trans. Hum.-Comput. Interact. **11**(3), 136–156 (2019). https://doi.org/10.17705/1thci.00117

31. Tarkkanen, K., Harkke, V., Reijonen, P.: Testing the unknown-value of usability testing for complex professional systems development. In: Abascal, J., Barbosa, S., Fetter, M., Gross, T., Palanque, P., Winckler, M. (eds.) Human-Computer Interaction - INTERACT 2015. Lecture Notes in Computer Science(), vol. 9297, pp. 300–314. Springer, Cham (2015). https://doi.org/10.1007/978-3-319-22668-2_24

32. Wiegers, K.E., Beatty, J.: Software Requirements. Pearson Education, London (2013)

33. Wu, J., Du, H.: Toward a better understanding of behavioral intention and system usage constructs. Eur. J. Inf. Syst. **21**(6), 680–698 (2012). https://doi.org/10.1057/ejis.2012.15

Governance-Focused Classification of Security and Privacy Requirements from Obligations in Software Engineering Contracts

Preethu Rose Anish[1]([⊠]), Aparna Verma[2], Sivanthy Venkatesan[2], Logamurugan V.[2], and Smita Ghaisas[1]

[1] TCS Research, Pune, India
{preethu.rose,smita.ghaisas}@tcs.com
[2] TATA Consultancy Services, Mumbai, India
{aparna.verma,sivanthy.v,logamurugan.v}@tcs.com

Abstract. **[Context and Motivation]** Security and Privacy (SP) compliance is an important aspect of running businesses successfully. Compliance with SP requirements by Software Engineering (SE) vendors, both in terms of the systems they implement and the practices they follow while implementing, gives customers an assurance that their data is accessed, stored, and processed securely. Failure to comply on the other hand, can entail heavy fines, lawsuits, and may even lead to loss of business through prohibition of those software in corresponding jurisdictions. SE contracts are known to be a useful source for deriving software requirements. **[Question/problem]** Mining any kind of information from contracts is a daunting task given that contracts are large and complex documents employing Legalese. **[Principal ideas/results]** We employ an exploratory study to come up with a model for a governance-focused classification of the SP requirements present in SE contracts for governance. Next, we report experiments conducted with Recurrent Neural Networks and Transformer-based models to automate this classification. Experiments conducted on 960 SE contracts received from a large vendor organization indicate that T5 performs best for both SP identification and classification tasks. With T5, we obtained an average F1 score of 0.90 each for identification of SP requirements. For the governance-focused classification, we obtained an average F1 score of 0.81 for the Security class and 0.80 for the Privacy class. **[Contribution]** Through an exploratory study, we present a model for a governance-focused classification of the SP requirements present in SE contracts. We further automate the extraction and the governance-focused classification of SP requirements by conducting experiments using 960 real-life SE contracts received from a large vendor organization.

Keywords: security · privacy · requirements · contracts · classification

1 Introduction

Recent laws such as European Union's Global Data Protection Regulation (GDPR), 2018, India's Personal Data Protection Bill, 2019; and The California Consumer Privacy Act, 2018 are bringing in sweeping changes in the Security and Privacy (SP) policies that

© The Author(s), under exclusive license to Springer Nature Switzerland AG 2024
D. Mendez and A. Moreira (Eds.): REFSQ 2024, LNCS 14588, pp. 92–108, 2024.
https://doi.org/10.1007/978-3-031-57327-9_6

govern and regulate the collection, use, retention, disclosure, and disposal of personal information. Despite these stringent regulations in place, the rate of cyberattacks, data breaches and unauthorized use of personal data is growing exponentially [1]. According to a recent report from IBM and the Ponemon Institute [2], the global average cost of a data breach in 2023 was USD 4.45 million, a 15% increase over 3 years. Meta's WhatsApp was ordered to pay a 225-million-euro ($266 million) penalty for failing to be transparent about how it handled personal information [3]. In 2021, Facebook agreed to pay a £500,000 (about $643,000) fine to U.K.'s information commissioner's office for its role in the Cambridge Analytica scandal [4]. These and similar examples elsewhere provide compelling grounds to consider SP requirements as a highly important subset of SE requirements. This is especially relevant today since we continue to expose more and more data to software systems embedded in various social media platforms that require us to create our personal and professional profiles, subscriptions to various apps incorporated in wearable devices, and user interfaces of several forums that require and store personal information. Compliance to SP requirements while handling financial data, health-related information and other personally identifiable information is an important responsibility that must be shouldered by software engineers and is therefore a concern that needs to be addressed at all stages of the Software Development Life Cycle (SDLC) [5]. An early detection of SP requirements is a precursor to ensuring such a compliance.

Contracts are known to be a source of software requirements and are a vital component of Information Technology (IT) outsourcing [6]. According to [7], 60–80% of business transactions are governed by contractual agreements. Non-compliance with contractual obligations can cause financial as well as reputational damage. SE contracts can help in the identification of high-level software requirements, thereby aiding the requirements elicitation phase of SDLC [6]. The SE requirements that emerge from contracts are mandatory to be delivered since they form parts of contractual obligations. Our interactions with the contract governance team in a large vendor organization revealed that up to 20% of clauses in any SE contract can contain SP requirements. As large organizations handle numerous contracts, each with many clauses, fulfilling these SP requirements is a significant part of contractual commitments. Due to the obligatory nature of the SP requirements present in contracts, it is crucial that they are identified early on for the benefit of all the stakeholders involved in different stages of the SDLC. Contracts, authored by legal experts on the customer and vendor side, can be difficult for software engineers to fully understand, yet they are responsible for implementing the commitments made by signing authorities. Our goal is to ensure that the contractual obligations pertinent to SP requirements are met with certainty. To achieve this goal, we conducted an exploratory study to understand from a contracts governance perspective, the SP-specific details present in SE contracts. For this, we formulated the following Research Question (RQ): *What details should we grasp to ensure effective governance of obligatory SP requirements present in software engineering contracts?*

Contract governance is the oversight of the contract and the relationship, partnered with the ongoing understanding of compliance and risk [8]. It covers everything from drafting to monitoring a contract's lifecycle. It involves continuous tracking and reporting on the fulfilment of contractual obligations. Further, governance includes assigning ownership of meeting the obligations to the right people/departments in an organization.

Through effective contract governance, all stakeholders gain clarity on their respective rights and responsibilities. This systematic approach establishes comprehensive guidelines that foster successful contract management within a business organization. Ultimately, it serves to proactively mitigate disputes, minimize risks, and optimize the overall value derived from contracts. Identifying the software requirements present in contractual clauses is an important first step in gathering the obligatory requirements of the customer and subsequently ensuring compliance. Works on identifying SE requirements at a high level exist. In [8], the authors identified 14 types of SE requirements found in contracts. While all the requirements committed to in the contract need governance, given the criticality of SP [1–5], in this paper we focus on SP requirements. For the effective governance of contracts, it is not only enough to identify the SP requirements present in contracts, but to classify them in a way that they can be owned and governed by the right people/departments in an organization. For instance, Security related requirements need to be differentiated between *Risk Management* and *Information Management,* because they entail different action items from people/departments responsible for those respective functions. The classification should help ensure that such nuances of a clause are brought out for the purpose of governance.

We used a grounded theory-based approach to first analyze the SP requirements present in SE contracts. We then created a model for classification of the SP requirements from a contracts governance perspective (henceforth referred to as governance-focused classification). Since contracts are large documents containing complex and ambiguous text, a manual extraction and/or classification of any kind of information from contracts is an arduous and time-consuming task. To circumvent this, we automated the extraction and classification of SP requirements from contracts. We performed experiments using deep learning models namely Recurrent Neural Network (RNN), Bidirectional Encoder Representation for Transformers (BERT) [9] and Text-to-Text Transfer Transformer model (T5) [10]. For the frequently encountered issue of class-imbalance, we employed GPT-3 [11] on one of the SP sub-types that had relatively low instance count in our dataset. We observed a significant enhancement in F1 scores using GPT-3 for this sub-type. To the best of our knowledge, this is the first attempt towards a governance-focused classification of SP requirements from SE contracts.

The contributions of this paper are as follows: (1) Through an exploratory study, we create a model for a governance-focused classification of SP requirements present in SE contracts, (2) We automate the extraction and governance-focused classification of these SP requirements.

The rest of the paper is organized as follows: Sect. 2 presents related work. Section 3 presents details of the exploratory study that we carried out to create a governance-focused classification model to classify SP requirements into various sub-types. Section 4 is on automating the extraction and classification of SP requirements. Section 5 presents results and discussion. Section 6 presents the experiments we conducted to validate the approach on publicly available contracts dataset. Section 7 presents threats to validity. Section 8 concludes the paper and provides directions for future work. In Sect. 9, we discuss our stand on Data Availability.

2 Related Work

Automating the extraction of SP requirements from natural language (NL) requirement artifacts has long been a research problem in the SE paradigm. Weber-Jahnke and Onabajo [12], mined security requirements from regulatory documents written in NL into a readymade ontology. Jindal et al. [13], extracted descriptions of security requirements present in software requirement documents (PROMISE) and classified the security requirements into *Authentication, Authorization, Access Control, Cryptography-Encryption* and *Data Integrity*. We focus on obligatory SP requirements present in SE contracts and present a fine-grained governance-focused classification for them. We identified 7 sub-types of security requirements and 3 sub-types of privacy requirements. Xiao et al. [14], proposed an approach to extract Access Control Policies from NL documents. Breaux and Anton [15] presented an approach for extracting access rights and obligations from regulation documents. Islam et al. [16], proposed a framework which aligns the security concepts used in laws and regulations with the terminology used in requirements engineering. Janpitak and Sathitwiriyawong [17] developed an automated compliance checking system that extracts compliance requirements from regulatory documents. Munaiah et al. [18], proposed a domain-independent model for identifying security requirements from requirement specifications. Farkhani and Razzazi [19], categorized security requirements into functional and non-functional requirements. Casillo et al. [5], presented an approach to decrease privacy risks during agile software development by automatically detecting privacy-related information in the context of user story requirements. They established that the application of transfer learning allows for considerable improvement in the accuracy of the predictions. Sainani et al. [6] established SE contracts as a potential source for mining SE requirements and identified 14 types of requirements from SE contracts, out of which two are Security and Privacy requirements. However, they did not sub-classify the SP requirements for governance. Jain et al. [20] proposed an approach for automatically identifying obligatory security and privacy requirements from SE contracts. However, just like [6], they did not present a sub-classification of the identified SP requirements that is crucial from governance perspective. Sharifi et al. [29] proposed a formal high-level specification language for contracts, called Symboleo, where contracts consist of collections of obligations and powers that define the legal contract's compliant executions. However, they did not include SP requirements and their sub-classification that is crucial from governance perspective.

We extract SP requirements from SE contracts and present a fine-grained governance-focused sub-classification for them. Further, we use state of the art classification techniques to automate this classification. To the best of our knowledge, there is no reported work on sub-classification of SP requirements from contracts.

3 Exploratory Study to Create a Classification Model for SP Requirements

In this section, we present the dataset details and the exploratory study we conducted to identify sub-types of SP requirements.

3.1 Dataset Details

Our dataset is comprised of 960 SE contracts received from the contract governance team in a large vendor organization. The vendor organization is a multinational IT services and consulting company that has been partnering with many of the world's largest businesses (henceforth referred to as the vendor organization). These 960 contracts were carefully chosen by the governance team of the vendor organization so that there is a fair representation of software project contracts spread across multiple business units, geographies, service lines, Master Service Agreements (MSAs), products and platforms. The dataset included SE contracts from nine application domains (healthcare, automotive, finance, banking, pharmaceuticals, telecom, utility, clothing-retail, and supermarket) with customers from ten different countries (Netherlands, United Kingdom, Australia, Germany, United States of America, Finland, Spain, India, United Arab Emirates and Sweden). There were 94,600 sentences in these 960 contracts with each contract document having a page count between 100 and 500 pages.

3.2 Exploratory Study Method

Three authors of this paper (henceforth referred to as participants) who have experience working with the legal and governance team within the vendor organization conducted the exploratory study. The exploratory study consisted of two parts. In part 1, the participants analyzed the 94,600 contractual sentences to identify SP requirements. In part 2, the participants further analyzed the identified SP requirements to identify governance-focused sub-types present in them.

Part 1 of the Exploratory Study. The participants manually analyzed the 94,600 sentences to identify SP requirements. Out of 94,600 sentences, 17,126 (~18%) sentences were identified as SP requirements - of which 11,731 belonged to Security and 5395 belonged to Privacy. This activity took 5 person days.

Part 2 of the Exploratory Study. To identify the sub-types of SP requirements present in SE contracts, the participants conducted an exploratory study by following the guidelines proposed in Hoda's Grounded Theory (GT) for SE [21]. GT is a research method that focuses on generating new theories through inductive analysis of data rather than from pre-existing theoretical frameworks. Such a method is useful in scenarios wherein the researcher does not have any preconceived notions about the phenomena under study. Nunes et al. [22] argued that while applying GT, knowledge of the context is a fundamental information resource in improving the researchers' understanding of activities, relationships, and stakeholders' thinking. This was clearly defended by Glaser and Strauss [23] when proposing that insight as well as theoretical sensitivity were the main components in the social scientist armory. Linking the emerging theory with the context makes the researcher's emerging theories denser, more complex, and more precise. Therefore, before analyzing the SP requirements, the participants had interactions with domain experts from the contract governance team within the vendor organization to understand the context of contracts governance. The domain experts from the contract governance team further gave the participants an overview of the company processes and organizational structure to which the obligations need to be mapped.

For identifying the governance-focused sub-types of SP requirements, the participants manually analyzed the 17,126 SP sentences. Each of the participants first individually analyzed each SP sentence and attached a coding word to a portion of the text – a phrase or a sentence, indicating which governance-focused facet of Security or Privacy requirement does the sentence address. It took the participants six working days to create these codes. The participants took three additional days to consolidate their individual codes and finally they worked together to jointly study the various coding words to arrive at high-level categories. The participants also discussed and clarified doubts/assumptions to arrive at a consensus regarding how and why particular concepts were important from the context of the problem being studied. The cases where a consensus could not be reached were forwarded to two domain experts from the contracts governance team in the vendor organization to determine the final label. The concepts finalized at the end of this process formed the governance-focused sub-types of SP requirements that exist in SE contracts. Figure 1 provides an example of the process of coding a text. We would like to highlight that focused coding was conducted to organize the codes into conceptual categories. More examples of coding are included at https://zenodo.org/records/105 84426.

Sanitized Examples of Contractual Clauses	Codes	Emering Categories as a result of focused coding	Emerging Concepts
Without limiting any of its other obligations under this Agreement, the Supplier shall ensure that if the <client_name> Data is placed on a portable electronic device (including laptops, memory sticks and back-up tapes) or transmitted electronically it is securely encrypted in accordance with <practice_name>.	supplier data	data security measures	Information Management
	secure data storage	encription practices	
	secure data transmission		
	secure encription		
	encription practice		

Fig. 1. Contractual Text, Codes and Conceptual Categories

We note that additional sub-types of SP requirements may emerge when the requirement analyst collects requirements from customers as a part of requirement elicitation phase of SDLC. However, in this work, we focus only on the **obligatory** SP requirements and their sub-types derived directly from contracts documentation. In the next sub-section, we define each of the identified sub-types along with an example.

3.3 Exploratory Study Results: SP Requirement Sub-Types in SE Contracts

Figure 2 presents the governance-focused sub-types of SP requirements that we have identified through our exploratory study. First, we present the definitions and examples of SP requirements found in SE contracts.

Security requirements found in SE contracts provide details pertinent to information access, retention and destruction of data, procedures to follow if a security incident occurs and maintaining business continuity and disaster recovery plan [6]. An example contractual obligation containing Security requirement: *The vendor undertakes that it has and shall continue to have in place and keep up to date business continuity plans and disaster recovery plans enough to minimize the possibility of any interruption in the services and allow the rapid restoration of the services should they, nonetheless, be disrupted by a disaster.*

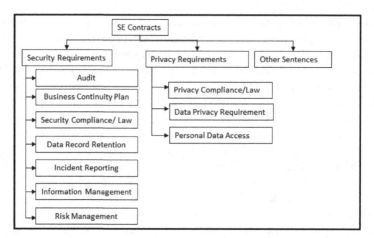

Fig. 2. Governance-focused Sub-types of SP Requirement in SE Contracts

Privacy requirements found in SE contracts provide details pertinent to various data privacy laws and requirements to adhere to data privacy policies of customer [6]. An example contractual obligation containing Privacy requirement: *Vendors are required to protect the personal and confidential information they receive about the customer and shall not transfer, store, access or otherwise process such information provided by customer without customer's prior written consent.*

Next, we provide definition and examples (obfuscated) of each sub-type of SP requirements that we identified through our exploratory study reported in Sect. 3.2. This provides an answer to our RQ. We first explain sub-types of Security requirements, followed by an exploration of sub-types of Privacy requirements.

Sub-types of Security Requirements in SE Contracts. We found the following sub-types of Security requirements in SE contracts:

Audit. This category includes obligations about the periodic audits that must be done for the deliverables and their various components such as the various security safeguards used for ensuring safety of the data. An example: *Service recipient retains the right to audit patch or firmware level of all service delivery systems including reference to any patches or firmware updates not deployed and the reason for non-deployment on an ad-hoc basis.*

Business Continuity Plan. This includes obligations related to detailed business continuity plans pertinent to the security of the software system, the need for their up-to-date documentation, their effective implementations and testing of robustness of procedures of business continuity. An example: *The Vendor shall provide disaster recovery services for the security of designated critical systems in accordance with the "Disaster Recovery" portion(s) of the applicable order as part of the services.*

Data Record Retention. This category is about defining security aspects around various types of data, their retention period, and conditions under which the data must be returned to the customer or destroyed. An example: *The Supplier shall ensure that if any data or electronic devices containing data is disposed of; such disposal takes place immediately in a secure manner such that the data is not recoverable.*

Incident Reporting. This category is about the need to inform the customer about a security or data breach as soon as the vendor becomes aware of it. It also outlines various procedures the vendor can take to mitigate the impact of the incident. An example: *The Supplier shall take all steps necessary to remedy the event and prevent its re-occurrence and provide a description of the measures taken or proposed to be taken to address the Security Breach.*

Information Management. This category includes specification about information disposal and stored information. This includes requirements about secure storage, transfer of customer data and destruction of data under various conditions as requested by the customer. An example: *All data on portable systems (including, but not limited to laptops, iPads, etc.) in the offshore development center must be encrypted.*

Risk Management. This category is about the need to have a process and to implement such a process to identify, assess and mitigate any current and possible future security risks. An example: *Supplier will undertake regular security risk assessments of infrastructure and components used to deliver the services and/or deliverables and maintain an up-to-date IT security risk register identifying the risks, assigning action items to appropriate personnel, and tracking the status of actions taken to mitigate or accept risks.*

Security Compliance/Law. This category is about various Industry standards and laws that must be complied with regarding security of the data and the deliverables. An example: *Without limiting any of its other obligations under this Agreement, the Supplier shall comply with PCI-DSS and PADSS standards.*

Sub-types of Privacy Requirements in SE Contracts. We found the following sub-types of Privacy requirements in SE contracts:

Privacy Compliance/Law. This category is about various Industry standards and laws that must be complied with respect to the privacy of data and deliverables. An example: *Supplier shall, and shall cause Supplier Agents to, comply with all applicable Laws regarding the handling, collection and transfer of personal information including the "Safe Harbor Principles" issued by the ⟨Country Name⟩ Commerce Department on ⟨Date⟩.*

Data Privacy Requirement. This category is about the policies and procedures to be followed to ensure privacy of certain data, such as Protected Health Information (PHI) or Personally Identifiable Information (PII), as defined in the contract. An example: *The Supplier shall inform < client_name > within forty-eight hours of receipt of any complaints about Personal Data and shall act in accordance with < client_name > reasonable instructions when dealing with that complaint.*

Personal Data Access. This category is about various conditions under which access to personal data of customer's customers is permissible to the vendor. An example: *Supplier will have access to PII data to include (name, address, DOB, phone numbers, emails, policy numbers, member numbers, SSN, etc.) in accordance with Exhibit ⟨Exhibit Number⟩, personal data access requirements in Agreement.*

Despite some sub-types having overlapped linguistic patterns, this fine-grained classification is essential for contract governance. It enables the accurate assignment of obligations to the right owners, ensuring accountability, particularly in cases of outsourced development where processes involve multiple stakeholders responsible for different steps. For instance, in a Privacy-specific SE requirement related to regulatory compliance, explicit mapping of stakeholders like the Legal Department is necessary to verify compliance with a law such as GDPR.

To summarize, we attempt to answer our RQ - *What governance-related details should we grasp to ensure satisfactory fulfillment of obligatory SP requirements in the SDLC?* Through an exploratory study, we explored the governance-focused details pertinent to SP requirements found in SE contracts. We identified 7 governance-focused sub-types of Security requirement and 3 governance-focused sub-types of Privacy requirements.

4   Automated Extraction and Classification of SP Requirements

In Sect. 3, we discovered governance-focused SP requirement sub-types through an exploratory study. Manual identification of these sub-types in large, complex SE contracts is time-consuming, error-prone, and requires heavy dependence on subject matter experts (SMEs). To address this, we automated the extraction and classification of SP requirements from SE contracts.

We divided the task of SP requirement extraction and classification into two stages. The first stage is *Level 1 Classification*, a ternary classification, wherein the entire dataset (94,600 instances) is classified into *Security*, *Privacy*, and *Others* class. The *Others* class comprises all sentences not in *Security* or *Privacy* class. *Level 2 Classification* involves two multi-class classification tasks on the *Security* and *Privacy* class, classifying them into their respective sub-types. We treat this as a multi-class classification problem due to the exclusive assignment of each clause to a single sub-type. The following subsections describe data pre-processing, model architecture, hyperparameter tuning, and experimental setup for both *Level 1* and *Level 2 classification*, using the labeled data from the exploratory study (reported in Sect. 3) as the ground truth.

In both *Level 1* and *Level 2 classification*, we used the Keras API with a TensorFlow backend in a Python 3 environment. The default Tokenizer class available in Keras has been used for tokenization purposes. The regular expression library in Python has been used to strip the contract clauses of all alphanumeric characters so that only English words are considered. The Natural Language Toolkit (NLTK) [24] library has been used for performing the basic text preprocessing tasks. We have removed all irrelevant stop-words from the clauses. Further, morphological analysis on the words has been performed using NLTK's Wordnet Lemmatizer, to remove the inflectional endings and only focus on the base English words.

For RNN-based experiments, we employed RNN with an attention mechanism. For BERT-based experiments, we utilized the BERT BASE model, as the BASE model is sufficient for solving multi-class classification problems. For T5 experiments, we used the T5-large model pretrained on the C4 data corpus. We fine-tuned the T5 model for text classification by providing input text and class labels. We conducted extensive hyperparameter tuning to find the optimal model, resulting in a learning rate of 5.6e-5, a batch size of 2, and 8 training epochs. Learning rate notably influenced model performance. Refer to Table 1 for the hyperparameter values used in fine-tuning BERT and T5.

Table 1. Hyperparameter values for fine-tuning BERT and T5

Hyperparameters	BERT	T5
train_batch_size	20	2
max_seq_length	100	NA
learning_rate	3.00E-05	5.60E-05
min_train_epochs	20	8

We utilized the Adam optimization algorithm for neural networks and employed the Categorical Cross Entropy loss function as a metric. We implemented a nested cross-validation technique, with Scikit-Learn's Random Search in the outer loop and stratified five-fold cross-validation in the inner loop to find optimal hyperparameter values. To tackle class imbalance, we used weighted categorical cross-entropy by assigning class weights as the reciprocal of the number of instances in each class.

5 Results and Discussion

The experimental results show that T5 outperforms BERT and the RNN models in both *Level 1* and *Level 2 classification*. Table 2 shows F1 scores for *Level 1 classification*. We get the best weighted average F1 score of 0.91 with T5 as compared to 0.90, 0.83, 0.84 with BERT, Bi-LSTM Attention and Bi-GRU Attention respectively.

Table 2. Results of level 1 classification (F1 score)

	Bi-LSTM Attention	Bi-Gru Attention	BERT	T5
Security	0.75	0.78	0.86	0.86
Privacy	0.67	0.69	0.80	0.87
Weighted Average	0.83	0.84	0.90	0.91

For *Level 2 Classification*, the class-wise Precision (P), Recall (R) and F1-Scores are shown in Tables 3 and 4. For Security class, we obtained the best average F1 score of 0.85 with T5 when compared to the other models. With T5, the highest average F1 score of 0.98 was obtained for *Business Continuity Plan*. For Privacy class, T5 has an average F1 score of 0.84 as compared to 0.81, 0.77 and 0.75 for BERT, Bi-LSTM Attention and Bi-GRU Attention respectively. With T5, the highest average F1 score of 0.89 was obtained for *Data Privacy Requirement* which is the most prominent sub-type within Privacy class. Most of the sub-types namely, *Audit, Data Record Retention, Incident Reporting* and *Security Compliance/Law* has an F1 score above 0.82. Only two sub-types from Security class namely *Information Management* and *Risk Management* and one from Privacy class namely *Personal Data Access* has a relatively low average F1 score. Low scores for these sub-types can be explained by (a) their rarity in our dataset, especially *Personal Data Access*, and (b) the high linguistic similarity between instances of *Information Management* and *Risk Management*, posing challenges even for state-of-the-art classification models such as T5. As fine-grained classification is crucial for SP governance, we are experimenting with the generative capabilities of GPT-3 to create synthetic sentences for the minority sub-types. We utilized the GPT-3 playground API with prompts for artificial sentence generation. While prompting GPT-3, we supplied natural language prompts that included a rationale and a corresponding example to infer the SP sub-type. An example prompt for generating *Personal Data Access* sentences and the corresponding GPT-3 output is presented in Fig. 3. Due to confidentiality agreement, we used publicly available contract examples in our prompts, ensuring their similarity in structure and complexity to the organizational dataset examples. Preliminary results from re-running the *Privacy* class experiment with augmented *Personal Data Access* instances generated by GPT-3 are promising. We employed 20 different prompts to create 200 synthetic *Personal Data Access* sentences (10 sentences per prompt) and observed a significant increase in the average F1 score (a rise from 0.30 to 0.71). To uphold the authenticity and reliability of the generated synthetic training data, we asked the domain experts from the vendor organization to review the generated synthetic samples and modify them if they deemed it necessary.

Table 3. Results of level 2 classification for security class

Security Sub-class	Bi-LSTM Attention			Bi-GRU Attention			BERT			T5		
	P	R	F1	P	R	F1	P	R	F1	P	R	F1
Audit	0.69	0.63	0.66	0.72	0.67	0.69	0.78	0.92	0.85	0.78	0.89	0.87
Business Continuity Plan	0.81	0.86	0.83	0.90	0.84	0.87	0.97	0.91	0.94	0.98	0.97	0.98
Data Record Retention	0.70	0.70	0.70	0.66	0.71	0.69	0.86	0.77	0.81	0.81	0.74	0.77
Incident Reporting	0.60	0.69	0.64	0.57	0.69	0.63	0.80	0.87	0.84	0.96	0.94	0.95
Information Management	0.42	0.45	0.44	0.50	0.61	0.55	0.53	0.61	0.57	0.54	0.57	0.55
Risk Management	0.33	0.14	0.20	1.00	0.43	0.60	0.67	0.75	0.71	0.80	0.80	0.80
Security Compliance/Law	0.79	0.80	0.79	0.82	0.76	0.79	0.90	0.85	0.88	0.88	0.87	0.88
Weighted Average	**0.72**	**0.73**	**0.72**	**0.75**	**0.73**	**0.74**	**0.85**	**0.83**	**0.84**	**0.85**	**0.85**	**0.85**

In addition to RNN and Transformer architectures, we trained traditional Machine Learning (ML) classifiers like Naive Bayes, Support Vector Machine (SVM), Logistic Regression (LR), Random Forest and AdaBoost classifiers on our dataset. For *Level 1*

Table 4. Results of level 2 classification for privacy class

Privacy Sub-classes	Bi-LSTM Attention			Bi-GRU Attention			BERT			T5		
	P	R	F1	P	R	F1	P	R	F1	P	R	F1
Privacy Compliance/Law	0.65	0.58	0.61	0.71	0.38	0.50	0.61	0.85	0.71	0.80	0.87	0.84
Data Privacy Requirement	0.83	0.94	0.88	0.81	0.97	0.88	0.89	0.91	0.90	0.87	0.91	0.89
Personal Data Access	0.00	0.00	0.00	0.00	0.00	0.00	1.00	0.05	0.09	1.00	0.22	0.31
Weighted Average	**0.74**	**0.80**	**0.77**	**0.73**	**0.80**	**0.75**	**0.85**	**0.83**	**0.81**	**0.87**	**0.85**	**0.84**

Input Prompt to GPT-3

A Personal Data Access clause is a clause that describes various conditions under which the access to personal data of customer's customers is permissible to vendors. For example, "*Supplier will have access to PII data to include (name, address, emails, etc.) in accordance with Exhibit, personal data access requirements in Agreement*". This is a Personal Data Access clause as it restricts Supplier access to Personal Identifier Information in accordance with an Exhibit. Can you generate 10 Personal Data Access clauses?

Generated Output

1.The Supplier shall only access Personal Identifiable Information (PII) of the customers as necessary for the performance of services under this Agreement and in accordance with Exhibit A.
2.The Supplier shall not use or disclose PII for any purpose other than as specifically authorized in writing by the Company.
3.The Supplier shall promptly notify the Company of any unauthorized access, use, or disclosure of PII.
4.The Supplier shall assist the Company in responding to any request from a customer for access to or deletion of their PII
5.This clause shall survive termination of this Agreement and the Supplier's duty to protect PII shall remain in effect indefinitely.
6. -------

Fig. 3. Synthetic data generation for Personal Data Access using GPT-3 prompting.

Classification, the average Precision scores range from 77% to 80%, average Recall range from 76% to 81% and average F1 scores vary between 73% and 79%. For classification of *Security* requirements, average Precision scores vary between 27% and 75%, average Recall range from 45% to 70% and average F1 scores range from 31% to 71%. Similarly, for classification of *Privacy* requirements, average Precision scores range between 57% to 70%, average Recall vary between 70% and 80% and average F1 scores vary between 65% and 75%. Traditional ML classifier performance varied significantly with respect to the training dataset. SVM excelled in *Security* class but struggled with *Privacy* class, while LR outperformed other classifiers in *Level 1* but faltered in *Level 2 Classification*. In some instances, certain ML models slightly outperformed RNN architectures, but overall, T5 consistently outperformed all other models.

While discussing model accuracies and false negatives with the contract governance team of the vendor organization, they recommended manual review for the missed sentences. They recognized that in such complex classification tasks requiring human cognition, the achieved accuracies would significantly reduce manual effort.

Nevertheless, we would also like to discuss the limitations on neural network models particularly in our context. We utilized neural network models (T5, BERT, and RNN) to automate the extraction and governance-focused classification of SP requirements from software engineering contracts. Although these models performed well, it is crucial to understand the challenges in classifying complex text such as contracts. One notable challenge with these models is the context length limit (context window), shorter context may result in missing important information, especially in the case of complex texts such as contracts. Since we are focusing specifically on individual clauses and not the entire contracts text, we mitigate this limitation of these neural network models. RNNs are helpful in sensing sequences, however, not very effective with long-range dependencies. For example, in case of security related requirements, the requirement for a supplier to comply with data protection policies may involve nuanced dependencies (Encryption Standards, Access Control Measures, Legal and Regulatory Compliance) that RNNs struggle to capture accurately, resulting in misclassification of the given requirement. Transformer-based models such as BERT and T5, even with their global context capturing abilities, still face challenges while dealing with complex text. For example, consider a security requirement in a contract where the term "encryption" is used. These models may interpret "encryption" with a global context of data protection, assuming it refers to safeguarding personal data during transmission and storage and therefore may classify it as a privacy requirement instead of security requirement. The black-box nature of these models makes it challenging to understand their predictions leading to misinterpretation and inaccurate classification. Addressing overfitting in these models is crucial to prevent memorization of specific instances, enabling the models to learn underlying patterns and nuanced dependencies. In Future, we plan to enhance these models and make architectural modifications to ensure precise understanding of subtle nuances in contractual text, ensuring more accurate classification.

6   Validation on Publicly Available Dataset

To comply with open science policies and to validate the generalizability of the proposed approach, we assessed it using a publicly available contracts dataset sample - Contracts Understanding Atticus Dataset (CUAD) [25]. CUAD consists of 510 commercial contracts. For the validation study, we picked 5000 sentences from CUAD using stratified sampling. We manually analyzed these sentences to label them as *Security* or *Privacy* or *Others*. After labeling, we had 750 sentences belonging to *Security* class, 370 sentences belonging to *Privacy* class and the remaining sentences belonging to *Others* class. 22% of sentences belonged to SP requirements in this subset of CUAD dataset. This closely matches the vendor organization's dataset, where 20% of sentences were SP requirements. Next, the sentences marked as *Security* or *Privacy* were analyzed further to classify them into governance-focused sub-types. One author (who also participated in the exploratory study reported in Sect. 3) and three software engineers from the vendor

organization who have experience working on contracts management projects, participated in this manual labeling. The three software engineers were given the definitions
of the sub-types of SP requirements identified through the exploratory study reported
in Sect. 3. They were then asked to classify the SP sentences into their respective governance specific sub-types. If they think that a sentence does not fit into any of the
sub-types identified in Sect. 3, they were asked to propose a new relevant sub-type to
accommodate the sentence. This activity was performed to check the generalizability
of the SP sub-types identified in Sect. 3. No new sub-types of SP requirements were
proposed indicating that the SP sub-types created by studying contracts from the vendor
organization are generalizable to other contract documents.

Next, we validated the generalizability of the T5 model that we trained on the contract
dataset from the vendor organization. For this, we first utilized the *Level 1 Classification* to classify the 5000 contractual sentences from CUAD dataset as *Security*, *Privacy*
and *Others*. For *Level 1 Classification*, we obtained an average F1 score of 0.93. We
then utilized the *Level 2 Classification* to classify the SP requirements into their respective sub-types. For *Level 2 Classification*, we obtained an average F1 score of 0.76 for
Security class and 0.73 for *Privacy* class. This indicates that T5-based model is largely
generalizable to unseen contracts.

7 Threats to Validity

The first possible threat to validity is the comprehensiveness of the SP requirement
sub-types identified in this study. We cannot claim that we have identified all possible
sub-types of obligatory SP requirements found in any possible SE contract or that the
sub-types identified in this study are universal and is therefore expected to be present in
every SE contract. However, since the exploratory study involved 960 large SE contracts
from nine application domains and 10 countries, it is reasonable to expect a similar set of
SP requirement sub-types in any set of SE contracts. This claim is further strengthened
by the validation on publicly available contracts dataset reported in Sect. 6.

The second threat is about the representativeness of our dataset - are the SP requirements in our dataset a good representation of SP requirements in any SE contract? SP
requirements are dynamic, requiring adjustments in response to changes in international
laws, standards, or client policies, such as GDPR implementation. Major security vul-
nerabilities like the Apache Log4J exploit have led to contract modifications related to
SP clauses. Hence, language models trained on existing SP requirement repositories
may require periodic retraining with new data to stay current in the ever-evolving security and privacy landscape. While addressing the practical challenges of dealing with
SP requirements, we enhance dataset representativeness by including SP requirements
from nine application domains across ten countries. We therefore expect our dataset
to be a good representation of SE contracts. Furthermore, we leverage the intentional
repetitiveness in contract language [26], which makes it reasonable to anticipate linguistic similarities across organizations, enhancing the generalizability of our findings.
Moreover, we employed 960 real-world SE contracts, totaling over 94,600 sentences,
with 17,126 sentences dedicated to SP requirements, adding a realistic dimension to our
experiments.

The third threat concerns whether the coding words from participants in the exploratory study truly represent the best way to categorize SP requirement sub-types in SE contracts. In our case, the participants had over 16 years of experience in SE, legal and governance roles within the vendor organization and had interacted with domain experts from the contract governance team. This combined expertise ensured that the coding words effectively represented the likely SP sub-types in any SE contract. This claim is further strengthened by the validation study reported in Sect. 6.

The fourth threat to validity concerns the utilization of Bi-LSTM Attention, Bi-Gru Attention, BERT and T5 model, despite more recent and better models like GPT-4 and Bard. We did not use GPT-4 or Bard as, (1) we are dealing with highly confidential contracts data and therefore we cannot expose this data to these recent models, and (2) the current token-based pricing model adopted by OpenAI, employing GPT-4 Turbo for our experiments are very expensive, rendering it unfeasible.

The fifth threat pertains to the reproducibility of experimental results. We validated the reproducibility of our approach by testing it on a publicly available contracts dataset (CUAD) consisting of 5000 contractual sentences. The results indicate that our best performing model T5 is largely generalizable to unseen contracts. However, due to the confidential nature of our dataset, we cannot make the T5 model trained on the contracts data from the vendor organization publicly available. As an alternative, we trained the best performing T5 model on 80% of the 5000 contractual clauses from the publicly available CUAD dataset and tested it on the remaining 20% data. For *Level 1 Classification*, we obtained an average F1 score of 0.97. For *Level 2 Classification*, we obtained an average F1 score of 0.69 for *Security* class and 0.61 for *Privacy* class. We have made this labeled data, trained model, and results publicly available. As is evident from the obtained F1 scores, the T5 model trained on CUAD data did not achieve the same classification accuracies as those trained on contracts data from the vendor organization. This difference is attributed to (a) relatively less training data from CUAD (4000 contractual clauses), and (2) the intricacies and nuances present in the vendor organization's contracts data, which primarily consists of large-scale outsourcing contracts.

8   Conclusion and Future Work

Contracts have been established as a valuable source in eliciting software requirements. The primary objective of this work is to explore the problem of identification and a governance-focused classification of SP requirements from SE contracts. Governance-focused classification of SP requirements aids in assigning ownership, promoting accountability, and ensuring compliance. T5 outperformed other classifiers in automating SP requirement identification and classification. We validated our approach using the publicly available CUAD dataset, confirming the generalizability of SP sub-types and T5 classification from vendor contracts to other unseen contracts.

Our future work involves conceptual and technical enhancements. Conceptually, we aim to broaden the contract governance model to encompass SP governance nuances such as risk detection related to information in SP clauses, and extraction of vital data like *Actor, Action, Condition, Frequency,* and *Lead time* from the SP clauses. On the technical front, we intend to tackle two major challenges faced during our solution development:

manual labeling of training data and addressing class imbalance. To alleviate the labeling challenge, we are exploring data programming [27] and self-training [28] techniques to reduce human effort. For class imbalance, we have initiated successful experiments with GPT-3 on one minority class. We plan to refine and extend this approach to generate synthetic data for the remaining minority classes to improve the classification accuracies.

9 Data Availability

The complete replication package of this work includes (a) the code (b) the labeled CUAD dataset, and (c) the contracts data received from the vendor organization that is used for training and testing the models. The first two artifacts in the replication package i.e. the code and the labeled CUAD data are made available publicly and can be accessed via https://zenodo.org/records/10584426. The third artifact i.e. the training and test dataset cannot be made publicly available as it includes confidential proprietary contracts document signed between the vendor organization and their different customers and carries Non-Disclosure Agreements (NDAs). The legal department of the vendor organization prohibits sharing of this confidential data.

However, to support reproducibility to the extent possible, we have retrained our original model on 80% of the labeled CUAD dataset and tested it on the remaining 20% data. This retrained model is also included in https://zenodo.org/records/10584426.

References

1. https://www.financierworldwide.com/data-privacy-and-cyber-security-the-importance-of-a-proactive-approach. Accessed 02 Nov 2023
2. IBM report: https://www.ibm.com/security/data-breach. Accessed 02 Nov 2023
3. Whatsapp case: https://www.bloomberg.com/news/articles/2021-09-02/whatsapp-fined-266-million-over-data-transparency-violations. Accessed 05 Nov 2023
4. https://www.bbc.com/news/technology-54722362. Accessed 02 Nov 2023
5. Casillo, F., Deufemia, V., Gravino, C.: Detecting privacy requirements from user stories with NLP transfer learning models. Inf. Softw. Technol., 106853 (2022)
6. Sainani, A., Anish, P.R., Joshi, V., Ghaisas, S.: Extracting and classifying requirements from software engineering contracts. In: 2020 IEEE 28th International Requirements Engineering Conference (RE) (pp. 147–157). IEEE (2020)
7. https://www.infosysbpm.com/offerings/functions/legal-process-outsourcing/white-papers/Documents/contract-process-helping-hurting.pdf. Accessed 02 Nov 2023
8. Contract Governance: https://www.linkedin.com/pulse/simple-keys-contract-governance-kelly-smith/. Accessed 02 Nov 2023
9. Devlin, J., Chang, M., Lee, K.: BERT: pre-training of deep bidirectional trans-formers for language understanding. In: Proceedings of the 2019 Conference of the North American Chapter of the Association for Computational Linguistics: Human Language Technologies, Volume 1 (Long and Short Papers), Minneapolis, MN, USA, pp. 4171–4186 (2019)
10. Raffel, C., et al.: Exploring the limits of transfer learning with a unified text-to-text transformer. J. Mach. Learn. Res. 21(1), 5485–5551 (2020)
11. GPT-3: Models - OpenAI API. Accessed 02 Nov 2023

12. Weber-Jahnke, J., Onabajo, A.: Mining and analysing security goal models in health information systems. In: Workshop on Software Engineering in Health Care, pp. 42–52. IEEE Computer Society (2009)
13. Jindal, R., Malhotra, R., Jain, A.: Automated classification of security requirements. In: 2016 International Conference on Advances in Computing, Communications and Informatics (ICACCI), pp. 2027–2033 (2016)
14. Xiao, X., Paradkar, A., Thummalapenta, S., Xie, T.: Automated extraction of security policies from natural-language software documents. In: Proceedings of the ACM SIGSOFT International Symposium on the Foundations of Software Engineering (FSE), pp. 12:1–12:11 (2012)
15. Breaux, T., Anton, A.: Analyzing regulatory rules for privacy and security requirements. IEEE Trans. Softw. Eng. **34**(1), 5–20 (2008)
16. Islam, S., Mouratidis, H., Wagner, S.: Towards a framework to elicit and manage security and privacy requirements from laws and regulations. In: Wieringa, R., Persson, A. (eds.) REFSQ 2010. LNCS, vol. 6182, pp. 255–261. Springer, Heidelberg (2010). https://doi.org/10.1007/978-3-642-14192-8_23
17. Janpitak, N., Sathitwiriyawong, C.: Information security requirement extraction from regulatory documents using GATE/ANNIC. In: 7th International Electrical Engineering Congress (iEECON) (2019)
18. Munaiah, N., Meneely, A., Murukannaiah, P.K.: A domain dependent model for identifying security requirements. In: Proceedings of the IEEE 25th International Requirements Engineering Conference (RE), Lisbon, pp. 506–511 (2017)
19. Farkhani, T.R., Razzazi, M.R.: Examination and classification of security requirements of software systems. Inf. Commun. Technol. **2**, 2778–2783 (2006)
20. Jain, C., Anish, P.R., Ghaisas, S.: Automated identification of security and privacy requirements from software engineering contracts. In: 2023 IEEE 31st International Requirements Engineering Conference Workshops (REW) (pp. 234–238) (2023)
21. Hoda, R.: Socio-Technical grounded theory for software engineering. IEEE Trans. Softw. Eng. (2021). https://doi.org/10.1109/TSE.2021.3106280
22. Nunes, J.M.B., Martins, J.T., Zhou, L., Alajamy, M., Al-Mamari, S.: Contextual sensitivity in grounded theory: The role of pilot studies. Electr. J. Bus. Res. Methods **8**(2), 73–84 (2010)
23. Glaser, B., Strauss, A.: The Discovery of Grounded Theory. Aldine, Chicago (1967)
24. Loper, E., Bird, S.: NLTK: the natural language toolkit. arXiv preprint cs/0205028 (2002)
25. CUAD dataset. https://www.atticusprojectai.org/cuad. Accessed 02 Nov 2023
26. Simonson, D., Broderick, D., Herr, J.: The extent of repetition in contract language. In: Proceedings of the Natural Legal Language Processing Workshop 2019 (pp. 21–30) (2019)
27. Ratner, A.J., De Sa, C.M., Wu, S., Selsam, D., Ré, C.: Data programming: creating large training sets, quickly. In: Advances in Neural Information Processing Systems 3567–3575 (2016)
28. Amini, M.-R., Feofanov, V., Pauletto, L., Devijver, E., Maximov, Y.: Self-training: a survey (2022)
29. Sharifi, S., Parvizimosaed, A., Amyot, D., Logrippo, L., Mylopoulos, J.: Symboleo: towards a specification language for legal contracts. In: 2020 IEEE 28th International Requirements Engineering Conference (RE), Zurich, Switzerland, pp. 364–369 (2020). https://doi.org/10.1109/RE48521.2020.00049

Explainability with and in Requirements Engineering

What Impact Do My Preferences Have?

A Framework for Explanation-Based Elicitation of Quality Objectives for Robotic Mission Planning

Rebekka Wohlrab[1]([envelope]) [ORCID], Michael Vierhauser[2] [ORCID], and Erik Nilsson[1]

[1] Chalmers University of Gothenburg, Gothenburg, Sweden
wohlrab@chalmers.se
[2] University of Innsbruck, Innsbruck, Austria
Michael.Vierhauser@uibk.ac.at

Abstract. *[Context and motivation]* Successful human-robot collaboration requires that humans can express their requirements and that they comprehend the decisions that robots make. Requirements in this context are often related to potentially conflicting quality objectives, such as performance, security, or safety. Humans tend to have preferences regarding how important different objectives are at different points in time. *[Question/problem]* Currently, preferences are often expressed based on assumptions of what importance level should be assigned to a quality objective at runtime. To assign meaningful preferences to quality objectives, it is important that humans understand the impact of these preferences on the behavior of a robot. To the best of our knowledge, there is yet no framework that supports the explanation-based elicitation of quality preferences. *[Principal ideas/results]* To address these needs, we have developed OBJUST, a framework that helps with the interactive elicitation of preferences for robot mission planning. *[Contribution]* The framework relies on the specification of human preferences and contrastive explanations. We evaluated our framework in a study with 7 participants. Our results indicate that the visual and textual explanations of the generated robotic mission plans help humans better understand the impact of their preferences, which can facilitate the elicitation process.

Keywords: quality attributes · elicitation · robot mission planning · contrastive explanation

1 Introduction

To enable human-robot collaboration, humans need to be able to *express* their (potentially changing) objectives. To support the elicitation of objectives, it is crucial that humans *understand* how robots work, what tasks they are performing, and why robots select particular actions that were considered optimal in a given situation [23]. In practice, multiple objectives are used in robot mission planning, many of which are quality attributes such as performance, energy consumption, safety, or security [28,34,35].

In certain situations, a quality objective might become more important (e.g., due to an arising safety hazard, where a robot should avoid a location), resulting in the need to replan missions [11]. However, it is not always obvious how different requirements result in different plans [45]. Previous work found that humans frequently struggle with understanding how their preferences of different quality objectives affect the automated planning process [2]. The need for explanations as a guiding tool in requirements engineering for self-adaptive systems has been raised [41]. To the best of our knowledge, there is no human-on-the-loop approach that helps humans to elicit changing preferences supported by explanations [29, 49]. It is not enough to elicit input once at the beginning, but the dynamic nature of run-time contexts and stakeholder preferences [41,49] may require recalculating a mission with new or updated references.

Stakeholders working with robots are not always people directly interacting with a robot, but may be supervisors who operate systems or observe them at a distance. For example, in a warehouse, there might be few humans who directly communicate with a robot. For these scenarios, it is beneficial to have a user interface that can help humans get an overview of the planning problem, indicate quality objectives, and understand the tradeoffs of a particular plan [13,19].

In this paper, we present the OBJUST framework that enables humans to *express their preferences* for robot mission planning, *provides explanations* of plans that come with different tradeoffs, and further helps humans *adjust preferences* if necessary. While many existing approaches rely on textual explanations only [45], our approach supports a combination of textual and visual explanations. OBJUST does not propose or prescribe a specific requirements elicitation technique, but provides a framework to help stakeholders assess the effect of different preferences on the expected outcome of a planning problem. In a preparatory step, requirements are collected from stakeholders, e.g., using brainstorming, interviews, or workshops [4,10,16]. The problem is that these requirements are often elicited based on assumptions about how systems will act at runtime and how preferences might impact the behavior of the final system. Making such assumptions is not feasible and desirable in practice. To mitigate this issue, once an initial set of requirements has been collected, OBJUST can be used to investigate the effect of different preferences on the planning results. We put a special emphasis on providing a comprehensive explanation of why certain paths were deemed optimal by the planning algorithm.

We claim three main contributions with our OBJUST framework. First, we provide a list of requirements for an elicitation and explanation framework. Second, we propose an architecture of our elicitation and explanation framework along with a domain model for the underlying knowledge base. To demonstrate the framework's applicability, we provide an open-source prototype implementation. Finally, we provide a user dashboard that supports user input and explainability.

Sect. 2 describes the background and related work. Sect. 3 presents our research method. In Sect. 4 we then present OBJUST. We describe our evaluation in Sect. 5. Sect. 6 presents the threats to validity. In Sect. 7, we discuss the results, limitations, and avenues for future research.

2 Background

Robotic systems need to consider different quality requirements, to ensure that the system can respond to changing conditions in the environment, such as changes in workload, resource availability, or emerging security threats [32]. Quality objectives are often encoded in *cost functions* representing the costs associated with a certain action or sequence of actions. For instance, the cost might represent the distance traveled, energy expended, or risk encountered [17,48]. In related work, these cost functions are often combined in a weighted sum [18,20].

However, it is not always obvious how the weights of such a global cost function should be set. Stakeholders require tools and decision-making techniques to assist their prioritization of quality objectives and reach consensus [41].

2.1 Motivating Example

We represent the mission planning problem as a graph, consisting of locations and edges. Fig. 1 depicts a map containing locations and edges that a robot can choose from to reach its destination, where the number over each edge is the distance between locations. In this example, we use three quality objectives: *travel time*, *safety*, and *privacy*. Safety is measured in terms of collisions and privacy in terms of the expected number of privacy intrusions.

Traversing a normal edge yields a safety cost of 0, a partially occluded edge costs 1, and an occluded edge costs 2. Passing a public location yields a privacy cost of 0, a semi-private location costs 1, and a private location costs 2.

To calculate the optimal path, we define the cost function for a plan σ as a weighted sum:

$$c(\sigma) = w_{tt} \cdot c_{tt}(\phi_{tt}(\sigma)) + w_{col} \cdot c_{col}(\phi_{col}(\sigma)) + w_{int} \cdot c_{int}(\phi_{int}(\sigma))$$

where $w_{tt}, w_{col}, w_{int} \in \mathbb{R}^+$, and $w_{tt} + w_{col} + w_{int} = 1$.

c_* is the local cost function for each quality attribute, $\phi_*(\sigma)$ is the total cost of each attribute in a path, and w_* is the weight of each quality attribute. Weights are used to encode stakeholders' preferences concerning the importance

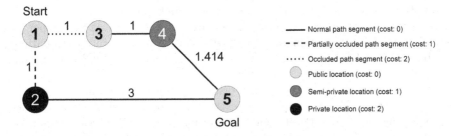

Fig. 1. A planning graph. The node colors represent 3 node types. The line styles (solid, dashed, dotted) represent normal, semi-occluded, and occluded edges.

Table 1. Cost function values of paths for different preferences.

w_{tt}	w_{col}	w_{int}	Locations	Optimal	ϕ_{tt}	ϕ_{col}	ϕ_{int}	c_{tt}	c_{col}	c_{int}	c
1	0	0	③,④	Yes	3.414	2	1	1	2	1	1
			②	No	4	1	2	1.172	1	2	1.172
0	1	0	③,④	No	3.414	2	1	1	2	1	2
			②	Yes	4	1	2	1.172	1	2	1
0.25	0.5	0.25	③,④	No	3.414	2	1	1	2	1	1.5
			②	Yes	4	1	2	1.172	1	2	1.293

of that quality attribute. The local cost function c_* is calculated by quantifying the local cost for a path in relation to the least expensive path's cost for the same attribute.

An automated planner may choose a path depending on the weighted preferences of quality objectives. Table 1 displays the attribute costs and cost function values for several sets of preferences. In the case where only travel time is relevant ($w_{tt} = 1$), the path ① → ③ → ④ → ⑤ is optimal. If the system only cares about safety ($w_{col} = 1$), then the path ① → ② → ⑤ is deemed optimal.

Fully prioritizing a single quality attribute is easy, but mixing weights does not always yield intuitive paths. E.g., for the preferences $w_{tt} = 0.25, w_{col} = 0.5, w_{int} = 0.25$, the path ① → ② → ⑤ is deemed optimal by the robot. However, why exactly was this path chosen? Even in this small graph, it is not always easy to calculate this by hand, or by intuition. For large maps, identifying the optimal path requires error-prone and time-consuming calculations. Therefore, solutions are needed that help humans to understand the consequences of their preferences.

2.2   Related Work

Priority Awareness: Samin et al. [39,40] have coined the concept of priority awareness with their Pri-AwaRE approach, to automatically adjust priorities to satisfy QoS requirements. Priorities are similar to preferences in OBJUST. Constraints can be expressed as well, but are not the focus of this paper. It appears promising to combine OBJUST with Pri-AwaRE, so that explanations can be given with respect to why certain requirements are fulfilled and what impact the adjusted priority values have.

Besides cost functions, there exist other ways of representing priorities of quality objectives. For example, some existing approaches select so-called knee solutions among a set of Pareto-optimal solutions [8,22]. OBJUST is similar in the sense that it selects Pareto-optimal solutions, takes the balanced points by default (if all priorities are the same, as in our example), but allows to deviate from them when quality objectives are reprioritized.

The need to dynamically adjust requirements for self-adaptive systems has been addressed, for example, by frameworks to enable goal model adaptation [25,34]. It focuses on run-time verification of adaptable goals. Similarly, approaches based on KAOS [1,9] have a general focus on functional goals and uncertainty. While goal modeling can capture quality-related aspects as well, our focus lies more on quality attributes and the impact that quality preferences have on generated plans. In this area, Bryl et al. [3] explored the use of goal modeling and requirements analysis, in conjunction with planning techniques using Tropos. Similar to our work, they describe a tool-supported approach for analyzing and exploring alternative requirements. Their work does not lie on the changing assignment of different weights to soft goals, whereas the assignment of preferences for different quality attributes is exactly the focus of OBJUST. Our previous work [52] has addressed the issue of run-time adaptation of quality attributes. We elaborated on the challenges and proposed steps towards quality attribute adaptation. Similarly, Li et al. developed a framework for preference adaptation and concluded that future works need to explain the impact of preferences to users [29]. This is the research gap we are addressing in this paper: making the impact of preferences explicit so that they can be better elicited.

Explainable Planning: In recent years, the area of explainable planning has received increased attention [24]. Contrastive explanations are among the most common forms of explanation proposed by related work [7,14,27,37]. They can help human users identify potential biases and errors in a robot's decision-making [33]. Contrastive explanations have been used to help humans understand why a robot mission plan was optimal for a given set of quality objectives [45]. Eifler et al. [14] developed a user interface that allows human users to iteratively explore the planning space, ask "why not" questions, and specify planning goals. Existing approaches often focus on in-situ explanations, in which a robot explains a current action it just took or is about to take [6,42]. OBJUST focuses on explanations of multiple planning alternatives for entire missions, to help humans give appropriate input during interactive robot mission planning.

Preference Elicitation: While traditional explainability approaches for robot mission planning have focused on describing why one plan is optimal or better than another, we focus on giving input to users who specify their (potentially changing) preferences to facilitate replanning at runtime.

Shaikh et al. [43] proposed a related approach that relies on a GUI. Similar to OBJUST, they also use sliders to indicate the importance of different quality objectives. Besides sliders, also a palette interface and a prism interface were implemented. Both the slider and the palette interfaces were found to be usable. OBJUST extends the use of interfaces for preference elicitation by providing a visualization and explanation component that can help humans when interactively exploring how their preference selection impacts the generated plans. In that way, preferences can be adjusted so that the desirable behavior is achieved.

3   Research Method

We applied design science [50] with several iterations of solution design and validation. In this paper, we focus on the framework for interactive elicitation and explanation as the design artifact. The goal of the research was to better *understand humans' needs* for explanation to guide the interactive elicitation of quality preferences, to *develop* the OBJUST framework supported by a prototype, and to *evaluate* to what extent the framework fulfills the needs of our participants. Our process consisted of (0) *identifying shortcomings* of existing solutions, (1) a *requirements elicitation phase* with human participants; (2) the *development* of the conceptual domain model and prototype implementation; and (3) an *evaluation phase* of the prototype. Interview material can be found on Figshare[1], and the implementation of our prototype is available online[2].

Table 2. Overview of the study participants, their occupations, and experiences.

Part	Occupation	Experience with technical subjects
1	Engineering manager	3–5 yrs
2	Graduating software development student	1–2 yrs
3	Graduating software engineering student	1–2 yrs
4	Software developer	6+ yrs
5	Backend developer	3–5 yrs
6	Cloud engineer/architect	6+ yrs
7	UX-design student	0 yrs
8	Consultant manager	6+ yrs
9	Software architect	6+ yrs
10	Product owner	6+ yrs

Table 2 shows an overview of the participants. All participants had experience with UX or technical subjects, with levels of experience ranging from senior students to practitioners who had worked in industry for more than 6 years.

Participation was voluntary, the participants were asked to give consent to participate in the study, and the procedures were explained. All participants were informed about their anonymity and assured that they could withdraw from the study at any point in time. No personally identifiable information was collected. The conducting researcher took notes and recorded the data from the survey.

After the interviews, the data was coded [15] using the QualCoder[3] tool for the thematic analysis. Codes were created in an iterative way and structured into categories of codes as a tree to arrive at our findings.

---

[1] https://doi.org/10.6084/m9.figshare.24006978.v1.
[2] https://github.com/SE-CPS/OBJUST_public.
[3] https://qualcoder.wordpress.com.

Requirements Elicitation Phase: We performed interviews with Part. 1–7 to investigate challenges when expressing preferences and understanding plans. Part. 8–10 were not available but participated in the evaluation phase instead.

One author performed the elicitation and was present in all sessions. The sessions relied on pen-and-paper calculations, in which the participants were asked what path would be optimal for a given set of quality objective weights. We encouraged the participants to think aloud. The participants were also asked questions about what features would be useful to alleviate such a task. The questions can be found in the supplementary materialsee footnote 3 and were mainly concerned with the participants' reasoning and perceived difficulties. Furthermore, participants were also asked for suggestions of features that might help to mitigate their challenges.

The data from the elicitation phase was then coded by the researcher who performed the interviews. Afterwards, the codes were grouped into themes and discussed in a data analysis workshop. We found that the challenges from the participants can be addressed by a number of requirements. As an outcome of this phase, we collected a set of six requirements/core features (cf. Sect. 4.1) serving as the input for the prototype implementation.

Development: We developed the framework based on the data from the elicitation phase. We systematically went through the requirements and understood what features were needed in a prototype that addressed the participants' challenges. We also developed a domain model by understanding the key concepts that were needed to reason about the robot mission planning domain.

Think-Aloud Study for Evaluation: Part. 1, 2, 5, and 7–10 were involved in this phase. The other participants were asked to participate but were not available. In the evaluation phase, participants worked on different tasks with the tool and were asked to fill out a short survey.

4   Framework for the Interactive Preference Elicitation

In this section, we introduce the elicited requirements for our framework (Sect. 4.1), the core implementation (Sect. 4.2), and visualization and explanation features (Sect. 4.3).

4.1   Requirements for Interactive Elicitation and Explanation

Based on our analysis, we arrived at a list of the following six functional requirements for an explanation visualization framework:

(R1) The system shall allow the user to prioritize quality objectives: R1 is concerned with the importance of eliciting the relevance of each quality objective, which in turn serves as an important input to automated planning.

(R2) The system shall display an optimal path on a map or graph, along with its costs: In the elicitation phase, we found that identifying the optimal path took a lot of time and it was difficult for participants to manage the complexity of planning problems quickly.

(R3) The system shall display alternative paths with their costs: R3 is relevant to compare a selected path to one that might seem optimal, but is not. This visual contrastive explanation was considered beneficial by the participants.

(R4) The system shall provide a textual explanation of why a path is optimal: During the elicitation, we found that participants struggled with the required calculations to understand what route was optimal given a cost function.

(R5) The system shall indicate important nodes that distinguish one path from other paths: Indicating differences between paths that are connected to key decision points in the map was considered crucial by the participants.

(R6) The system shall support traceability between quality (input) data and generated plans: R6 ensures that different paths can be easily traced back to the corresponding input requirements.

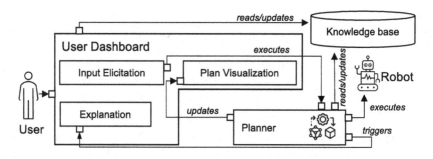

Fig. 2. Overview of the elicitation and explanation framework

4.2   Core Implementation

Fig. 2 provides an overview of the framework, which is described in the following.

Knowledge Base: Since it is often useful to explicitly capture concepts and relevant elements in a domain model [31], we have created a model for the adaptive planning of quality objectives. It supports traceability between quality data and the generated plans (R6). The model is used as the basis for the knowledge base and instantiated for each plan generation. For the initial instantiation, we reuse graph data from a public repository[4].

Fig. 3 provides an overview of the three main parts of the domain model. First, the *Qualities* part (top) describes the quality input provided by stakeholders, i.e., their constraints and preferences. Stakeholders can indicate how

---

[4] https://github.com/cmu-able/explainable-planning.

important a quality objective is and define constraints, for example, to restrict the value of a specific quality objective measure to a certain range. The quality objectives and preferences are then combined in a single *Cost Function*, e.g., as a weighted sum. Second, the *Environment* part (bottom) captures the structure of the *Map*, i.e., *Edges* (that may be occluded) and *Locations* with their privacy levels. For edges, a probability of an edge being successfully traversed can be specified. Finally, the *Planning Output* part (middle) represents the result of the automated planning. The Qualities part and the Environment part are input by humans, or given by the planning context. The Planning Output part (i.e., the set of locations that shall be visited) is fed into the robot for plan execution.

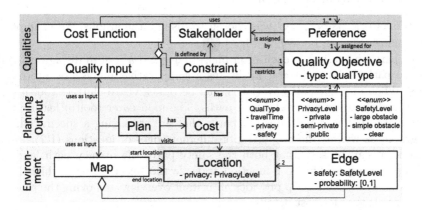

Fig. 3. Domain model for the underlying knowledge base

Input Elicitation: Fig. 4 shows an overview the user dashboard. The form on the left is used to elicit the users' preferences ❶ for quality objectives (R1). For the input elicitation, the map, start and end locations, and preferences are specified. For the specification of preferences, the Analytic Hierarchy Process [38] is used. Users can use sliders to perform pairwise comparisons between quality objectives and indicate how much they prefer one quality objective over another. Based on this input, values between 0 and 1 can be computed to set the weights of different objectives. This approach has been applied to elicit preferences for different objectives before [30,51] and we considered it useful for this framework as well.

Planner: In our implementation, we opted for Dijkstra's path-planning algorithm [12,47]. Particularly for robotic applications, a wide variety of mission planning and path planning algorithms have been proposed [26,46]. OBJUST provides a flexible component-based framework that allows to easily exchange the planning component and use a different algorithm. The only requirement is for the algorithm to support multi-objective optimization, and that it can output an optimal path, along with the resulting costs of different quality objectives. For instance, A* [21] can be considered, as it is superior in time efficiency compared to Dijkstra [5,44]. For the heuristics, travel time can be approximated by using the Manhattan distance. For privacy or safety, it is not obvious what the admissible heuristics should be.

All quality objectives in the cost function are normalized to ensure that an objective with a generally higher cost cannot dominate others. We normalize objectives by comparing the cost for each quality objective in the current path to the lowest possible cost of that objective.

4.3   Visualization and Explanation

The framework *vis-network*[5] was chosen to display the graphs. An example of the visualization can be seen in Fig. 4 **4**.

All participants in the elicitation phase requested a tool to display multiple optimal paths at once since it facilitates a comparative analysis of alternative paths. When multiple paths overlap, it can become a visual clutter. Tooltips are useful in distinguishing overlapping paths. The mouseover tooltip feature that can be seen in Fig. 4 indicates the paths that are traversing a specific edge in the graph, as well as the properties of different edges or locations (R2). In the example, it can be seen that both the safety path, the privacy path, and the path optimizing for the cost function traverse the selected edge, which has a safety cost of 0. The tooltip provides an instant overview, removing the need for a separate legend for cost details.

Displaying the optimal path was considered the most crucial feature in the elicitation phase. The interviewees did not only state that they would like the framework to visualize the selected path, but also the optimal paths if you optimized for only one quality objective. Therefore, the tool indicates both the plan optimized for the global cost function (where the quality objectives' weights from the input elicitation are factored in), as well as the plans optimized for each individual quality objective. The various alternative paths are highlighted with different colors, so that users can easily distinguish what objectives a path was optimized for (R3). Examples can be seen in Fig. 4 in **2**, **3**, and **4**.

Table with Cost Overview: One feature indicates the costs of different paths in a table-like structure **2**. In the elicitation phase, several participants (Interviewees 1, 2, and 5) asked for a detailed table where the entries would contain different weights and the corresponding paths. The table color-codes the costs associated with each optimal path for easier identification. In Fig. 4, a few cost items in the list are highlighted in bold, implying that they represent the lowest cost for a specific quality objective. This feature aims to mainly combat the complexity that interviewees perceived during their tasks in the elicitation phase, removing the need to manually calculate each path's cost.

---

[5] https://visjs.github.io/vis-network/docs.

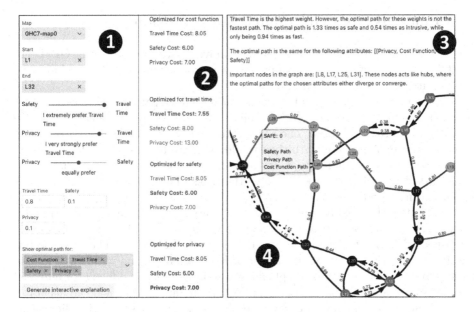

Fig. 4. Screenshot of the user dashboard, with input elicitation (1), cost overview (2), textual explanation (3), and plan visualization (4)

Textual Explanation: OBJUST provides sentence-based feedback to clarify visual information ❸ (R4). It is achieved using three features:

(i) Descriptive Text, clarifying why a specific path was chosen. In Fig. 4, it explains that even though *travel time* had the highest weight, the algorithm chose to optimize a path for other quality objectives because the difference in cost for safety and privacy was larger than the difference in *travel time*.

(ii) Equivalent Paths: This feature shows which objectives have the same costs for a path. It saves the users time when analyzing paths, especially in larger graphs.

(iii) Important Nodes (R5): This feature lists important nodes in the graph. These nodes act like hubs, where the optimal paths for the chosen objectives diverge or converge. The feature helps to reduce the complexity of larger graphs by segmenting them, enabling users to focus on a smaller portion of the graph.

5   Findings from Our Think-Aloud Sessions

The goal of our think-aloud sessions was to investigate to what extent OBJUST fulfills the needs of our participants, with a particular focus on the explanation capabilities and interactive dashboard.

We conducted think-aloud sessions with 7 participants, performed by the same researcher as in the elicitation phase. We worked with a smaller map (6 locations) and then a larger one (37 locations). The researcher asked the participants to explain what path they would estimate to be optimal for different combinations of weights for quality objectives. Subsequently, the participants were asked to utilize the tool to solve the tasks. The participants also ranked each feature's usefulness.

Regarding the perceived usefulness of the prototype, all participants strongly agreed that *"The tool allows me to accomplish my tasks"*. The participants were satisfied with the tool and considered it to save them time. The most useful features were the explainability features. Even when dealing with small maps, the complexity of robot mission planning is so high that it is difficult for humans to identify an optimal path manually. No participant could provide an accurate or satisfactory explanation of why a specific path was chosen in the elicitation phase without the aid of the prototype, except if only a single quality objective was prioritized. With the tool, participants could give increasingly better explanations after each task.

Reducing the Need for Time-Consuming Calculations: Without OBJUST, participants had to calculate multiple paths and their costs manually to understand which path was optimal. The operations were mainly a combination of sums and products. Still, many participants struggled with the calculations and finding an optimal path quickly. When evaluating the prototype, both the text-based features and the visualization of paths were considered helpful. For example, Part. 2 stated that *"displaying multiple paths was really nice to quickly get an overview, instead of having to calculate them by hand"*.

The text features were perceived as particularly helpful by participants with limited knowledge of algorithms. To complement the text features, the visual features provided a quick and simple explanation of what the outcome was, removing the need for a separate legend. Part. 7 stated that *"it was quick and easy to find the optimal path, and I could then use the other features to understand why it was chosen"*. The tooltips indicating equivalent paths were considered beneficial as well, as they reduced the need to examine multiple individual paths and helped participants focus on groups of paths instead.

Reducing Complexity: In the elicitation phase, we found that the complexity associated with finding the optimal path was challenging for our participants. Fig. 5 depicts that in the eyes of the participants, several features were helpful in reducing the perceived complexity of the planning problem. The participants deemed it crucial to display the optimal path in large graphs, as they couldn't see themselves calculating it by hand, no matter the time limit: *"There is no way I could ever find the optimal path in the large maps by myself, let alone multiple optimal paths."* (Part. 1).

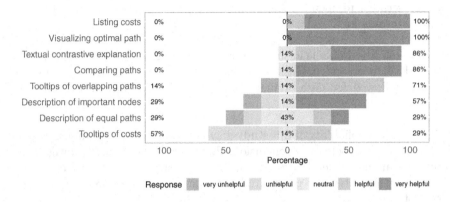

Fig. 5. Responses on features' *helpfulness in reducing the perceived complexity.*

The cost list and descriptive text were found useful in reducing the complexity: *"I found the cost list super useful because it allows me to easily compare the costs no matter how complex the paths are."* (Part. 8).

Part. 9 liked that *"the graph gets larger, but the descriptive text explanation remains concise"*. When asked which features were their favorites, every single participant mentioned the cost list. Part. 7 stated that they liked the *"display of costs for each path. It is good with the color differentiation and clear language"* .

6   Threats to Validity

External Validity: The number of study participants was limited, which constitutes a threat to external validity. Further workshops and more participants are needed to strengthen the findings. With a larger number of participants, we will be able to perform experiments in a controlled setting and draw statistically significant conclusions about the usefulness and usability of OBJUST.

Another issue is that different levels of experience might influence how useful the participants think the framework is. To mitigate this, participants with a mix of experience were chosen. The external validity is also compromised by the focus on an example with a fixed set of quality objectives. Furthermore, we have only investigated the problem in one example domain, with a single implementation using one path planning algorithm (Dijkstra). To further confirm the applicability of our framework in a broader context, additional evaluation is necessary, including a comparison with other mission planning algorithms. However, the main focus of this paper was not the use of mission planning algorithms, but rather developing an approach to improve understandability and reduce the perceived complexity for humans.

Construct Validity: The quality attributes used in our prototype were easy to understand for all participants. However, the participants in our study might not have the same interpretation of words such as "quality objective", "plan", or "preference". This could have led to issues in our qualitative analysis. It was therefore crucial to spend a few minutes with each participant to establish a common terminology.

Internal Validity: Possible misunderstandings in the interviews and analysis may have led to incorrect conclusions. Asking participants to answer Likert-scale questions allowed for data triangulation with the data from the think-aloud sessions.

Reliability: The reliability of this study may be influenced by our interpretations. They may have affected the conclusions drawn from the data. To mitigate this threat, we aimed to clearly describe our methods and keep a transparent chain of evidence. All interview guides and questionnaire answers were made public (See footnote 3).

7   Discussion and Future Work

We presented OBJUST, a framework for the interactive elicitation and explanation of quality-oriented mission planning. Our evaluation has indicated that the framework is useful for eliciting quality preferences, supported by explanations of their impact on generated plans. To the best of our knowledge, there is no approach that focuses on this gap and combines both visual and textual explanations.

In the following, we discuss three major findings:

Use of Contrastive Explanations: Notably helpful features were the concurrent display of optimal paths for different quality objectives, and their comparison using a list of their costs. Our findings confirm previous works about the usefulness of contrastive explanations [36,45].

Use of Elicitation Techniques: In previous works on elicitation, it was found that it is non-trivial for humans to understand the impact of a set of priorities on robot mission plans [41,51]. Therefore, OBJUST includes both an elicitation and an explanation component. We found that without a clear explanation, it is extremely difficult to understand what plan a given set of preferences leads to.

Visual and Textual Elements: Our participants stated that the visual features were useful to very quickly grasp the optimal path, compared to manually calculating it. Visualizations are useful for explaining *what* plan is deemed optimal. However, visual features are harder to generalize and require more development time compared to text-based features. To apply OBJUST to other systems, it would be necessary to design appropriate visualizations that are domain-specific.

Text-based features are great at explaining *why* a plan was deemed optimal. They are also highly generalizable for different types of robotic systems. The domain model/vocabulary used in the textual explanation can be adjusted, so that it is easy to generate explanations for another domain and system.

The presented evaluation is only a pilot study. In the future, we plan to conduct a study involving more participants with both technical and non-technical backgrounds. We would like to involve more practitioners and preferably no students. Such an evaluation would help us to assess whether the approach is applicable in practice and how much the explanations help end users. We envision different versions of OBJUST, depending on the system at hand and the concrete setup. The GUI of OBJUST can be used for users to monitor the real-time behavior of a robot, intervene when necessary, and specify different preferences depending on what is desired in a given context. We expect that the general conceptual framework presented in this paper can be reused and then tailored to specific contexts and systems. For contexts with many quality objectives and many possible locations, mechanisms are needed to hide and display relevant information, so as not to overwhelm the user.

Acknowledgments. This work was partially supported by the Wallenberg AI, Autonomous Systems and Software Program (WASP) funded by the Knut and Alice Wallenberg Foundation.

References

1. Baresi, L., Pasquale, L., Spoletini, P.: Fuzzy goals for requirements-driven adaptation. In: Proceedings of the 18th International Requirements Engineering Conference, pp. 125–134 (2010)
2. Bowers, K.M., Fredericks, E.M., Cheng, B.H.C.: Automated optimization of weighted non-functional objectives in self-adaptive systems. In: Colanzi, T.E., McMinn, P. (eds.) SSBSE 2018. LNCS, vol. 11036, pp. 182–197. Springer, Cham (2018). https://doi.org/10.1007/978-3-319-99241-9_9
3. Bryl, V., Giorgini, P., Mylopoulos, J.: Supporting requirements analysis in Tropos: a planning-based approach. In: Ghose, A., Governatori, G., Sadananda, R. (eds.) PRIMA 2007. LNCS (LNAI), vol. 5044, pp. 243–254. Springer, Heidelberg (2009). https://doi.org/10.1007/978-3-642-01639-4_21
4. Byrne, J.G., Barlow, T.: Structured brainstorming: a method for collecting user requirements. In: Proceedings of the Human Factors and Ergonomics Society Annual Meeting, vol. 37, pp. 427–431. SAGE Publications Sage CA: Los Angeles, CA (1993)
5. Candra, A., Budiman, M.A., Hartanto, K.: Dijkstra's and a-star in finding the shortest path: a tutorial. In: Proceedings of the 2020 International Conference on Data Science, Artificial Intelligence, and Business Analytics (DATABIA), pp. 28–32. IEEE (2020)
6. Chakraborti, T., Sreedharan, S., Grover, S., Kambhampati, S.: Plan explanations as model reconciliation. In: Proceedings of the ACM/IEEE International Conference on Human-Robot Interaction (HRI), vol. 2019-March, pp. 258–266. IEEE (2019)

7. Chen, S., Boggess, K., Feng, L.: Towards transparent robotic planning via contrastive explanations. In: Proceedings of the IEEE/RSJ International on Intelligent Robots and Systems (IROS), pp. 6593–6598 (2020)
8. Chen, T., Li, K., Bahsoon, R., Yao, X.: FEMOSAA: feature-guided and knee-driven multi-objective optimization for self-adaptive software. ACM Trans. Soft. Eng. Methodol. **27**(2), 1–50 (2018)
9. Cheng, B.H.C., Sawyer, P., Bencomo, N., Whittle, J.: A goal-based modeling approach to develop requirements of an adaptive system with environmental uncertainty. In: Schürr, A., Selic, B. (eds.) MODELS 2009. LNCS, vol. 5795, pp. 468–483. Springer, Heidelberg (2009). https://doi.org/10.1007/978-3-642-04425-0_36
10. Coulin, C., Zowghi, D., Sahraoui, A.E.K.: A situational method engineering approach to requirements elicitation workshops in the software development process. Softw. Process Improv. Pract. **11**(5), 451–464 (2006)
11. Dennis Ding, X., Englot, B., Pinto, A., Speranzon, A., Surana, A.: Hierarchical multi-objective planning: from mission specifications to contingency management. In: Proceedings of the International Conference on Robotics and Automation, pp. 3735–3742 (2014)
12. Dijkstra, E.W.: A note on two problems in connexion with graphs. Numer. Math. **1**(1), 269–271 (1959)
13. Driewer, F., Sauer, M., Schilling, K.: Discussion of challenges for user interfaces in human-robot teams. In: Proceedings of the 3rd European Conference on Mobile Robots (2007)
14. Eifler, R., Brandao, M., Coles, A., Frank, J., Hoffmann, J.: Evaluating plan-property dependencies: a web-based platform and user study. In: Proceedings of the International Conference Automated Planning and Scheduling, vol. 32, pp. 687–691 (2022)
15. Flick, U.: The SAGE Handbook of Qualitative Data Collection. SAGE Publications Ltd, Thousand Oaks (2017)
16. Franco, Á.J., Assar, S.: Leveraging creativity techniques in requirements elicitation: a literature review. Requirements Eng. Mag. **2016**(02), 1–15 (2016)
17. Garcia, M.P., Montiel, O., Castillo, O., Sepulveda, R., Melin, P.: Path planning for autonomous mobile robot navigation with ant colony optimization and fuzzy cost function evaluation. Appl. Soft Comput. **9**(3), 1102–1110 (2009)
18. Gasparetto, A., Zanotto, V.: Optimal trajectory planning for industrial robots. Adv. Eng. Softw. **41**(4), 548–556 (2010)
19. Goodrich, M.A., Schultz, A.C., et al.: Human-robot interaction: a survey. Found. Trends Hum.-Comput. Interact. **1**(3), 203–275 (2008)
20. Gulati, S., Jhurani, C., Kuipers, B., Longoria, R.: A framework for planning comfortable and customizable motion of an assistive mobile robot. In: Proceedings of the International Conference on Intelligent Robots and Systems, pp. 4253–4260. IEEE (2009)
21. Hart, P., Nilsson, N., Raphael, B.: A formal basis for the heuristic determination of minimum cost paths. IEEE Trans. Syst. Sci. Cybern. **4**(2), 100–107 (1968)
22. Hassan, S., Bencomo, N., Bahsoon, R.: Minimizing nasty surprises with better informed decision-making in self-adaptive systems. In: Proceedings of the 10th International Symposium on Software Engineering for Adaptive and Self-Managing Systems, pp. 134–145 (2015)
23. Hellström, T., Bensch, S.: Understandable robots-what, why, and how. Paladyn, J. Behav. Robot. **9**(1), 110–123 (2018)

24. Hoffmann, J., Magazzeni, D.: Explainable AI planning (XAIP): overview and the case of contrastive explanation (Extended Abstract). In: Krötzsch, M., Stepanova, D. (eds.) Reasoning Web. Explainable Artificial Intelligence. LNCS, vol. 11810, pp. 277–282. Springer, Cham (2019). https://doi.org/10.1007/978-3-030-31423-1_9
25. Iftikhar, M.U., Weyns, D.: ActivFORMS: Active formal models for self-adaptation. In: Proceedings of the 9th International Symposium on Software Engineering for Adaptive and Self-Managing Systems, pp. 125–134. ACM, New York (2014)
26. Karur, K., Sharma, N., Dharmatti, C., Siegel, J.E.: A survey of path planning algorithms for mobile robots. Vehicles 3(3), 448–468 (2021)
27. Krarup, B., Cashmore, M., Magazzeni, D., Miller, T.: Model-based contrastive explanations for explainable planning. In: Proceedings of the 29th International Conference on Automated Planning and Scheduling (2019)
28. Lera, F.J.R., Llamas, C.F., Guerrero, Á.M., Olivera, V.M.: Cybersecurity of robotics and autonomous systems: Privacy and safety. Robotics - Legal, Ethical and Socioeconomic Impacts (2017)
29. Li, N., Zhang, M., Li, J., Kang, E., Tei, K.: Preference adaptation: user satisfaction is all you need! In: Proceedings of the 18th Symposium on Software Engineering for Adaptive and Self-Managing Systems, pp. 133–144 (2023)
30. Liaskos, S., Jalman, R., Aranda, J.: On eliciting contribution measures in goal models. In: Proceedings of the 20th IEEE International Requirements Engineering Conference, pp. 221–230 (2012)
31. Lim, W.C.: Managing Software Reuse: a Comprehensive Guide to Strategically Reengineering the Organization for Reusable Components. Prentice-Hall, Inc., Hoboken (1998)
32. Mahdavi-Hezavehi, S., Durelli, V.H., Weyns, D., Avgeriou, P.: A systematic literature review on methods that handle multiple quality attributes in architecture-based self-adaptive systems. Inf. Softw. Technol. 90, 1–26 (2017)
33. Miller, T.: Explanation in artificial intelligence: insights from the social sciences. Artif. Intell. 267, 1–38 (2019)
34. Morandini, M., Penserini, L., Perini, A.: Towards goal-oriented development of self-adaptive systems. In: Proceedings of the 2008 International Workshop on Software Eng. for Adaptive and Self-Managing Systems, pp. 9–16 (2008)
35. Paucar, L.H.G., Bencomo, N.: Re-pref: support for reassessment of preferences for non-functional requirements for better decision-making in self-adaptive systems. In: Proceedings of the 24th International Requirements Engineering Conference, pp. 411–414. IEEE (2016)
36. Prabhushankar, M., Kwon, G., Temel, D., AlRegib, G.: Contrastive explanations in neural networks. In: Proceedings of the IEEE International Conference on Image Processing (ICIP), pp. 3289–3293. IEEE (2020)
37. Reynolds, O., García-Domínguez, A., Bencomo, N.: Automated provenance graphs for models@ run. time. In: Proceedings of the 23rd International Conference on Model Driven Engineering Languages and Systems: Companion Proceedings, pp. 1–10 (2020)
38. Saaty, R.: The analytic hierarchy process-what it is and how it is used. Math. Model. 9(3), 161–176 (1987)
39. Samin, H., Bencomo, N., Sawyer, P.: Pri-AwaRE: Tool support for priority-aware decision-making under uncertainty. In: Proceedings of the 29th Int'l Requirements Engineering Conference, pp. 450–451. IEEE (2021)
40. Samin, H., Bencomo, N., Sawyer, P.: Decision-making under uncertainty: be aware of your priorities. Softw. Syst. Model. 1–30 (2022). https://doi.org/10.1007/s10270-021-00956-0

41. Sawyer, P., Bencomo, N., Whittle, J., Letier, E., Finkelstein, A.: Requirements-aware systems: a research agenda for RE for self-adaptive systems. In: Proceedings of the 2010 18th IEEE International Requirements Engineering Conference, pp. 95–103. IEEE (2010)

42. Setchi, R., Dehkordi, M.B., Khan, J.S.: Explainable robotics in human-robot interactions. Procedia Comput. Sci. **176**, 3057–3066 (2020)

43. Shaikh, M.T., Goodrich, M.A.: Design and evaluation of adverb palette: A gui for selecting tradeoffs in multi-objective optimization problems. In: Proceedings of the ACM/IEEE International Conference on Human-Robot Interaction, pp. 389–397 (2017)

44. Soltani, A.R., Tawfik, H., Goulermas, J.Y., Fernando, T.: Path planning in construction sites: performance evaluation of the Dijkstra, A*, and GA search algorithms. Adv. Eng. Inform. **16**(4), 291–303 (2002)

45. Sukkerd, R., Simmons, R., Garlan, D.: Tradeoff-focused contrastive explanation for MDP planning. In: Proceedings of the 29th IEEE International Conference on Robot and Human Interactive Communication (RO-MAN), pp. 1041–1048. IEEE (2020)

46. Tipaldi, M., Glielmo, L.: A survey on model-based mission planning and execution for autonomous spacecraft. IEEE Syst. J. **12**(4), 3893–3905 (2017)

47. Wang, H., Yu, Y., Yuan, Q.: Application of Dijkstra algorithm in robot path-planning. In: Proceedings of the International Conference on Mechanic Automation and Control Engineering, pp. 1067–1069. IEEE (2011)

48. Wang, X., Liu, Z., Liu, J.: Mobile robot path planning based on an improved A* algorithm. In: Proceedings of the International Conference on Computer Graphics, Artificial Intelligence, and Data Processing, vol. 12604, pp. 1093–1098. SPIE (2023)

49. Weyns, D., Gerostathopoulos, I., et al.: Preliminary results of a survey on the use of self-adaptation in industry. In: Proceedings of the 2022 International Symposium on Software Engineering for Adaptive and Self-Managing Systems, pp. 70–76 (2022)

50. Wieringa, R.J.: Design Science Methodology for Information Systems and Software Engineering. Springer, Berlin (2014). https://doi.org/10.1007/978-3-662-43839-8

51. Wohlrab, R., Garlan, D.: A negotiation support system for defining utility functions for multi-stakeholder self-adaptive systems. Requirements Eng. **28**, 3–22 (2021)

52. Wohlrab, R., Meira-Góes, R., Vierhauser, M.: Run-time adaptation of quality attributes for automated planning. In: Proceedings of the 17th Symposium on Software Engineering for Adaptive and Self-Managing Systems for Adaptive and Self-Managing Systems, pp. 98–105 (2022)

Candidate Solutions for Defining Explainability Requirements of AI Systems

Nagadivya Balasubramaniam[✉], Marjo Kauppinen, Hong-Linh Truong, and Sari Kujala

Department of Computer Science, Aalto University, Espoo, Finland
{nagadivya.balasubramaniam,marjo.kauppinen,linh.truong, sari.kujala}@aalto.fi

Abstract. **[Context and Motivation]** Many recent studies highlight explainability as an important requirement that supports in building transparent, trustworthy, and responsible AI systems. As a result, there is an increasing number of solutions that researchers have developed to assist in the definition of explainability requirements. **[Question]** We conducted a literature study to analyze what kind of candidate solutions are proposed for defining the explainability requirements of AI systems. The focus of this literature review is especially on the field of requirements engineering (RE). **[Results]** The proposed solutions for defining explainability requirements such as approaches, frameworks, and models are comprehensive. They can be used not only for RE activities but also for testing and evaluating the explainability of AI systems. In addition to the comprehensive solutions, we identified 30 practices that support the development of explainable AI systems. The literature study also revealed that most of the proposed solutions have not been evaluated in real projects, and there is a need for empirical studies. **[Contribution]** For researchers, the study provides an overview of the candidate solutions and describes research gaps. For practitioners, the paper summarizes potential practices that can help them define and evaluate the explainability requirements of AI systems.

Keywords: Explainability Requirements · Explainable AI · Explainability Practices · AI Systems

1 Introduction

Several studies in the field of requirements engineering (RE) have indicated that explainability is a key quality requirement that must be defined when developing AI systems [6, 16, 17, 24, 25]. For example, the interview study of data scientists highlighted explainability as a new important requirement that focuses on explaining the decisions and models of machine-learning-based systems [24]. In addition, the analysis of the ethical guidelines indicated that companies and public organizations consider explainability to be essential part of the transparency of AI systems [6]. The need of explanations and establishing transparency are also considered as a vital part of autonomous car development [18, 19].

D. Mendez and A. Moreira (Eds.): REFSQ 2024, LNCS 14588, pp. 129–146, 2024.
https://doi.org/10.1007/978-3-031-57327-9_8

In addition to transparency, the definition of explainability requirements can support the development of trustworthy and responsible AI. For example, Langer et al. [20] highlight the role of explainability in building trustworthy systems and proposes techniques to audit the explanations from four crucial perspectives: technical, psychological, legal, and ethical. Two studies on the development of ethical AI systems report explainability as an important requirement when developing responsible AI [14, 15].

A systematic mapping study on RE for AI systems found limited empirical studies on explainability and it emphasizes the need for further research on explainability [25]. The goal of this paper is to analyze *what kind of candidate solutions are proposed for defining explainability requirements of AI systems*. We conducted a literature study using an experience-based approach [21]. This literature review is a part of our longitudinal study for which we have chosen the Technology Transfer Research Methodology (TTRM). One of the research activities of TTRM is to conduct literature studies in order to find candidate solutions for industrial challenges [23]. In this literature study, we found 12 publications that propose solutions for defining explainability requirements. The main contributions of this literature study are:

- We identified different types of candidate solutions, including frameworks, approaches, models, and catalogues for defining explainability requirements. This paper provides an overview of these proposed solutions by analyzing the content, basis, and empirical evaluation level of them. The literature study also revealed significant research gaps, such as the lack of empirical evaluations of the candidate solutions and the need for analyzing the current state of explainability in practice.
- Based on the analysis of 12 publications, we identified 30 practices that can support the development of explainable AI systems. We classified these practices to assist the elicitation, analysis, modelling, and validation activities of RE. We also identified practices for testing explainability requirements and explanations.

This paper is organized as follows. Section 2 presents the definition of explainability requirements and relevant literature reviews. In Sect. 3, we describe the research method of the literature study. Section 4 presents the results of the analysis of the candidate solutions and good RE practices for defining explainability requirements. Finally, we discuss our results and validity threats, and conclude the paper.

2 Related Work

In what follows, we first cover the definition of explainability requirements and provide an overview of the RE activities for defining explainability requirements. Finally, we describe how our study complements the existing literature studies related to explainability requirements.

Based on the definitions of explainability requirements [1] and explainability [10], we define the explainability requirements using the following four questions:

- Who are the relevant target groups that need explanations?
- What aspects of the AI system need to be explained to the target groups?
- In which context or situation, do the aspects need to be explained?
- Who explains the aspects to the target groups?

The definition of explainability requirements can follow a typical RE process. Requirements definition is often divided into the following four RE activities: elicitation, analysis, modelling, and validation [7, 24, 25]. The main purpose of elicitation is to discover the needs of users and other stakeholders [29]. Analysis is an activity in which the identified needs of stakeholders are analyzed and prioritized, and potential conflicts are resolved [11]. The purpose of modelling is to represent individual requirements in a systematic way and to visualize requirements relationships [11]. In validation, it is checked that the requirements are clear and feasible and that they meet the needs of the stakeholders [7].

We identified three relevant literature reviews on the explainability of AI systems in the RE field. The first study is a mapping study that identified 43 existing publications to specify and model requirements of AI systems [25]. This work delineated the potential challenges and limitations during AI system development and proposed explainability as one of the new requirements for building AI systems. The study highlights that explainability is one of the areas that need further research. Our literature study focuses especially on explainability requirements of AI systems and also includes publications that were not part of the mapping study. The mapping study covered papers from 2010-mid 2021. Our study includes publications from 2010 to Aug 2023.

The second literature study analyzed 79 papers and proposes six activities for developing explainable systems [13]. In addition, the study recommends 14 practices to support the six activities that are vision definition, stakeholder analysis, back-end analysis, trade-off analysis, explainability design and evaluation. In our literature study we especially focus on the RE activities and systematic definition of explainability requirements of AI systems. We also analyze this publication in detail, and it is one of the candidate solutions of this study.

The third literature study analyzed 58 papers and focused on identifying the characteristics of explanations and the evaluation methods to measure the impact of explanations on system quality [12]. This work proposed a quality framework that aims to facilitate the analysis, operationalization, and evaluation of explainability requirements. Our study complements this paper by suggesting a set of practices for defining the explainability requirements of AI systems. This is one of the candidate solutions that we analyzed in more detail in this study.

3   Research Methodology

The goal of the study is to analyze *what kind of candidate solutions are proposed for defining explainability requirements of AI systems*. To answer this high-level research question, we conducted a literature study using an experience-based approach. We selected this approach because it provides practical guidelines for the data collection and selection procedures [21]. Another reason for using this approach was that it supports young researchers in conducting their first literature study [21].

This literature study supports our long-term goal to do collaboration with practitioners using the Technology Transfer Research Methodology (TTRM). The core of TTRM is close industry-academia collaboration, the development of candidate solutions for industrial challenges, and the iterative empirical evaluation of candidate solutions

[23]. One of the research activities of TTRM focuses on identifying potential solutions for industrial challenges [23]. The industrial challenge of our study is to investigate how explainability requirements can be defined in practice. To tackle this challenge, we analyzed the candidate solutions using the following research questions (RQs):

- RQ1: Which candidate solutions have been proposed for defining the explainability requirements of AI systems?
- RQ2: What knowledge is the development of candidate solutions based on?
- RQ3: How has the candidate solution been evaluated?
- RQ4: Which RE practices does the candidate solution contain?

According to Kuhrmann et al. [21], the experience-based study design includes three phases: 1) preparation, 2) data collection and data cleaning, and 3) study selection. In the preparation phase, we defined the research questions, search queries, and the inclusion and exclusion criteria of the literature study.

The first author of the paper was responsible for the preparation phase and the second author reviewed the research questions, search strings, and inclusion/exclusion criteria. The first author tested two search strings: 'explainability requirement' or 'requirement AND explainability'. We decided to focus the search process on publications that specifically deal with explainability requirements and to use the search term 'explainability requirement' in this literature study. The inclusion criteria were: (I1) The paper introduces a new candidate solution or proposes an existing solution for defining explainability requirements; and (I2): The paper contains results related to RE and the explainability of AI systems. The exclusion criteria of this study were: (E1) The paper is not a peer-reviewed publication; and (E2): The paper does not focus explicitly on explainability requirements.

The second phase of this study was data collection and it consisted of both manual and automatic search. As recommended by Kuhrmann et al. [21], the first step of data collection was to identify appropriate data sources. Because the focus of this study was on requirements engineering (RE), we selected the following RE publication forums: RE journal, RE and REFSQ conferences, and workshops (AIRE and AI4ES) in the RE and REFSQ conferences. The first author of the paper performed a manual search on all these publication forums to identify papers related to explainability requirements. The first author read the title and abstract of all RE publications from Jan 2010 to Aug 2023, identified 26 papers related to explainability, and read the whole papers. After using the inclusion and exclusion criteria, seven papers were selected from the manual search.

We also performed automatic searches using the search term 'explainability requirement' in three standard libraries: IEEE Xplore, SpringerLink, and ScienceDirect. The first author identified 1783 articles and read the titles and abstracts of them to identify publications related to explainability requirements. Then, the author read the full papers using the inclusion/exclusion criteria. Based on this automatic search, five new papers were selected. This meant that a total of twelve publications were selected from the manual and automatic search processes.

In the analysis of the selected articles, we performed a thematic analysis using the research questions of this study. First, the line-by-line coding technique [9] was used for analyzing data related to the first three research questions (RQ1-RQ3). The first author performed this coding process for all 12 publications. Based on the coded publications, the first author wrote summaries of the main content of the candidate solution (RQ1), the basis of the candidate solution (RQ2), and the empirical evaluation level of the solution (RQ3). The second author reviewed the summaries based on the coded publications. The results of this analysis process are in Sect. 4.1.

The final analysis of the literature study focused on the practice-related codes (RQ4). The first author identified the practices for defining explainability requirements and categorized them together with the second author. Then, the first and second authors analyzed the codes/papers together to understand the context of these practices. After that, they mapped each practice to the RE activities, which are elicitation, analysis, modelling, validation, design, and evaluation of explainability requirements. The results of this analysis are in Sect. 4.2.

4 Results

4.1 Candidate Solutions for Defining Explainability Requirements

Our study aims to identify and analyze potential candidate solutions that can support practitioners in defining the explainability requirements of AI systems. Table 1 provides an overview of the candidate solutions proposed in the 12 publications. There are different types of candidate solutions, such as approaches, frameworks, models, principles, templates, and practices. Almost all these candidate solutions are based on the existing literature.

The proposed candidate solutions are based on the analysis of literature from several fields, such as user-centered design, human-computer interaction, philosophy, psychology, and RE. For example, Chazette et al. [10] performed an interdisciplinary literature study and used knowledge from the fields of computer science, philosophy, and psychology. The quality framework for explainability [12] is based on the analysis of explainability and human-computer interaction (HCI) literature. Moreover, the activities and practices to develop explainable systems [13] draws knowledge from the disciplines of RE and HCI. Two studies [3, 8] utilized user-centered design (UCD) literature to suggest candidate solutions to capture explainability needs.

There are two candidate solutions [2, 27] that bring in the knowledge from machine learning (ML) and Explainable AI (XAI) fields to define explainability requirements. Likewise, Phillips et al. [28], Habiba et al. [26], and Jovanovic and Schmitz [5] used XAI literature to propose their candidate solutions.

Only two studies [6, 8] included the analysis of the current state-of-the-practice when the candidate solutions were developed. Chazette and Schneider [8] conducted a survey with 107 end users to discover their opinions on the need of explanations and perceived advantages and disadvantages of explanations. The study indicated that explainability is a quality requirement that is closely associated to transparency [8]. Furthermore, Balasubramaniam et al. [6] analyzed the ethical guidelines of AI in 16 organizations and proposed a model and template that assist in defining explainability requirements of AI systems. The analysis of the ethical guidelines also highlighted that the organizations view explainability as a key part of transparency [6].

Only four studies [3, 6, 12, 27] report results on the evaluation of the candidate solutions. To assess the usefulness of the proposed solutions, three studies included

Table 1. Overview of the candidate solutions

Candidate solution	Basis of the candidate solution	Empirical evaluation
Köhl et al., 2019 [1] **Definitions** and **practices** to define and evaluate explainability requirements	Existing literature and conceptual analysis	An example – automated hiring system
Chazette and Schneider., 2020 [8] **User-centered design activities** to elicit and analyze explainability of a system	Existing literature	-
Nguyen et al., 2021 [2] **Holistic explainability approach** for end-to-end ML development	Existing literature and authors' knowledge	An example – base transceiver system
Phillips et al., 2021 [28] **Principles** for building explainable AI systems	Existing literature and authors' knowledge	-
Ramos et al., 2021 [3] **Approach** to capture users' explainability needs using personas	Existing literature	*Questionnaires with 60 users and 38 designers*
Chazette et al., 2022 [10] **Definition, conceptual model,** and **knowledge catalogue** for explainability	Interdisciplinary SLR of 229 papers and two workshops	An example – navigation system for autonomous cars
Chazette et al., 2022 [12] **Quality framework** for explainability	Literature study of 58 papers	*Case study – navigation app*
Chazette et al., 2022 [13] **Activities** and **practices** to develop explainable systems	Literature study of 79 papers and interviews of 19 participants	-
Habiba et al., 2022 [26] **Preliminary framework** to identify explainability needs of stakeholders	Existing literature and authors' knowledge	-
Jovanovic and Schmitz, 2022 [5] **Explainability-related attributes** to guide design and development	Existing literature and authors' knowledge	Examples of attributes in different domains
Balasubramaniam et al., 2023 [6] **Model** of explainability components, **template**, and **practices** to define explainability requirements	Ethical guidelines of 16 organizations, interviews of 3 practitioners and workshop with 7 practitioners	*A small-scale study - recruitment system*
Li and Han, 2023 [27] **Framework** to select explainability methods of ML systems	Existing literature and authors' knowledge	*Questionnaire and interviews of 10 students*

a light-weight evaluation: 1) a workshop with practitioners [6], 2) questionnaires with users and designers [3], and 3) questionnaire and interviews with students [27]. Chazette et al. [12] conducted a case study to evaluate the quality framework. They applied the framework to define the explainability requirements of an existing navigation system and to incorporate the explanations into the system. They also conducted an experiment to evaluate the quality and effect of the explanations. In the following part, we summarize the candidate solution for defining explainability requirements.

Köhl et al. [1] provide a starting point to the elicitation, modelling, and evaluation of explainability requirements. The study proposes *definitions* of three explainability concepts: 1) an explanation, 2) an explainable system, and 3) an explainability requirement. This work served as a pioneer study of explainability requirements for two candidate solutions, [6] and [10].

Chazette et al. [8] recommend *four iterative user-centered design (UCD) -based activities*: 1) discovering explainability needs, 2) creating different personas for explanations, 3) performing requirement negotiations and trade-off analysis, and 4) building prototypes and testing. They also propose concrete practices to be used in these activities. This paper brings in knowledge from the UCD field to support defining explainability requirements.

Nguyen et al. [2] propose a *holistic explainability approach* for end-to-end ML development that contains three key steps. The first step aims to identify the stakeholders who impact or would be impacted by the explainability aspects of ML service. The second step focuses on identifying the ML tasks and associated explainability aspects. In the third step, a template is used for collecting and documenting explainability requirements. This paper pointed out the important ML tasks and their associated explainability aspects such as quality of data, model selection, feature attribution, and training configuration.

Philips et al. [28] introduce *four principles* that are foundational for building explainable AI systems. They are: 1) the system gives explanations, 2) the explanations are meaningful to humans, 3) the explanations accurately reflect the system's process, and 4) the system communicates its knowledge limitations. In addition, the work proposes two essential characteristics of explanations, purpose, and style. The purpose of the explanation delineates why a person needs an explanation and what question the explanation intends to answer. The style of explanations helps in deciding how the explanations are delivered. The three key attributes that support in deciding explanation style are: level of detail, degree of human-machine interaction, and explanation format. Further, risk analysis, explanation trade-off analysis, and compliance with regulatory and legal requirements are emphasized as significant activities in building explainable AI systems. Moreover, the work indicates that it is important to evaluate the explainable AI algorithms with respect to the four principles proposed in this study.

Ramos et al. [3] investigate the process of creating personas to understand the users' explainability needs. In this work, the authors propose *a five-step approach to cover users' explainability needs*. As the first step, questionnaires are used to collect users' perceptions. Then, empathy maps are created based on the responses obtained from the questionnaire. Further, similar empathy maps are clustered together and then personas are created for modelling different user groups. Finally, the personas are validated with target users and designers. The feasibility of the approach was evaluated by an empirical study with 61 users and 38 designers. The results of the empirical study indicate that the proposed approach is good for creating explainability personas.

Chazette et al. [10] propose three candidate solutions to define explainability during the RE process and design. The researchers performed a systematic literature review (SLR) of 229 papers and two workshops to propose the solutions. The *definition of explainability* summarizes important variables such as aspects of a system that should be explained, context in which to explain, explainers that do the explaining, and addressees that receive the explanation. The *conceptual model* aims to visualize the impact of explainability on four potential quality dimensions: 1) users' needs, 2) cultural values, laws and norms, 3) domain aspects and corporate values, and 4) project constraints and system aspects. The *knowledge catalogue* summarizes whether the 57 quality aspects in the conceptual model are positively or negatively influenced by explainability. The catalogue highlights the importance of analyzing conflicts between explainability and other quality requirements.

Chazette et al. [12] present *a quality framework* that aims to facilitate the analysis, operationalization, and evaluation of explainability requirements. The framework has three key categories. The first category focuses on capturing the business goals, users' perceptions, and context of explainability. Defining the context helps in understanding how the system supports the users' goals. The second category covers three key characteristics that impact the system features: 1) when the explanations are given, 2) what aspects need to be explained and in what detail, and 3) how the explanations are delivered. The third category captures three key details that are required for a successful evaluation of explainability, they are: the method of evaluation, level of evaluation, and metrics of evaluation.

The study [12] evaluated the proposed quality framework in a navigation app for smartphones with a workshop with seven participants during the elicitation and design stage and with four participants during the evaluation stage. The study reveals the usefulness of the quality framework for the elicitation of explainability requirements. Moreover, the work suggests the need for a thorough design process for explanations to tackle design trade-offs.

Chazette et al. [13] recommend *six core activities and a set of practices* for the development of explainable systems. Based on the literature review, the six identified activities were mapped to requirements engineering, design and implementation, and validation and testing phases of software lifecycle development [13]. The first four activities vision definition, stakeholder analysis, back-end analysis, and trade-off analysis were mapped to the RE phase of explainable system development. Explainability design and evaluation activities were mapped to design and testing phases respectively. In addition, this work suggests a set of qualitative practices that support these six activities from the perspective literature and practitioners.

Habiba et al. [26] propose a ***preliminary five-step framework*** to define the explainability requirements of stakeholders. The purpose of the first step is to identify the relevant stakeholders who need explanations. The second step focuses on eliciting and documenting explainability requirements. Establishing a common vocabulary to improve the understandability of explainability is the third step. After that, the trade-off and uncertain requirements are negotiated and validated as an iterative process. Finally, the requirements are classified based on the stakeholders and the level of explanations needed.

Jovanovic and Schmitz [5] propose ***four explainability-related attributes*** which assist in delineating the application domains and to guide the design and development of explanation systems. These attributes are risk, user, timeline, and automation. The purpose of risk is to capture the negative consequences that could occur because of AI decisions. The user attribute specifies the primary users' domain knowledge and skills. The timeline attribute defines whether AI decision-making is done in real-time or with a flexible response time. The automation attribute focuses on identifying the level of autonomy in AI decision-making. This study highlights the importance of user requirements when defining the explainability of AI systems.

Balasubramaniam et al. [6] propose ***a model of explainability components***, a ***template***, and ***six good practices*** for defining the explainability requirements of AI systems. The model of explainability components assists in identifying the four key components of explainability, which are addressees (to whom to explain?), aspects (what to explain?), contexts (in what kind of situation to explain?), and explainers (who explains?). In addition, the model contains examples of each component. To model individual explainability requirements in a structured way, the paper proposes a simple template based on user stories. The lessons learned by applying the model and the template in practice are presented as six good practices for defining explainability requirements [6]. The practices emphasize the importance to understand the user and organizational processes and the definition of the purpose of the AI systems. This work used technology transfer research methodology (TTRM) to transfer the academic findings into industry. In this study, the model was tested with seven practitioners who are potential users of a recruitment game system to define the explainability requirements. The study reported that the model and the template were useful when defining the explainability requirements of AI systems.

Li and Han [27] propose ***a framework*** to choose explainability methods of ML systems systematically and automatically. The framework has three parts. In the first part, the problems are represented as explainability requirements, and they are modelled based on the goal-based approach. The second part focuses on modelling the explainability methods and their evaluation indicators. Next, the contexts of the explainability methods are modelled in relation to the real-world situation. The two types of contexts for choosing the explainability methods are data contexts and method contexts.

4.2 Practices for Defining Explainability Requirements

In this section, we describe the set of practices relating to explainability requirements that are identified from the candidate solutions of this study. Figure 1 shows practices that supports the definition of explainability requirements.

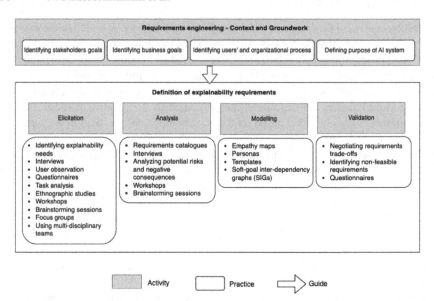

Fig. 1. Practices for defining explainability requirements

We identified four RE practices that are prerequisites for the definition of explainability requirements. The first practice is *identifying stakeholders' goals* captures who the stakeholders are and what are their explainability needs [2, 8]. The second practice of *identifying business goals* is from organization's viewpoint about how explainability could support their quality goals [12]. Next, *identifying users' and organizational process* supports in creating basis and context for the definition of explainability requirements [6]. The fourth practice suggests *defining the purpose of the AI system* from both users' and other stakeholders' perspective brings in the roles of different stakeholders in the AI system [6, 28].

Conducting interviews [2, 6, 8, 13] assist in the elicitation and analysis of explainability requirements. Chazette et al. [8] suggest interviews as a technique to discover the explanation needs and expectations of the users. In addition, interviews can assist in analyzing aspects such as when explanations are needed and how to design the explanations [2, 13].

Conducting user observation [13], *task analysis* [8, 28], *and ethnographic studies* [8] are helpful in the definition of explainability requirements in the elicitation and evaluation phases. Practitioners suggest end-user observation to identify the explanation needs and expectations of the users [13]. Similarly, task analysis and ethnographic studies are useful in discovering the users' needs and to gain a deeper understanding of the explainability context and the domain [8].

Using questionnaires [2, 3, 8, 13] aids the elicitation, validation, testing, and evaluation of explainability requirements. The studies highlight using questionnaires to collect the users' needs and requirements of explainability [2, 3]. Ramos et al. [3] propose to use questionnaires when validating the users' perceptions.

Conducting workshops [6, 8], *brainstorming sessions* [13], *and using multi-disciplinary teams* [6, 8] can help in understanding the behaviors of the system and to clarify technical, business, and environmental contexts of the explanations during the elicitation and analysis phases. One of the studies highlights that workshop with multi-disciplinary teams brings in both organizational-level and system-level perspective on explainability [6]. Practitioners suggest brainstorming to specify the explainability approach when explaining algorithms and to enhance communication between the team members [13]. Furthermore, *focus groups* can be used during the elicitation and evaluation of explainability requirements [13].

Creating requirements catalogues [1, 8, 10, 13] for requirements analysis helps to identify conflicts between explainability and other quality requirements such as privacy, security, and usability. Each company can develop its own requirements catalogues summarizing its own important requirements conflicts based on Chazette et al.'s [10] extensive requirements catalogue on explainability. In addition, requirements catalogs can support other practices such as creating soft-goal interdependency graphs (SIGs) [1, 8]. Further, requirements catalogues can be useful for developers to recognize quality-related conflicts and to acquire domain knowledge during explainability definition [13].

Analyzing potential risks and negative consequences [6, 28] helps in identifying stakeholders' fears and negative perceptions on using AI during the explainability requirements analysis.

Creating empathy maps [3] *and personas* [3, 13] are used for modelling the explainability needs. Empathy maps help in modelling six user emotions: says, does, sees, hears, feels, and thinks. Therefore, empathy maps can support in modelling the users' feelings and attitudes on explainability [3]. Personas help in representing target audiences and their explainability needs as fictional characters [3, 13].

Using templates [2, 6, 12] assists in representing individual explainability requirements of AI systems. Nguyen et al. [2] and Balasubramaniam et al. [6] in their work propose templates that assist in representing individual explainability requirements. In addition, Chazette et al. [12] used Volere's template to represent the explainability requirements of their case AI system.

Creating soft-goal interdependency graphs (SIGs) [1, 27] is essential to identify conflicts between the requirements. The SIGs aim to represent and document the explainability design and reasoning process. Furthermore, the soft goals help to clarify the interrelationships between the explainability requirements and to identify conflicting quality requirements [1].

Negotiating requirements trade-offs [26, 28] and *identifying non-feasible requirements* [26] are useful practices when validating explainability requirements. These practices are part of an iterative process which is important in validating especially when there is uncertainty and tradeoffs in the requirements. Furthermore, requirements catalogues can assist the requirements trade-off analysis [1].

4.3 Practices for Designing, Testing, and Evaluating Explainability

In this section, we classify the practices relating to the design, implementation, testing, and evaluation of explainability. Figure 2 highlights the practices that support in the design and evaluation of explainability.

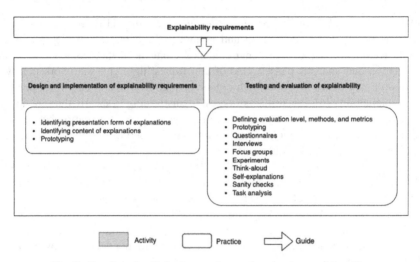

Fig. 2. Practices for designing, testing, and evaluating explainability

Identifying presentation form of explanations [10, 12, 28] is a practice that supports the design decision process when implementing explainability. The purpose of this practice is to specify how the explanations are given to users, for example, via text, audio, or visual. In addition, the tone of explanations i.e., casual tone or formal tone is also decided [10, 12].

Identifying content of explanations [12, 28] supports in clarifying what information is conveyed in explanations. This practice is useful in deciding whether the information in the explanations is generic or system-level, static or dynamic, abstract, or detailed [12].

Prototyping [8, 13] is used during the design, implementation, testing, and evaluation of the explanations and explainability requirements. Chazette et al. [13] highlight low-fidelity prototyping and high-fidelity prototyping to design various forms of explanations and to evaluate their effectiveness. Prototyping is also useful to understand the impact of explainability on user experience [8].

Defining evaluation levels, evaluation methods, and metrics [10, 12] is a starting point to evaluate the customers' overall explainability goals and the quality of explanations. It is important to decide whether to do the evaluation on the system level or on the explanation level. In addition, *questionnaires, interviews,* and *experiments* can be used as a method to gain end-user feedback on explanations. The evaluation metrics are defined depending on the contexts of explanations, e.g., soundness, completeness, and usefulness [10, 12]. *Questionnaires* [8, 12] can also help in the evaluation of explanations which by revealing whether the users understood the explanations given to them by the system [8, 12] and then assessing the quality of explanations [12].

Using think-aloud [1] and *self-explanations* [1] are useful for understanding users' perceptions of given explanations. Self-explanations reveal how well the users understand the given explanations [1]. In addition, *applying sanity checks* [28] with humans can be used to evaluate the accuracy of explanations.

Task analysis [28] can be used to evaluate the meaningfulness of explanations by asking humans to perform a task. Then, it is also possible to measure the time taken to complete the task and decision accuracy on the performed task [28].

5 Discussion

5.1 Candidate Solutions for Explainability Requirements of AI Systems

The goal of this literature study was to analyze what kind of candidate solutions have been proposed for defining explainability requirements of AI systems. The results show that researchers have already suggested a considerable number of different solutions for defining and testing explainability requirements. These solutions include approaches, frameworks, models, catalogs, and practices. Most of the proposed practices are existing RE practices and UCD methods that focus on a specific RE activity (e.g. elicitation, analysis, modelling, and validation of explainability requirements). The analyzed publications also contain more comprehensive solutions that can be used not only for RE activities but also for designing, testing, and evaluating the explainability of AI systems (e.g. [2, 12, 27]).

One important motivation of this literature study was to identify candidate solutions that practitioners can use when defining explainability requirements. Based on the longitudinal action research on RE process improvement [4], we recommend that practitioners focus first on a small set of existing RE and UCD practices when they start improving the definition of explainability requirements. Starting with existing good practices and tailoring them for explainability requirements can make their adoption easier in AI projects.

Recommendation 1: *Identifying explainability needs* [6, 10] is one of the key practices, as it forms the basis for the systematic definition of explainability requirements. The first step of this practice includes the discovery of target groups who need explanations, the aspects of the AI systems that need to be explained, and contexts in which target groups need explanations [6, 10]. In addition, one needs to define explainers who give explanations to the target groups [6, 10].

Recommendation 2: *Conducting interviews* and *workshops* with *multi-disciplinary teams* can assist in the identification of explainability needs. Likewise, Nuseibeh and Easterbrook suggest interviews as a elicitation technique [29]. Maalej et al. also recommend conducting interviews and workshops with stakeholders for responsible AI [15].

Recommendation 3: *Using requirements catalogues* during the trade-off analysis to identify the conflicts between explainability and other quality requirements such as privacy, security, and usability [1, 8, 10, 13].

Recommendation 4: *Using a simple template similar to user* stories helps representing the individual explainability requirements in a structured way [6]. Similarly, Inayat et al. [30] suggest user stories for agile RE, and Maalej et al. [15] for responsible AI.

Recommendation 5: *Using reviews and prototypes* can help the validation of explainability requirements. Inayat et al. also propose reviews and prototyping as good practices for agile RE [30].

Recommendation 6: *Prototyping* combined with *think-aloud* and *interviews* can be used for testing the explainability requirements and explanations Systematically defined explainability requirements form an important foundation for testing and evaluation of explanations.

5.2 Research Gaps

In addition to the recommendations for practitioners, the analysis of the candidate solutions revealed significant research gaps:

Research Gap 1: *Lack of empirical evaluations of candidate solutions.* This literature study revealed that most of the candidate solutions have not been evaluated in a real setting. This can lead to technology transfer problems where researchers develop new solutions, but organizations do not adopt these solutions. To support the industrial adoption of new technologies, researchers need to evaluate them in a real setting with real users and applications [22]. Wohlin and Runeson propose three research methodologies to ensure successful industry-academia research collaboration in software engineering [23]. They are Design Science, Action Research, and the Technology Transfer Research Methodology [23]. Empirical evaluations of candidate solutions can also contribute to the understanding of the current state-of-the-practice on explainability.

Research Gap 2: *Need for more knowledge about the current state-of-the-practice on explainability.* It is critical to understand the current state-of-the-practice to discover practitioners' viewpoints on explainability and developing explainable systems. For instance, two studies [24, 31] discussed important quality requirements for ML-based system from the perspective of practice. Vogelsang et al. [24] interviewed data scientists and reported the importance of explainability when developing ML systems. On the other hand, the study results in [31] revealed that explainability is not a critical quality requirement from the perspective of practitioners. These conflicting results can be a starting point for recognizing the perception of explainability in different organizations and the need for further empirical studies on explainability.

Research Gap 3: *Lack of understanding of the role of explainability requirements in the design, implementation, testing, and evaluation of explainability of AI systems.* We identified three solutions that integrate explainability requirements with design and evaluation. Nguyen et al. [2] propose a holistic approach that highlights the importance of linking explainability requirements to the ML process. In addition, Li and Han [27] suggest an explainability requirements analysis framework for automatic selection of explainability methods for ML systems. For an advanced evaluation of explainability, Chazette et al. propose a quality framework [12]. It is important to continue to investigate how explainability requirements guide the design and evaluation of AI systems.

Research Gap 4: *Need to better understand the current state-of-the-art of explainability.* In addition to the RE field, explainability is actively studied in the disciplines of XAI and HCI. It is also important to integrate knowledge from other fields into RE, which can support the implementation of explainability requirements in AI projects.

5.3   Threats to Validity and Limitations

There are three main validity threats in this literature study. Researcher bias in the selection process is one of the main threats to validity because the selection process was performed mostly by one researcher. It is possible that relevant publications were excluded during the selection process, although we tried to define the inclusion and exclusion criteria clearly based on the main research question.

The second validity threat relates to researcher bias in the analysis process. To reduce this bias, we used researcher triangulation. The first author performed the analysis based on the detailed research questions. To reduce bias, the first and second authors organized regular meetings throughout the analysis process and discussed which data should be used, how to analyze the data, and how to present the analysis results. The background of the first and second authors is in the field of requirements engineering, which can lead to unconscious bias. Therefore, they also organized meetings with the third author who brought a technical perspective to the analysis process.

Thirdly, the major concern for construct validity is that the number of the analyzed publications is only twelve. We note that we focused on publications that relate to explainability requirements or explainability in the RE field. This means that there can be potential solutions for the definition of explainability requirements and for the testing and evaluation of explainability of AI systems outside our focus area, for example, in the field of explainable AI (XAI). However, we see that defining explainability requirements has gained attention among RE researchers and hence bringing the current state-of-the-art is valuable to structure future research in the RE field.

The scope of the search process is the main limitation of our study. This literature study focused on the main RE publication forums and explainability requirements specifically. This means that our search process was not comprehensive. However, we believe that this study provides a starting point for systematic literature reviews. The publications of this literature study can be the start set for snowballing [32] and to finding out whether more candidate solutions for defining explainability requirements have been published outside the discipline of RE.

6 Conclusions

The goal of this study was to investigate what kind of candidate solutions are proposed for defining the explainability requirements of AI systems. The results show that researchers have already developed various types of solutions, including approaches, models, and frameworks to define explainability requirements. Furthermore, our paper suggests that there are existing RE and UCD practices that can be tailored for the systematic definition and evaluation of explainability requirements. The literature study also revealed that most of the proposed solutions have not been evaluated in real contexts. This leads to the conclusion that there is an urgent need to conduct empirical studies in which the solutions are used for defining and evaluating the explainability requirements of AI systems.

In the future, we aim to collaborate with companies to study the current state of the practice in defining the explainability requirements of AI systems. We also want to evaluate potential solutions with practitioners to identify benefits and challenges in implementing the solutions. Our long-term goal is to conduct empirical studies on the definition of explainability requirements in real contexts and to investigate which practices are useful in different stages of the development of explainable AI systems.

References

1. Köhl, M.A., et al.: Explainability as a non-functional requirement. In: International Requirements Engineering Conference, pp. 363–368 (2019)
2. Nguyen, M.-L., Phung, T., Ly, D.-H., Truong, H.-L.: Holistic explainability requirements for end-to-end machine learning in IoT cloud systems. In: International Requirements Engineering Conference Workshops (REW), pp. 188–194 (2021)
3. Ramos, H., Fonseca, M., Ponciano, L.: Modeling and evaluating personas with software explainability requirements. In: Ruiz, P.H., Agredo-Delgado, V., Kawamoto, A.L.S. (eds.) Human-Computer Interaction: 7th Iberoamerican Workshop, HCI-COLLAB 2021, Sao Paulo, Brazil, September 8–10, 2021, Proceedings, pp. 136–149. Springer, Cham (2021). https://doi.org/10.1007/978-3-030-92325-9_11
4. Kauppinen, M., Vartiainen, M., Kontio, J., Kujala, S., Sulonen, R.: Implementing requirements engineering processes throughout organizations: success factors and challenges. Inf. Softw. Technol. 46(14), 937–953 (2004)
5. Jovanovic, M., Schmitz, M.: Explainability as a user requirement for artificial intelligence systems. Computer 55(2), 90–94 (2022)
6. Balasubramaniam, N., Kauppinen, M., Rannisto, A., Hiekkanen, K., Kujala, S.: Transparency and explainability of AI systems: from ethical guidelines to requirements. Inf. Softw. Technol. 159, 107197 (2023)
7. Kotonya, G., Sommerville, I.: Requirements Engineering. Wiley, Hoboken (1998)
8. Chazette, L., Schneider, K.: Explainability as a non-functional requirement: challenges and recommendations. Requirements Eng. 25(4), 493–514 (2020)
9. Charmaz, K.: Constructing Grounded Theory, 2nd edn. Sage publications (2014)
10. Chazette, L., Brunotte, W., Speith, T.: Explainable software systems: from requirements analysis to system evaluation. Requirements Eng. 27(4), 457–487 (2022)
11. Sommerville, I., Sawyer, P.: Requirements Engineering: A Good Practice Guide. Wiley, Hoboken (2004)
12. Chazette, L., Klos, V., Herzog, F., Schneider, K.: Requirements on explanations: a quality framework for explainability. In: Requirements Engineering Conference, pp. 140–152 (2022)

13. Chazette, L., Klünder, J., Balci, M., Schneider, K.: How can we develop explainable systems? Insights from a literature review and an interview study. In: International Conference on Software and System Processes and International Conference on Global Software Engineering, pp. 1–12 (2022)
14. Arrieta, A.B., et al.: Explainable Artificial Intelligence (XAI): concepts, taxonomies, opportunities and challenges toward responsible AI. Inf. Fus. **58**, 82–115 (2020)
15. Maalej, W., Pham, Y.D., Chazette, L.: Tailoring requirements engineering for responsible AI. IEEE Comput. **56**(4), 18–27 (2023)
16. Habibullah, K.M., Horkoff, J.: Non-functional requirements for machine learning: understanding current use and challenges in industry. In: International Requirements Engineering Conference, pp. 13–23 (2021)
17. Balasubramaniam, N., Kauppinen, M., Kujala, S., Hiekkanen, K.: Ethical Guidelines for solving ethical issues and developing AI systems. In: Morisio, M., Torchiano, M., Jedlitschka, A. (eds.) PROFES 2020. LNCS, vol. 12562, pp. 331–346. Springer, Cham (2020). https://doi.org/10.1007/978-3-030-64148-1_21
18. Koopmann, B., et al.: Challenges in achieving explainability for cooperative transportation systems. In: International Requirements Engineering Conference Workshops (REW), pp. 114–119 (2022)
19. Cysneiros, L.M., Raffi, M., Leite, J.C.S.P.: Software transparency as a key requirement for self-driving cars. In: International Requirements Engineering Conference (RE), pp. 382–387 (2018)
20. Langer, M., Baum, K., Hartmann, K., Hessel, S., Speith, T., Wahl, J.: Explainability auditing for intelligent systems: a rationale for multi-disciplinary perspectives. In: Requirements Engineering Conference Workshops (REW), pp. 164–168 (2021)
21. Kuhrmann, M., Fernández, D.M., Daneva, M.: On the pragmatic design of literature studies in software engineering: an experience-based guideline. Empir. Softw. Eng. **22**(6), 2852–2891 (2017)
22. Ivarsson, M., Gorschek, T.: Technology transfer decision support in requirements engineering research: a systematic review of REj. Requirements Eng. **14**, 155–175 (2009)
23. Wohlin, C., Runeson, P.: Guiding the selection of research methodology in industry–academia collaboration in software engineering. Inf. Softw. Technol. **140**, 106678 (2021)
24. Vogelsang, A., Borg, M.: Requirements engineering for machine learning: perspectives from data scientists. In: Requirements Engineering Conference Workshops (REW), pp. 245–251 (2019)
25. Ahmad, K., Abdelrazek, M., Arora, C., Bano, M., Grundy, J.: Requirements engineering for artificial intelligence systems: a systematic mapping study. Inf. Softw. Technol. **158**, 107176 (2023)
26. Habiba, U.-E., Bogner, J., Wagner, S.: Can requirements engineering support explainable artificial intelligence? Towards a user-centric approach for explainability requirements. In: International Requirements Engineering Conference Workshops (REW), pp. 162–165 (2022)
27. Li, T., Han, L.: Dealing with explainability requirements for machine learning systems. In: Computers, Software, and Applications Conference (COMPSAC), pp. 1203–1208 (2023)
28. Phillips, P.J., et al.: Four principles of explainable artificial intelligence. National Institute of Standards and Technology (U.S.), NIST IR 8312 (2021)
29. Nuseibeh, B., Easterbrook, S.: Requirements engineering: a roadmap. In: International Conference on Software Engineering, pp. 37–46 (2000)
30. Inayat, I., Salim, S., Marczak, S., Daneva, M., Shamshirband, S.: A systematic literature review on agile requirements engineering practices and challenges. Comput. Hum. Behav. **51**(27), 915–929 (2015)

31. Habibullah, K.M., Gay, G., Horkoff, J.: Non-functional requirements for machine learning: understanding current use and challenges among practitioners. Requirements Eng. **28**(2), 283–316 (2023)
32. Wohlin, C., Kalinowski, M., Felizardo, K.R., Mendes, E.: Successful combination of database search and snowballing for identification of primary studies in systematic literature studies. Inf. Softw. Technol. **147**, 106908 (2022)

Artificial Intelligence for Requirements Engineering

Opportunities and Limitations of AI in Human-Centered Design a Research Preview

Anne Hess[1]([✉]) [ID], Thomas Immich[2], Jill Tamanini[1] [ID], Mario Biedenbach[1] [ID], and Matthias Koch[1] [ID]

[1] Fraunhofer IESE, 67663 Kaiserslautern, Germany
{anne.hess,jill.tamanini,mario.biedenbach,
matthias.koch}@iese.fraunhofer.de
[2] Centigrade GmbH, 66123 Saarbrücken, Germany
thomas.immich@centigrade.de

Abstract. *[Context and motivation]* AI has significantly increased its capabilities and popularity since the emergence of Large Language Models. Generative AI, in particular, shows potential to support a variety of RE activities. *[Question/problem]* While the opportunities of AI in RE are being discussed, there is little reflection on the limitations and concerns regarding the use of AI. Moreover, holistic investigations of these aspects within the software engineering lifecycle are sparse. *[Principal ideas/results]* We propose a research agenda that aims to systematically investigate the potential of AI within a human-centered design (HCD) process to derive meaningful application scenarios and recommendations for AI. *[Contribution]* In this research preview, we share initial results of workshop sessions conducted with RE and UX experts to determine opportunities and limitations of AI within the HCD process and provide insights into ongoing research activities on the example of "persona agents".

Keywords: Human-centered design · Artificial intelligence · Research agenda

1 Introduction

Artificial Intelligence (AI) is a mega trend, and many approaches and solutions have emerged that illustrate the potential of AI. In the RE community, too, various papers have been published that report on successful applications and solution ideas to support RE-related activities with the help of AI and related subfields such as machine learning (ML) [1–3]. The intended support provided by these techniques mainly focuses on activities that are cumbersome for humans because large amounts of information need to be kept in mind and/or processed. Besides the elicitation of requirements [4, 5], subsequent analysis and specification activities have been investigated with respect to the potential of AI support, such as prioritization of requirements [6, 7] or quality assurance of requirements [8, 9]. With the recent breakthroughs in the realm of Large Language Models (LLMs), in particular the ChatGPT versions publicly available since late 2022, new horizons and discussions have emerged that aim to explore the potential of these technologies for RE-related activities [10–12].

D. Mendez and A. Moreira (Eds.): REFSQ 2024, LNCS 14588, pp. 149–158, 2024.
https://doi.org/10.1007/978-3-031-57327-9_9

While we see many publications about using AI for certain RE activities, we claim that systematic and holistic investigations of the opportunities of AI (respectively LLM-based models) within the software engineering lifecycle as well as of limitations and concerns regarding the application of AI (such as [10]) are rare. Such investigations could serve as a valuable guideline, especially for practitioners, and be helpful for SMEs that observe the ongoing development and discussion but lack the necessary resources to explore AI technologies for their individual development processes.

We take the widely known human-centered design (HCD) approach as a baseline for our investigations [13]. It involves the human's – in particular the user's – perspective at every stage of the development process to enhance the usability and user experience (UX) of interactive systems. We follow this approach in our research and industry projects and have been experiencing a variety of challenges, which motivated us to define a research agenda as part of our research-industry collaboration that aims to explore the opportunities and limitations of applying AI within the HCD lifecycle. We envision consolidating the results and lessons learned in our research activities and incorporating them into concrete guidelines illustrating meaningful application scenarios and recommendations to help practitioners choose appropriate AI-based support for their respective HCD activities, and to motivate future research.

The remainder of this paper is structured as follows: Sect. 2 takes a deep-dive into our vision of using generative AI to create and interact with "persona agents" as a motivating example of a potential AI-based application scenario in HCD that we are currently investigating as part of our research agenda. In Sect. 3, we provide insights into the aforementioned research agenda and related research questions, including initial results from first explorative studies. The paper concludes with a summary and an outlook on next steps in Sect. 4.

2   Persona Agents – a Motivating Application Scenario

This section provides insights into ongoing research activities on the example of persona agents, which we recently initiated as part of our research agenda (cp. Section 3).

2.1   Challenges from an Industry Perspective

Personas, as descriptive models of real users, are used to act as user proxies to compensate for the lack of real user involvement within the software engineering process [14]. In practice, personas face many challenges, not least the time it takes to create and maintain them. In addition, manual persona development is criticized because it is often based on small samples, one-time data collection, and non-algorithmic methods [15]. In addition, personas are heavily dependent on their creators, who often create stereotypes or mirrors of themselves [16]. This may be reflected in their usage, as they are mainly used by those who produced them. Their potential breadth of use is thus not being exploited.

At the same time, personas are often underutilized or even neglected due to skepticism about them, lack of user access, or lack of available time. Skepticism often arises from the "non-algorithmic" nature of the method. Engineers criticize its subjective nature and potential lack of empirical rigor. They typically favor methodologies grounded in quantitative analysis and empirical data [17]. And even if personas *are* being used, their utilization is much higher in the early phases of a project, while at later stages, they are often "forgotten" [17]. From an RE and UX perspective, this is unfortunate, as the factual resolution of a user requirement cannot be proven without matching quantitative usage data analytics and qualitative usability testing insights with personas created based on assumptions and data gained during early user research activities. We conclude that despite their potential for closing such "requirements loops", personas are "dying" along the HCD process, as their value in relation to the effort being spent to maintain them seems to decrease over time.

2.2 Solution Ideas

Even though many challenges can be observed, the continuous usage of personas is a valuable tool throughout the HCD process as it enhances empathy towards future users [18]. This motivated us to elaborate concrete solution ideas for improving the utilization of personas as part of our collaboration, which brings together the view of professional UX service providers requiring UX methods to be economically feasible and the scientific and goal-oriented view of researchers regarding the definition and execution of research and evaluation activities. The first core research question that we aim to investigate can be stated as: *To what extent can AI help to expand the utilization of personas over all activities of the HCD process (cp. Figure 1)?*

We hypothesize that AI, respectively LLMs, have great potential to overcome the aforementioned challenges. We envision that development team members could "get in touch" with personas by leveraging generative agents [19] that can be chatted or video-conferenced with (we call such generative agents that represent a specific persona "persona agents").

To increase the longevity and value of personas we have elaborated an initial (still hypothetical) set of guidelines/requirements that persona agents, respectively conversations with persona agents, should fulfill. These include:

- Persona agents should be easily accessible to all team members, just as if they were real people (e.g., through a standard address book or via MS Teams chat/call).
- Persona agents should have background stories and personality traits so that they are engaging for team members and stakeholders to communicate with.
- Persona agents' utterances should be marked as being certain or risky, depending on whether they are backed up by user research or based on human assumptions or generated by an LLM. In any case, available user research data should be made accessible in the respective communication context.
- During the conversation with a persona agent, a neutral assistant should indicate which user research methods could best serve to remove any risky responses by a persona agent.

2.3 Research Activities

As a next step, we will perform empirical investigations in real project settings in which persona agents are actively used by development teams. Thereby, we aim to assess and evaluate the influence of the persona agents and related guidelines/requirements regarding the utilization of personas throughout a complete HCD lifecycle. Using suitable data collection measures (such as observations, data logging, interviews/surveys), we aim to investigate metrics such as:

- *Frequency of visits*: How often is a persona agent being visited or contacted by a team member?
- *Number of user research activities*: How many user research activities have been triggered because the persona agent conversation made their necessity obvious?
- *Number of design decisions*: How many design decisions are based on the conversation insights with the persona agent?
- *Frequency of reference*: How often is the persona agent referred to by the team (e.g., how often does a team member use the persona agent's name during a conversation or forward the agent's contact information to another team member)?

We are aware that generative persona agents acting like "real people" might introduce ethical risks (e.g., crossing of human boundaries, emotional dependencies) and limitations (see also Sect. 3). This motivates the investigation of further research questions, which we aim to investigate in the future, such as: what is the effect of emotionally engaging the AI persona agent? To what extent should the persona agent be authentic, i.e., act exactly like the user (e.g., not admitting mistakes, hiding the truth)?

To differentiate our research from published work on AI-based personas (e.g., [20, 21]), we see a core contribution in the utilization of LLMs to interact with "persona agents" rather than "temporary personas". That is, creating "ad-hoc" persona agents in real-world projects that can be communicated with, that evolve by augmenting user research insights over time, making use of their long-term memory and their capability to plan, execute, and even interact with themselves autonomously [22], offers new possibilities to investigate the aforementioned metrics across a process in the longer term. With generative AI enabling personas to 'come to life', we anticipate that our research will lead to groundbreaking developments at this new intersection.

3   AI-Supported Human-Centered Design – Objectives and Preliminary Results

The idea of persona agents introduced in the previous section is part of a larger research agenda that looks into possible AI support for activities within the HCD process. The core research objective underlying this research agenda aims to *investigate opportunities for AI as well as possible limitations and concerns from the viewpoint of team members (e.g., requirements engineers, designers, software architects, developers, testers) as users of AI-supported solutions, respectively as consumers of AI-generated results.* Specifically, we want to find answers to the following research questions (RQ):

- RQ_1: Which challenges exist in HCD activities that could be streamlined or simplified by involving AI-based solutions?
- RQ_2: Which opportunities do AI-based solutions offer to tackle the challenges?
- RQ_3: What limitations, concerns, or risks exist from the viewpoint of team members with regard to the usage of AI-based solutions in HCD? And how to mitigate them?

As a first starting point of our research activities towards these goals and questions, we ran moderated discussion sessions at Fraunhofer IESE with a total of 14 RE/UX experts to explore and discuss challenges within typical HCD activities (see Fig. 1) experienced by these experts in their projects. In addition, the experts were invited to reflect on opportunities, limitations, and concerns regarding AI support within the corresponding activities. In addition, one of the authors co-organized a workshop at the "Mensch und Computer" conference in 2023, with the objective of identifying opportunities and limitations of AI in HCD activities from the viewpoint of RE/UX experts from both research and industry. In the following, we share the first insights we gained during the aforementioned workshop sessions.

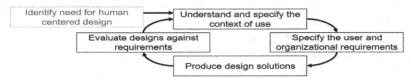

Fig. 1. Human-centered design process according to ISO 9241–210:2019 (based on [13])

3.1 AI Support for Usage Context Analysis

The objective of this HCD activity is to understand the conditions under which a software is or will be used. This includes the users, tasks, resources, as well as physical and social usage environments. To support this activity, various methods and techniques are available both for the elicitation of relevant information (e.g., interviews, surveys, observations) and for the documentation of the elicited data, such as personas, empathy maps, user journey maps, task descriptions, physical models, etc.

Challenges in the Analysis of Contexts of Use:

- The elicitation of relevant information (via surveys, interviews, fieldwork, etc.) is highly dependent on the qualifications of the moderator, consumes a lot of time and financial means (e.g., travel expenses), and is therefore often not done adequately.
- It is often difficult to recruit not only enough stakeholders/users, but also the right ones, who are motivated and/or available to participate in elicitation activities.
- The creation of personas as a valuable artifact in HCD is time-consuming, and personas are not used sufficiently (cp. Section 2).

Opportunities for Tackling the Challenges with the Help of AI:

- Free-text fields in surveys, respectively any data/notes elicited during interview sessions, field observations/contextual inquiries, etc. could be quickly analyzed and clustered with the help of AI (*increased efficiency*),
- AI could provide support in training skills regarding questionnaire design and/or conducting interviews (respectively in validating questionnaires/interview guidelines) to eliminate any biases (*improved quality of elicited data*).

- In the event of insufficient access to relevant stakeholders, AI could serve as a proxy to collect relevant knowledge (e.g., about relevant user groups, domain knowledge, workflows, pain points) (*compensate for missing data*).

Limitations and Concerns: While the elicitation of usage context information via AI might bring benefits, it also raised concerns during the workshop discussions. These are reflected in the following questions, which we aim to incorporate into future research activities:

- How reliable is AI-generated data?
- Is there a risk of drawing false conclusions through "hallucinations" (i.e., responses generated by an AI that contain false or misleading information presented as fact)?
- Does the elicited data represent "real" needs and information that "human" stakeholders would have stated/shown?
- How "believable" are AI-based artifacts from the viewpoint of their consumers (e.g., interaction designers, developers, testers)?
- To what extent is it possible to feel empathy towards the represented user groups based on AI-generated artifacts or "persona agents" (cp. Section 2)?

3.2 AI Support for Requirements Specification

The objective of the second activity is to derive, analyze, and specify (non-)functional requirements and communicate these requirements to the team members. A typical outcome of this activity is a requirements specification comprising various RE-related artifacts (such as stakeholders, goals, vision, task and use case descriptions, system functions, quality requirements, etc.).

Challenges in the Specification of User Requirements:

- The derivation and specification of requirements including their quality assurance (e.g., to avoid conflicts, inconsistencies, ambiguities, incompleteness, etc.) is a complex and time-consuming task that also requires skill and experience.
- Establishing a shared understanding of requirements, respectively requirements-related information (such as vision, goals, needs, etc.), in the team is difficult due to different role-specific information needs, insufficient communication of changes, inappropriate or too many different communication channels, insufficient onboarding of new team members, and social issues within the team.
- Ideation activities aiming to identify novel ideas and requirements are time-consuming (e.g., organizing, running, as well as preparing, analyzing, and documenting the results of creativity workshops requires several days, if not weeks).
- Creating variations and evaluating their potentials according to project goals is time-consuming, subjective, and requires a lot of experience.

Opportunities for Tackling the Challenges with the Help of AI:

- With the help of AI, requirements engineers could benefit during the preparation of requirements-related information (e.g., structuring information consistently, generating role-specific views, addressing team members' information needs) (*increased quality/acceptance of requirements specifications*).
- Quality assurance activities can be supported by AI, making it easier to trace them and maintain an overview (*increased quality of requirements specifications*).
- Initial experiences have also shown that AI has the potential to support the identification of (novel) solution ideas for given problems (*solution-finding*) as well as assessment and opportunity evaluation for relevant markets/user groups.
- New forms of requirements artifacts such as "living" personas could be more engaging for team members and motivate them to consume requirements-related information more frequently/intensively (*increase in engagement and motivation*).
- In the event of insufficient access to relevant stakeholders, AI could serve as a proxy to identify/specify requirements (*compensate for missing data*).

Limitations and Concerns: While the experts see a lot of potential in supporting requirements specification and quality assurance with the help of AI (not least because of the availability of already proven AI-based solutions for these activities), the discussions of limitations and concerns were mainly related to the reliability and "believability" of AI-generated requirements and raised questions similar to those outlined in Sect. 3.1.

3.3 AI Support to Produce and Evaluate Design Solutions

The last two activities of the HCD lifecycle focus on the development of design solutions (third activity) and their continuous evaluation (fourth activity) according to the previously specified requirements respectively to new emerging requirements in the space of agile development. All team members and stakeholders constantly work together and communicate with each other to fulfill all necessary tasks. Both activities are highly interrelated as the design solutions typically evolve in various iterations from low-fidelity prototypes to high-fidelity solutions based on results and feedback gained during the evaluation activities.

Challenges in the Design and Evaluation Activities:

- Communicating solution concepts and interaction flows to the team and to stakeholders at various abstraction levels in order to synchronize / validate requirements and get (early) feedback is very time-consuming and prone to errors (e.g., as requirements might have changed or been misunderstood).
- Designs as a visual means must still be documented in a textual manner to explain their meaning, inner logic, and interplay. The handover process (from designers to developers) is therefore time-consuming, provides friction, and is prone to error.
- Accessibility, both in terms of visual design (sufficient contrast, font sizes, or color settings, etc.) and comprehensibility (simple text), is not only becoming increasingly important, but is also required by law in many countries. There is also a great need for more inclusiveness to represent marginalized groups within solutions that are aimed

mostly at binary norm-typical users only. However, many standards are unfamiliar, and implementation of related requirements is time-consuming.

- Following HCD principles, design solutions should be consistently evaluated from the perspective of future users starting from early concepts to final solutions. However, it is also just as hard to recruit suitable representatives of user groups (or usability experts as proxies) for evaluation activities as it is for the elicitation activities (see Sect. 3.1). Similarly, the analysis of evaluation data is time-consuming.

Opportunities for Tackling the Challenges with the Help of AI:

- AI offers the possibility of ad-hoc generation of visualizations in meetings or workshops, so everyone can contribute to the solution (*supports co-design*). By creating a concrete image, ambiguities are eliminated (*increased shared understanding*). AI can transform a hand-drawn scribble into a polished design and thereby reduce effort to a minimum (*increased efficiency*). In fact, co-design with AI is another way to *reduce redundant work* in a collaborative manner, e.g., changes to the design resulting from new findings can be easily initiated and updated at all necessary points. The capabilities of AI to read images and understand requirements, could be used to map designs to their targeted requirements, thereby highlighting discrepancies (*improved consistency, conformity to requirements*)
- Established patterns, best practices and guidelines, e.g., for accessibility and inclusiveness, could be utilized by the AI to suggest/create coherent designs (*conformity to standards*). Moreover, AI can be a powerful tool in creating copy for the targeted audience within the available space in no time (*increased efficiency*). With AI, the interface can be specifically adapted to the specific user groups and their needs (*efficient adaptation to marginalized user groups*).
- AI-based personas (or persona agents as introduced in Sect. 2) could be invited into meetings where features are discussed and evaluated regarding their potential impact. The proxies' feedback facilitates decision-making, especially in later stages when real usage data can be matched and synchronized with proxies. This allows assumptions about the product or its users to be validated constantly (*continuous evaluation, compensate for missing access to user representatives*).

Limitations and Concerns: In addition to the limitations stated in Sect. 3.1, the experts see the challenge of finding the right level of closeness to the proxy personas and defining the desired depth and type of contact with them. Even a warning that the proxy is not a real person can be forgotten if the proxy performs convincingly. There could be too much empathy, subjecting the team to excessive pressure.

4   Summary and Next Steps

In this paper, we provided insights into the objectives and initial results of ongoing research activities that aim to investigate opportunities, limitations, and concerns regarding AI-supported HCD as part of our research-industry collaboration.

Besides first insights gained during workshop sessions conducted with RE/UX experts, we shared our vision of using generative AI to create and interact with "persona agents" as a motivating and promising example of an AI-based application scenario in HCD. We claim that the capabilities of "persona agents" have great potential to increase the longevity and value of personas and to provide a novel and valuable contribution within the landscape of AI-based persona approaches.

We are aware that in the literature promising reflections respectively AI-based solutions can be found that directly contribute to our research questions. To consolidate the existing knowledge and experience, we plan to conduct a systematic literature review to further explore such synergies and to map these findings to our research questions. Moreover, we plan to run empirical studies such as case studies and/or experimental investigations in the context of our projects in order to further explore existing solutions and to draw conclusions regarding potentials, limitations, concerns, and risks in HCD activities. We envision that the insights and experiences gained during our research activities will ultimately be consolidated and incorporated into a publicly available guideline that could serve as a valuable source of information both for researchers to contribute to the research agenda and for practitioners aiming to learn about the potentials and limitations of AI within HCD processes. We believe that the application of AI in practical projects will tremendously improve the level of user centricity and UX maturity of teams and organizations.

References

1. Dalpiaz, F., Niu, N.: Requirements engineering in the days of artificial intelligence. IEEE Softw. **37**(4), 7–10 (2020). https://doi.org/10.1109/MS.2020.2986047
2. Zhao, L., et al.: Natural language processing for requirements engineering: a systematic mapping study. ACM Comput. Surv. **54**(3), 1–41 (2022). https://doi.org/10.1145/3444689
3. Liu, K., Reddivari, S., Reddivari, K.: Artificial intelligence in software requirements engineering: state-of-the-art. In: IEEE 23rd International Conference on Information Reuse and Integration for Data Science (IRI), San Diego, CA, USA, pp. 106–111 (2022)
4. Sharma, S., Pandey, S.K.: Integrating AI techniques in requirements elicitation. In: Proceedings of International Conference on Advancements in Computing & Management (ICACM) 2019. https://doi.org/10.2139/ssrn.3462954
5. Lim, S., Henriksson, A., Zdravkovic, J.: Data-driven requirements elicitation: a systematic literature review. SN Comput. Sci. **2**(1) (2021). https://doi.org/10.1007/s42979-020-00416-4
6. Lunarejo, M.I.L.: Requirements prioritization based on multiple criteria using artificial intelligence techniques. In IEEE 29th International Requirements Engineering Conference (RE), Notre Dame, IN, USA, pp. 480–485 (2021)
7. Qayyum, S., Qureshi, A.: A survey on machine learning based requirement prioritization techniques. In: Proceedings of the 2018 International Conference on Computational Intelligence and Intelligent Systems, Phuket Thailand, pp. 51–55 (2018)
8. Hayes, J.H., Payne, J., Leppelmeier, M.: Toward improved artificial intelligence in requirements engineering: metadata for tracing datasets. In: IEEE 27th International Requirements Engineering Conference Workshops (REW), Jeju Island, Korea (South), pp. 256–262 (2019)
9. Sinpang, J.S., Sulaiman, S., Idris, N.: Detecting ambiguity in requirements analysis using mamdani fuzzy inference. J. Telecommun. Electr. Comput. Eng. (JTEC) **9**, (3–4), 157–162 (2017). https://jtec.utem.edu.my/jtec/article/view/2936

10. Arora, C., Grundy, J., Abdelrazek, M.: Advancing requirements engineering through generative AI: assessing the role of LLMs (2023). https://arxiv.org/abs/2310.13976
11. White, J., Hays, S., Fu, Q., Spencer-Smith, J., Schmidt, D.C.: ChatGPT prompt patterns for improving code quality, refactoring, requirements elicitation, and software design (2023). https://arxiv.org/abs/2303.07839
12. Zhang, J., Chen, Y., Niu, N., Wang, Y., Liu, C.: Empirical evaluation of ChatGPT on requirements information retrieval under zero-shot setting (2023). https://arxiv.org/abs/2304.12562
13. International Organization for Standardization: ISO 9241–210:2019 Ergonomics of human-system interaction: Part 210: Human-centred design for interactive systems. Standard (2019)
14. Karolita, D., McIntosh, J., Kanij, T., Grundy, J., Obie, H.O.: Use of personas in requirements engineering: a systematic mapping study. Inf. Softw. Technol. **162**, 107264 (2023). https://doi.org/10.1016/j.infsof.2023.107264
15. Xu, W.: AI in HCI Design and User Experience (2023). https://arxiv.org/abs/2301.00987
16. Emmanuel, G.S., Polito, F.: How related are designers to the personas they create? In: Soares, M.M., Rosenzweig, E., Marcus, A., Eds., Lecture Notes in Computer Science, Design, User Experience, and Usability: Design Thinking and Practice in Contemporary and Emerging Technologies. Springer International Publishing, pp. 3–13 (2022). https://doi.org/10.1007/978-3-031-05906-3_1
17. Salminen, J., Jansen, B.J., An, J., Kwak, H., Jung, S.-G.: Are personas done? Evaluating their usefulness in the age of digital analytics. Persona Stud. **4**(2), 47–65 (2018). https://doi.org/10.21153/psj2018vol4no2art737
18. Salminen, J., Guan, K., Jung, S.-G., Jansen, B.J.: A survey of 15 years of data-driven persona development. Int. J. Hum.-Comput. Interact. **37**(18), 1685–1708 (2021). https://doi.org/10.1080/10447318.2021.1908670
19. Park, J.S., O'Brien, J.C., Cai, C.J., Morris, M.R., Liang, P., Bernstein, M.S.: Generative agents: interactive simulacra of human behavior (2023). https://arxiv.org/abs/2304.03442
20. Zhang, X., et al.: PersonaGen: a tool for generating personas from user feedback (2023). http://arxiv.org/pdf/2307.00390v2
21. Kocaballi, A.B.: Conversational AI-Powered Design: ChatGPT as designer, user, and product (2023). http://arxiv.org/pdf/2302.07406v1
22. Qian, C., et al.: Communicative agents for software development (2023). http://arxiv.org/pdf/2307.07924v4

A Tertiary Study on AI for Requirements Engineering

Ali Mehraj$^{(\boxtimes)}$, Zheying Zhang , and Kari Systä

Tampere University, Tampere, Finland
{ali.mehraj,zheying.zhang,kari.systa}@tuni.fi

Abstract. Context and Motivation: Rapid advancements in Artificial Intelligence (AI) have significantly influenced requirements engineering (RE) practices. **Problem:** While many recent secondary studies have explored AI's role in RE, a thorough understanding of the use of AI for RE (AI4RE) and its inherent challenges remains in its early stages. **Principal Ideas:** To fill this knowledge gap, we conducted a tertiary review on understanding how AI assists RE practices. **Contribution:** We analyzed 28 secondary studies from 2017 to September 2023 about using AI in RE tasks such as elicitation, classification, analysis, specification, management, and tracing. Our study reveals a trend of combining natural language process techniques with machine learning models like Latent Dirichlet Allocation (LDA) and Naive Bayes, and a surge in using large language models (LLMs) for RE. The study also identified challenges of AI4RE related to ambiguity, language, data, algorithm, and evaluation. The study gives topics for future research, particularly for researchers who want to start new research in this field.

Keywords: Tertiary study · Requirements engineering · Artificial intelligence · Machine learning · Natural language processing

1 Introduction

Requirements engineering (RE) plays a crucial role in software development to ensure that the final software product meets its stakeholders' needs and expectations and delivers value to them [14]. It is characterized by challenges such as obtaining accurate inputs, analyzing often ambiguous and conflicting information, and integrating RE processes with broader software product development activities [5,14]. With the rapid advancements in artificial intelligence (AI), various AI techniques are increasingly used to improve the efficiency of software engineering (SE) tasks, including RE. This integration of AI for SE (AI4SE) or AI for RE (AI4RE) has been explored by researchers, particularly in identifying and analyzing requirements.

In the context of our ongoing task for the project on 6G software[1], which focuses on extremely distributed and heterogeneous massive networks of connected devices, the role of AI in RE tasks emerges as one key challenge. One

---

[1] http://tinyurl.com/3y86t8ne.

© The Author(s), under exclusive license to Springer Nature Switzerland AG 2024
D. Mendez and A. Moreira (Eds.): REFSQ 2024, LNCS 14588, pp. 159–177, 2024.
https://doi.org/10.1007/978-3-031-57327-9_10

task in this project funded by Business Finland and part of 6G Bridge[2] program, aims to leverage AI to address challenges in RE. To initiate this endeavor, we conducted a comprehensive analysis of the current state of AI applications in RE. Our objective is to understand the specific RE tasks where AI is used, identify the AI techniques in use, and assess their impacts. To achieve the goals, we formulated 3 research questions (RQs).

- RQ1. What are the RE tasks where AI is applied?
- RQ2. What AI techniques are used in RE tasks?
- RQ3. What are the challenges for implementing AI for RE?

We conducted a tertiary study, analyzing existing secondary studies in the field to present a broad perspective on AI's role in RE. The structure of the paper is as follows. Section 2 describes the related work - mainly other tertiary studies in the area. The research method - including the research questions are described in Sect. 3. The results are given in Sect. 4 and their implications are discussed in Sect. 5. The threats to validity are discussed in Sect. 6 and finally the conclusions are given in Sect. 7.

2   Related Work

There are several tertiary studies investigating the intersection of RE and the application of AI in SE tasks. Kotti et al. [11] conducted a comprehensive review of 83 secondary studies, focusing on the state of ML techniques in SE. Similarly, Cico et al. [3] reviewed 11 studies published between 2009 and 2020, aiming to identify software development tasks where ML techniques have been applied. Both studies [3,11] were broader in scope, encompassing SE at large, with only part of their studies dedicated to RE and related tasks. In the field of RE, Bano et al. [1] presented the first tertiary study to categorize RE topics reported in secondary studies. The study reviewed 53 papers published from 2006 to 2014, evaluating their quality and identifying potential research gaps. Additionally, Kudo et al. [12] examined four secondary studies on requirements patterns, published between 2011 and 2018, and proposed a preliminary research agenda on this topic. However, these early studies [1,12] did not cover topics related to AI techniques.

Moreover, there are other tertiary studies in the literature, each with a different focus such as the method for conducting a systematic review in SE [9] and the quality concerns in the secondary studies [2]. Yet, none of these studies specifically explored the state-of-the-art in AI to assist RE practices. Our work presents the first tertiary study with this focus, aiming to identify the use of AI techniques in RE and the current challenges in implementing AI4RE.

---

[2] https://tinyurl.com/mr3yssfk.

3   Research Method

In this section we describe the literature search strategy and data extraction from the selected studies. We adopted comprehensive guidelines by Kitchenham et al. [7,8] to establish our search strategies and the review protocol before conducting the tertiary study.

Weekly meetings were held with all authors to review the progress of the research and to discuss further actions. A shared spreadsheet was used to facilitate collaboration among the authors.

3.1   The Search and Selection Process

We conducted our search between 01.09.2023 and 19.09.2023 in four databases: SCOPUS, ACM Digital Library, IEEE Xplore, and Web of Science. All the searches were based on title, keywords and abstract except for ACM Digital Library where we included search on full text for a few keywords. We further conducted a free search on Google Scholar and arXiv to find relevant papers.

All selected studies were filtered using the inclusion and exclusion criteria set by the authors. The studies were then reviewed by reading the full text and the studies that fulfilled the quality assessment criteria set by the authors were selected.

Define Search String. The search string was developed iteratively, drawing from keywords in the research questions, and expanded with synonymous terms. The string comprises three parts: a) keywords restricted to secondary studies, b) keywords related to AI and ML, and c) keywords related to RE.

We used the logical operator "AND" to connect these parts, ensuring precise search results. Within each part, initially, primary keywords like "literature review" and "mapping studies" for Part a, "AI", "NLP", etc. for Part b, and "requirements engineering" and "requirements management" for Part c were selected. To broaden the search scope, we included synonyms, abbreviations, and related terms, using the wildcard symbol (*) for variations of specific keywords. These terms within each part were connected with "OR" to maximize relevance in the search results. To ensure both the consistency and comprehensiveness of our search string, we refined it by examining terms from papers from initial search results. Keywords selected for generating our search string are listed in Table 1. We also adjusted the string's syntax to match the syntax requirements of different databases.

Identify Inclusion and Exclusion Criteria. With the inclusion and exclusion criteria, we ensured the relevance of the selected secondary studies. To be included, a study must address the use of AI techniques to assist RE activities and meet at least one other from criteria 2–5. Our exclusion criteria rule out non-English articles, duplicates, and entire conference volumes. We also exclude studies not directly related to RE, those presenting requirements analysis for AI

Table 1. Keywords for the search string

Keyword type	Keywords
Secondary study related	SLR, systematic literature review, literature review, mapping stud*, secondary stud*, meta-analy*
AI related	artificial intelligence, AI, machine learning, ML, NLP, Natural language processing, LLM, Large language model
RE related	requirements engineering, requirements specification, requirements identification, requirements elicitation, requirements analysis, requirements prioritiz*, requirements validation, requirements acquisition, requirements extraction, requirements classification, requirements change, requirements volatility, requirements tracing, requirements management, requirements provision, requirements quality, requirements review, requirements dependency, user stor*, user feedback*, user opinion*, user review

systems, and those that discuss AI in contexts unrelated to RE or SE. Detailed inclusion and exclusion criteria are presented in Table 2.

Table 2. Inclusion and Exclusion Criteria

Inclusion	1. Secondary studies on AI applied to RE tasks
	2. Papers analyzing specific AI techniques and tools in RE
	3. Papers on specific methods or approaches for integrating AI in RE
	4. Papers assessing the quantity or quality effects of AI on RE
	5. Papers addressing challenges of implementing AI for RE
Exclusion	1. Paper not written in English
	2. Paper not focusing on the subject of RE
	3. Paper focusing on RE for AI
	4. Paper presenting AI in a context unrelated to RE
	5. Duplicate papers
	6. Entire conference proceedings books or volumes

Selection of Studies. After defining the inclusion and exclusion criteria, we conducted our first search on SCOPUS. This yielded a total of 122 results. The first author then independently screened the search results, applying the inclusion and exclusion criteria. Whenever there was ambiguity regarding a paper's relevance, the first author reviewed the full paper and discussed it with the second author to reach a consensus on its inclusion. From SCOPUS, 33 studies were included.

Subsequent searches were conducted on ACM Digital Library, IEEE Xplore, and Web of Science, resulting in 16, 22, and 39 papers respectively. Because the SCOPUS search had already identified most of the relevant studies, removal of the duplicates left 36 unique papers.

Each of these 36 papers underwent both forward and backward snowballing. The forward snowballing found 460 papers, out of which 3 met our criteria. In backward snowballing only one paper from 1808 references was considered relevant, bringing our total to 40 papers.

Recognizing the emerging interests in trending topic on generative AI in SE, we expanded our search to include Google Scholar and pre-publication archive arXiv. The former added one paper and the latter 3. This resulted in a final selection of 44 papers for our review. The selection of studies from different sources is illustrated in Fig. 1.

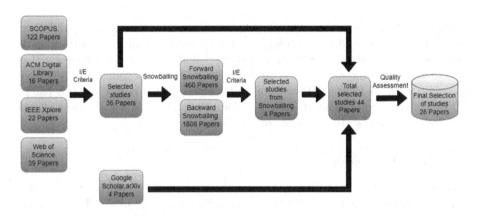

Fig. 1. Study Search and Selection

Assessment of Quality. We ensured the quality of the selected papers with quality assessment step. Assessment questions are listed in Table 3. Questions QA1 and QA2 assess the studies' relevance. The other questions evaluate the research methods and the validity and reliability of the review findings.

The assessment of the studies was based on points and the maximum number of points is 8. Studies that have at least 6 points were selected as primary studies for this research. To emphasize the importance of a paper's relevance to our research topic, questions focusing on this aspect, i.e. QA1 and QA2, are assigned a weight of 2 points each. The other quality assessment questions focus on the quality of the studies. Therefore, QA3, QA4, QA5, and QA6 are assigned a standard weight of 1 point each.

Conducting the initial quality assessment process resulted in the selection of 32 studies. The first author then identified the initially selected studies with SP identifier. After going through the papers thoroughly with all authors, we further adjusted the quality assessment scores for SP18 due to inaccessible primary

Table 3. Grading scales for QA1-2: 0 = irrelevant, 1 = partially relevant, 2 = completely relevant; QA3-6: 0 = undefined, 0.5 = partially defined, 1 = well defined

No	Question
QA1	Does the secondary study clearly focus on AI4SE or SE activities?
QA2	Does the secondary study clearly focus on AI4RE or RE activities?
QA3	Are the inclusion or exclusion criteria described and appropriate? Is the search strategy clear enough?
QA4	Was the methodology of the study clearly outlined, consistently applied, and does it adhere to recognized standards or guidelines for systematic reviews?
QA5	Were the findings from primary studies synthesized systematically, with potential biases or heterogeneity among those studies adequately discussed?
QA6	Did the authors discuss potential biases or limitations in their review process, and were there efforts made to minimize them?

studies and SP22 due to the study being a tertiary review itself with a different focus. Thus, these studies were removed. Furthermore, we discovered that there were two sets of articles (SP2 and SP8, SP9 and SP13) with different titles but shared the same authors, research objectives, and questions. We removed the older versions (SP8, SP13) as duplicates and kept the newer versions (SP2, SP9). Thus, the final count of our selected studies was 28. The assessment scores and selected studies are accessible in the shared spreadsheet[3].

3.2 Data Extraction

Data extraction was conducted by reading the full text of the selected studies. We collected statistical data for each paper, including publication year, data sources, timeframe of selected primary studies, type of study, guidelines, and number of primary studies. Alongside collecting statistical data, we collected data related to our research questions. We initially collected data on AI techniques reported in studies on RE and their relevance to our research questions. We soon realized that the extracted data should specify AI for particular RE tasks, and added an additional data item of specific RE task(s). The data extraction template used in our study is represented in Table 4.

Table 4. Data extraction template

Publication details	Title, Authors, Publisher, Publication Year
Research Context	Research aims/objectives, Research questions, Type of study, Databases used for search, Number of primary studies, Dataset years, Secondary study guideline, RE tasks, AI techniques, Results, Relevance to research questions
Insights	Insight for researchers, Insight for practitioners, Future work

---

[3] https://doi.org/10.6084/m9.figshare.24564829.v2.

4   Results

The secondary studies selected for this research are listed in Appendix A and cited as SP1-SP32 in the following discussion. Notably, studies SP8, SP13, SP18, and SP22 were excluded from the list in the quality assessment phase. The final selection comprises 28 studies, published between 2017 and 2023. The selection consists 8 systematic mapping studies and 20 systematic literature reviews. Figure 2. illustrates their distribution over time, with most published in 2020 and later. Of these, 16 papers were published in journals, 9 in conferences, and 3 as pre-publications on arXiv.

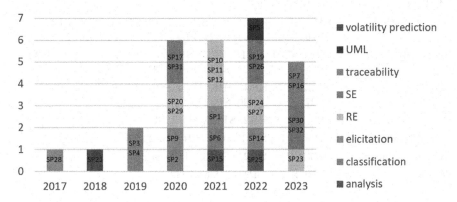

Fig. 2. Year-wise distribution of the selected studies

The selected studies were grouped into eight categories based on their research topics inferred from the title and abstract of the paper, as shown in Fig. 2. Among the 28 papers, 8 (29%) are dedicated to studies on AI4RE and 6 (21%) on AI4SE. The remaining 14 papers (50%) review the studies on specific RE tasks, such as requirements elicitation, classification, analysis, traceability, modeling, and requirements volatility prediction.

Notably, more than two secondary studies have specifically reviewed the topic of AI for requirements elicitation or analysis. To ensure accuracy and mitigate double-counting issues when synthesizing the studies' findings, we used a Corrected Covered Area (CCA) matrix [2,13] to evaluate primary study overlaps in these secondary studies. The CCA value was calculated as 2 for studies on requirements elicitation (i.e., SP1, SP9, SP14, and SP28) and 1.8 for studies on requirements classification (SP2, SP3, SP4, SP6), indicating only slight overlap - a value over 10 would indicate high overlap. This minimal overlap implies a very low risk of double-counting bias, ensuring the reliability of the aggregated data to address RQ1 and RQ2.

4.1   RQ1. What Are the RE Tasks Where AI Is Applied?

Analyzing the selected studies, we found that requirements elicitation, analysis, and classification emerged as the three most reviewed RE tasks where the application of AI techniques is most frequently investigated. As shown in Table 5, studies with identifiers in bold represent reviews concentrating on a distinct RE task, while the other identifiers are broader reviews on AI4RE or AI4SE which contain relevant discussions on these RE tasks.

Table 5. AI applied RE tasks discussed in the selected studies

RE Task	#	Study Identifier
Req. elicitation	17	**SP1**, SP4, **SP9**, SP10, SP11, SP12, **SP14**, SP17, SP19, SP20, SP23, SP24, SP25, SP27, **SP28**, SP29, SP32
Req. analysis	16	SP10, SP11, **SP15**, SP17, SP19, SP20, SP23, SP24, **SP25**, SP26, SP27, SP28, SP29, SP30, SP31, SP32
Req. classification	13	**SP2, SP3, SP4, SP6**, SP10, SP12, SP20, SP24, SP25, SP27, SP29, SP30, SP32
Req. traceability	9	SP7, SP10, SP12, **SP16**, SP19, SP24, SP27, SP29, SP32
Req. specification	9	**SP5**, SP10, SP11, SP12 SP17, SP23, SP24, SP25, SP27
Req. management	6	**SP7**, SP10, **SP21**, SP23, SP24, SP26
Others	5	SP10, SP19, SP24, SP30, SP32

Requirements elicitation is a critical process in understanding the context of a software product and identifying the stakeholders' needs, expectations, and constraints as requirements to be fulfilled. AI plays an important role in automatically analyzing large volumes of user feedback across various platforms. Systematic review SP1 of 68 studies emphasizes AI's role in data-driven requirements elicitation, while another review SP14 of 86 studies focuses on ML techniques. These studies also identified common data sources for AI-assisted requirements elicitation, including user reviews, discussion forums, social media, and crowdsourcing platforms. For instance, SP9 analyzed 12 studies on ML's use in identifying software requirements on Stack Overflow, while SP28 reviewed 24 studies on opinion mining from app store reviews and ratings. Other data sources include software repositories, product descriptions, and system-user interaction logs. Additionally, specialized datasets like the DePaul's NFRs corpus [4], PURE [6], and SecReq [10] are widely used in research on training ML models to effectively differentiate and classify requirements.

In AI-assisted requirements analysis, the primary studies have reviewed and identified several important tasks, including ambiguity detection (SP10, SP20, SP24, SP27, SP32), which resolves lexical, syntactic, semantic, and pragmatic ambiguities in requirements documents; quality analysis (SP11, SP20, SP26, SP27, SP30), which identifies requirements completeness and template adherence; user review analysis (SP15, SP24, SP25, SP27, SP28, SP31, SP32), which extracts insights from user feedback on satisfaction, preferences, and product expectations; security and vulnerability analysis (SP10, SP19, SP24), which identify potential system threats and weaknesses to ensure security requirements are adequately defined and addressed; and requirements verification (SP10, SP26, SP17, SP11), which confirms that requirements meet the necessary standards and conditions.

Requirements classification organizes requirements information into distinct groups based on specific criteria such as functionality, priority, and quality attributes. This assists in effectively managing requirements to ensure that all requirements are appropriately addressed throughout the software development lifecycle. However, this process can be laborious and error-prone. AI, particularly NLP and ML, provides enhancement by automating classification tasks. The review studies indicate these techniques effectively prioritize requirements (SP3, SP10, SP20, SP25, SP27) and distinguish requirements in various categorizations (SP2, SP3, SP4, SP6, SP10, SP12, SP20, SP24, SP25, SP27, SP29, SP30, SP32) such as including quality attribute categories, functionality, security requirements, and business rules. Notably, study SP4 reviews the research on the use of AI to analyze user feedback into categories such as sentiments, intentions, user experiences, and other relevant topics, assisting in identifying emerging requirements, preferred features, and change requests, among others.

Requirements traceability studies primarily address three distinct tasks: traceability link generation (SP12, SP16, SP7, SP24, SP27, SP32), traceability link recovery (SP10, SP12, SP16, SP19, SP32), and bug localization (SP16, SP32). Traceability link generation creates and manages relationships between requirements and other software artifacts like issues, user stories, commits, source code, test cases, or regulations, aiding product development and project management. Traceability link recovery focuses on re-establishing lost or unclear connections between requirements and associated software artifacts. Bug localization aims at the process of identifying the specific source code files, functions, or lines of code that may be responsible for a reported bug or defect. It typically involves analyzing bug reports or issue descriptions provided by users or testers and mapping them to the corresponding segments of the source code. AI improves traceability link accuracy, reduces human effort, provides recovery information, and enhances trustworthiness, especially in large, complex software projects with vast repositories, diverse artifacts, and extensive codebases.

Requirement specification involves documenting descriptions of the required behavior and quality properties of the software. AI assistance in requirement specification primarily focuses on textual requirements documentation (SP10, SP11, SP17, SP23, SP25) and model generation (SP5, SP10, SP12, SP27). The process of model generation and transformation typically includes identifying modeling concepts from textual descriptions to construct conceptual models. Specifically, SP5 systematically reviews 39 selected studies that applied various NLP techniques to generate UML diagrams from textual specifications.

Requirements management, as addressed in studies SP7, SP10, SP23, and SP26, involves organizing and managing software requirements. This focus area encompasses the documentation, organization, and control of requirements for a software project or system throughout its entire life cycle. AI assistance in change management was reviewed in studies such as SP7, SP10, and SP24, covering the management and control of changes to the software project or system requirements. Requirement volatility prediction is the subject of the review study SP21, focusing on forecasting or estimation of the likelihood and extent of changes or modifications to the software project requirements over time.

Some studies (SP19, SP24, SP30, SP32) addressed RE broadly or specified the RE tasks as 'other'. SP24 reviewed AI's use in requirements reuse, leveraging existing requirements artifacts to expedite new software system development.

4.2   RQ2. What AI Techniques Are Used in RE?

In our analysis of the selected studies, we identified a diverse range of AI techniques applied to various RE tasks. These techniques are divided into three categories: Natural Language Processing (NLP), Machine Learning (ML), and Deep learning (DL) models. As illustrated in Table 6, the first column lists the RE tasks, highlighted in bold. In each task, the non-bolded items represent specific sub-tasks. Every technique in the second column is referenced alongside the selected papers, which also indicates the frequency of its mention across the selected papers. This notation not only identifies the techniques applied for various RE tasks, but also provides insight into its prevalence in the selected papers.

Latent Dirichlet Allocation (LDA), a topic modeling method that can be used to classify texts to particular topics, was the most prevalent ML technique for requirements elicitation, identified in all 4 review studies on AI for requirements elicitation (SP1, SP9, SP14, SP28). LDA was used in app reviews (SP1, SP25, SP28), user stories (SP15), and documents (SP2, SP6, SP9, SP14, SP27, SP28) to identify hidden or latent topics. LDA models were also used to determine a word's weight or value in relation to the topic (SP2, SP27).

Another widely used ML technique was Naïve Bayes (NB), which was identified as the most popular technique in studies on requirements classification (SP2, SP6). NB is a classifier generally used for text classification. The use of NB spanned various tasks including requirements elicitation (SP1, SP14, SP28), analysis (SP10, SP25), and traceability identification (SP16). Additionally, techniques such as Support Vector Machine (SVM), Decision tree, Random forest, and Logistic regression were identified for their implementation in a wide range of RE tasks, encompassing requirements elicitation, classification, analysis, and management.

NLP techniques have shown dominance in RE tasks to enhance the ability to process, understand, and manipulate textual data. They were reported in approximately 93% of the primary studies in SP1 for requirements elicitation. Moreover, the majority of the primary studies dedicated to requirements specification have applied NLP techniques, as reported in studies of SP11, SP5, and SP26. Among the various NLP techniques, POS-tagging was the most popular NLP technique alongside preprocessing, Tokenizing, TF-IDF, Bag-of-words

(BoW), and Vector Space Model (VSM) were notably prevalent in the selected studies. Additionally, other NLP techniques such as syntatic analysis, sentiment analysis, stemming, text-chunking, text-parsing, and n-grams were frequently used for identifying ambiguity in quality analysis.

Table 6. AI technologies in RE tasks

RE Tasks	AI Techniques
Elicitation	**NLP:** POS Tagging (SP1, SP28), Syntatic analysis (SP1, SP14), Tokenizing (SP1, SP14), Bag-of-words (SP1, SP14), TF-IDF (SP1, SP14), N-grams (SP1, SP14), Sentiment analysis (SP1), Word2vec (SP14), FastText (SP14); **ML:** LDA (SP1, SP9, SP14, SP28), NB (SP1, SP14, SP28), SVM (SP1, SP9, SP14), logistic regression (SP1, SP14, SP28), Decision tree (DT) (SP1, SP14, SP28), Random forest (RF) (SP1, SP14), ASUM (SP14, SP28), multinomial Naïve Bayes (SP1), Biterm Topic Model (BTM) (SP1, SP14), TSA (SP14), K-medoidds (SP14), X-means (SP14), K-means (SP14), K nearest neighbor (SP14), Regression (SP14), J48 (SP28), CART (SP28); **DL:** BERT (SP14), CNN (SP14), RNN (SP14), FNN (SP14)
Classification	**NLP:** Sentiment Analysis (SP4); **ML:** NB (SP2, SP6), J48 (SP2, SP6), Artificial Neural Networks (SP2, SP6), SVM (SP2, SP6), K Nearest Neighbor (SP2, SP6), RF (SP2, SP6), LDA (SP2, SP6), BTM (SP2, SP6), Bagging (SP2), Sequential Minimum Optimizer SMO (SP2), Bayesian Network (SP2), K-Means (SP2), Adaptive Boost (SP2), Extra tree (SP2), Gradient Boosting (SP2), Stocastic Gradiet Descent Classifier (SP2), DT (SP2, SP6), Hierarchic (SP6), Hierarchial K-means (SP6), K-means (SP6), Random K-label Naive Bayes (RKNB) (SP6), Logic Rules (SP6), Logistic Regression (SP6); **DL:** BERT (SP6), CNN (SP2)
Analysis	
User review analysis	**NLP:** POS tagging (SP15, SP25), vector space model (SP15, SP25), Preprocessing (SP15, SP25), named-entity recognizer (SP15), bag-of-words (SP15, SP25), clustering (SP15), TF-IDF (SP15), WuP similarity (SP15), Lemmatization (SP15), Semnatic role labeling (SP15), Skip-gram (SP15), Similarity matrix (SP15), Fuzzy set theory (SP15), Collocation (SP25), Syntatic parse tree (SP15); **ML:** LDA (SP15, SP25), Logistic Regression (SP15, SP25), NB (SP25), SVM (SP25), DT (SP25), RF (SP25), Neural Network (SP25), Linear Regression(SP25), K-nearest neighbor (SP25), K-Means (SP25)
Quality Analysis	**NLP:** POS-tagging (SP11, SP20), Preprocessing (SP11), text-parsing (SP11), text-classification (SP11), lemmatization (SP11), tokenization (SP11, SP20), n-grams (SP11), TF-IDF (SP11, SP20), stemming (SP11), text-parsing (SP11, SP20), text-clustering (SP11), text-chuking (SP11, SP20), semantic analysis (SP11), Ontology (SP24); **ML:** K-NN classifier (SP10), Sequential Minimum Optimizer (SMO) (SP10), NB (SP10), Binary Logistic Regression (BLR) (SP10), Stochastic Gradient Descent (SGD) (SP10), Logistic Regression (SP10), K-fold cross-validation (SP10); **DL:** BERT (SP23), DistilBERT(SP23), RoBERTa(SP23)

(continued)

Table 6. (*continued*)

RE Tasks	AI Techniques
Validation	**ML:** K-means algorithm (SP17), Binary Logistic Regression (BLR) (SP10), Stochastic Gradient Descent (SGD) (SP10), Logistic Regression (SP10)
Specification	
Textual specification	**NLP:** POS-tagging (SP11), Preprocessing (SP11), text-parsing (SP11), text-classification (SP11), lemmatization (SP11), tokenization (SP11), n-grams (SP11), TF-IDF (SP11), stemming (SP11), text-parsing (SP11), text-clustering (SP11)), text-chuking (SP11), semantic analysis (SP11); **ML:** Artificial Neural Networks (SP23), Neural Networks (SP23); **DL:** CNN (SP23)
Modeling & Model Transformation	**NLP:** POS-tagging (SP5, SP26), text-parsing (SP5, SP26), Tokenizer (SP26), lemmatization (SP5, SP26), stemming (SP5), Correlation and Dependencies Analysis (SP26)
Management	
Baseline maintenance	**ML:** Bayesian networks (SP7), SVM (SP7), Laplacian feature mapping (SP7), RF (SP7), K- nearest neigbours (KNN) (SP7), Logistics regression (SP7), DT (SP7), Demand-dependant extraction (RD-AL) (SP7); **DL:** FNN (SP7), RNN (SP7), Deep Neural Networks (DNN) (SP7), CNN (SP7)
Change management	**ML:** Backward propagation of errors neural network (BPNN) (SP7), K-means (SP7), Neural Networks (SP7), Logistic Regression (SP7), Reinforcement learning (SP7), RE-STROM (SP7), SVM (SP7), RF (SP7), Bayesian Networks (SP7), Decision Tree (SP7), ME (SP7), Multi-layer perceptron classifier (MLPC) (SP7), RaM-POMDP (SP7), Multi label learning (MLL) (SP7)
Volatility/ vulnerabilities prediction	**ML:** Logistic Regression (SP21, SP10), Fuzzification/Defuzzification (SP21), Binary Classifier (SP21), Linear Regression (SP21), Artificial Neural Network (SP21), Bayesian Network (SP21), Binary Logistic Regression (BLR) (SP10), SGD (SP10)
Traceability	**NLP:** POS-tagging (SP24), Tokenizing (SP24), Parsing (SP24), Sentiment Analysis (SP24), VSM (SP16, SP24); **ML:** RF (SP7, SP16), DT (SP7, SP16), LDA (SP16, SP27), SVM (SP7, SP27), Reinforced learning (SP7), Bayesian Networks (SP7), K nearest neighbor(SP7), ATLaS (SP7), DCTracVis (SP7), TLR-ELtoR (SP7), ALCATRAL (SP7), Logistic regression (SP7), LTR (SP7), Onto-Req (SP7), S2Trace (SP7), LSI (SP16), SZZ (SP16), Naive Bayes (SP16), J48 (SP16), LSA (SP27), K-fold cross-validation (SP10); **DL:** CNN (SP7, SP16), Long short-term memory (LSTM) (SP7, SP16), BERT (SP16, SP27), BI-LSTM (SP7), BI-GRU (SP7, SP27), RNN (SP24, SP27), T-BERT (SP7, SP32), LLM (SP32)
Others	
Reuse	**NLP:** text-parsing (SP24); **ML:** DT (SP24), Collaborative Filtering-based recommendation (SP24)
Recollection	**NLP:** POS-tagging (SP26), Tokenizer (SP26), Text-parsing (SP26), semantic analysis (SP26)

Recent studies on AI4RE have increasingly reported the use of Deep Learning (DL) Models. The Convolutional Neural Network (CNN) was the most frequently used model, with its applications in requirements elicitation (SP14), classification (SP2), specification (SP23), management(SP7), and traceability (SP7, SP16). Additionally, other models like Recurrent Neural Network (RNN) and Feedforward neural network (FNN) have been implemented for requirements elicitation (SP14) and management (SP7). The Bidirectional Encoder Representations from Transformers (BERT) family has been applied for requirements elicitation (SP14), classification (SP6), analysis (SP23), and traceability (SP16, SP27). This includes variations like DistilBERT (SP23) and RoBERTa (SP23) for requirements analysis, and T-BERT (SP7, SP32) specifically for requirements traceability identification. Notably, a recent study (SP32) also addressed the implementation of Large language models (LLM) in RE tasks which could be a future trend for applying NLP approaches.

Our analysis also revealed that NLP was a popular approach, especially in requirements elicitation, analysis, specification, and model generation and transformation. These studies also highlighted a trend of preprocessing text documents, user stories, and user reviews using NLP techniques before applying other ML or DL techniques to assist RE tasks. A recent study (SP32) also addressed the implementation of LLM in RE tasks which could be a future trend for applying NLP approaches. It was further discovered that a wide range of deep learning models were implemented for requirements elicitation and traceability compared to other RE tasks, indicating a trend.

4.3   RQ3. What Are the Challenges of Implementing AI for RE?

Studies SP15, SP25, and SP28 highlighted the challenges in using NLP for analyzing app reviews and user stories. SP15 and SP28 reported that results generated by NLP for user story analysis are context- or domain-dependent, limiting their broader applicability. Additionally, SP15 notes the necessity of human intervention in automating data mining from user stories due to the inherent ambiguity. SP5 addressed the limitation of NLP in processing long or complex sentences. SP25 raises concerns about the quality of app review analysis and sentiment analysis. Furthermore, SP28 addresses the limitations and lack of maturity of existing automated solutions for identifying fake and spam reviews. However, a recent advancement is noted in SP32, where LLMs, particularly those addressing anaphoric ambiguity, show potential in resolving ambiguity issues. This suggests a promising direction for overcoming some of the current limitations in NLP for analyzing complex requirements information.

The challenges in datasets for AI4RE are multifaceted, involving issues of accessibility, quality, relevance, and the need for better data management and sharing practices. SP1, SP9, and SP10 emphasized the need to investigate the relevance and structure of textual data and its volume, especially when applying advanced algorithms like deep learning. This highlights the necessity of understanding how various types of data and their complexities impact on successful AI strategies and model training. Studies such as SP1, SP5, SP10, and SP25 advocated for the sharing of evaluation datasets and mining tools to facilitate replication of experiments, indicating a broader challenge in the AI4RE community regarding open access to data and tools, as well as the need for scalable and efficient data processing methods, as addressed in SP3.

The challenges in evaluating AI solutions revolve around the need for more efficient, robust, and standardized evaluation methods, along with ensuring consistency in research quality. While most research primarily investigated the effectiveness of AI methods in terms of precision and recall, there is a notable lack of emphasis on efficiency. This is particularly crucial when dealing with large datasets, as addressed in SP3, SP7, SP17, and SP26. Additionally, SP15 noted that precision evaluation produced unexpected results even though the recall evaluation results were as expected. Addressing these challenges is essential for ensuring the practical applicability of AI solutions in RE - another challenge addressed in SP2.

5   Discussions

The integration of AI in RE is a paradigm shift to enhance the efficiency and quality of the software development process. Our tertiary study on AI4RE sheds light on the evolving trends in both research and practice.

The growing interest in the use of AI to assist RE is a transformative shift, moving from traditional methods like interviews and workshops to leveraging AI for tasks such as identifying, extracting, and managing requirements, using diverse information sources like user reviews, documentation, and community discussions. However, we did not find empirical studies to proof real-world effectiveness of these AI techniques. Such studies should be done in industry settings.

The research indicates a notable trend towards hybrid approaches, combining ML with other AI techniques like NLP. This synergy is particularly effective in handling multidimensional and multi-variety data in dynamic or uncertain environments. Preliminary research shows promising results when ML methods are integrated with techniques like IR, domain ontology, and optimization algorithms. In addition, The use of information retrieval and ML techniques in creating and recovering trace links between issue reports and other software artifacts is gaining attention, highlighting the practical application of these hybrid AI approaches in managing complex SE tasks. This evolving nature of AI4RE underscores the importance of context-specific solutions and the need for practitioners to adapt to the rapidly changing landscape of AI technologies.

Challenges such as the maturity of existing automated solutions, evaluation methods, and the availability and quality of source requirements information provide a roadmap for future advancements in AI4RE. Addressing these challenges is crucial in both research and practice. For researchers, these challenges offer direction, pointing towards areas that can lead to more robust and efficient solutions, ultimately enhancing the quality of requirements and the overall software development process. Practitioners, on the other hand, can view these challenges as a checklist. By proactively acknowledging and addressing these challenges, we can enhance the quality of requirements and the overall software development process.

6   Threats to Validity

To ensure the validity of our study, we designed search strategies and a review protocol grounded in established guidelines [7–9]. We also adhered to the categories recommended in [15] to address the potential threats.

Our search string included keywords extracted from our research questions, which were further diversified by adding synonyms. The search was conducted in major citation databases such as Scopus and Web of Science, and digital libraries of IEEE Xplore and ACM Digital Library. To enhance the coverage of studies, snowballing was conducted as a supplementary. Recognizing the emerging trend of generative AI methods, such as the use of LLMs in RE and SE, we were aware of the potential lack of recent secondary studies in the digital libraries and expanded our search in Google Scholar and pre-publication archives like arXiv, which resulted in 4 additional recent studies. While there is a potential risk of omitting relevant studies, we are confident that our study represents a comprehensive overview of secondary studies within our defined scope, and any threats to the internal and external validity are minimal.

To ensure transparency in our study selection, quality assessment, and data extraction, we shared all of our data and findings in a spreadsheet[4] that we used throughout the research process. The 'overview' sheet of the spreadsheet gives information on the contents of all the sheets in the file, making navigation easier for the reader.

To address the threat to construct validity, our methods from search strategy to data extraction process were entirely based on established guidelines. Since a tertiary study is a systematic review of secondary studies, this adds abstractness and increases the risks of double-counting and the consequence of biases in data interpretation, which threats the internal and conclusion validities [2,13]. To evaluate the risk of double-counting bias, we applied the CCA matrix [13] to analyze overlaps of primary studies reported in the secondary studies which have a similar topic. The resulting low overlap score implies a very low risk of double-counting risk and allows us to aggregate findings when addressing RQ1 and RQ2. Additionally, our synthesized results were cross-referenced with secondary studies on a broader topic like AI4RE or AI4SE.

7   Conclusion

In this tertiary study, we analyzed a total of 28 relevant studies to investigate the state-of-the-art in AI4RE. The findings of this tertiary study indicate the wide use of AI techniques for different RE tasks. The popularity of NLP techniques and ML techniques, such as LDA and Naïve Bayes, was addressed for various RE tasks. The research also highlights the implementation of deep learning models and LLMs for RE tasks. Furthermore, we have addressed the challenges of implementing AI4RE related to NLP methods, the use of datasets, and the AI solution evaluation.

---

[4] https://doi.org/10.6084/m9.figshare.24564829.v2.

In addition, the research explores potential avenues for future investigations. Further research can be conducted on the implementation of specific techniques or investigate the state-of-the-art applications of AI for a specific RE task. The challenges addressed in this study can be avenues for further research as well. Our research can further help identify current trends and the findings could be helpful for researchers who want to start new research in this field.

Acknowledgements. This work has been supported by Business Finland (project 6GSoft, 8548/31/2022)

A   List of Primary Studies

SP1. Lim, Sachiko, Aron Henriksson, and Jelena Zdravkovic. "Data-Driven Requirements Elicitation: A Systematic Literature Review." SN computer science 2.1 (2021)

SP2. Perez-Verdejo, J. Manuel, Angel J. Sanchez-Garcia, and Jorge Octavio Ocharan-Hernandez. "A Systematic Literature Review on Machine Learning for Automated Requirements Classification." 2020 8th International Conference in Software Engineering Research and Innovation (CONISOFT). IEEE, 2020. 21–28.

SP3. Ijaz, Khush Bakht, Irum Inayat, and Faiza Allah Bukhsh. "Non-Functional Requirements Prioritization: A Systematic Literature Review." 2019 45th Euromicro Conference on Software Engineering and Advanced Applications (SEAA). IEEE, 2019. 379–386.

SP4. Santos, Rubens, Eduard C Groen, and Karina Villela. "A Taxonomy for User Feedback Classifications." N.p., 2019

SP5. Ahmed, Sharif, Arif Ahmed, and Nasir U. Eisty. "Automatic Transformation of Natural to Unified Modeling Language: A Systematic Review." 2022 IEEE/ACIS 20th International Conference on Software Engineering Research, Management and Applications (SERA). Ithaca: IEEE, 2022. 112–119.

SP6. Lopez-Hernandez, Delmer Alejandro et al. "Automatic Classification of Software Requirements Using Artificial Neural Networks: A Systematic Literature Review." 2021 9th International Conference in Software Engineering Research and Innovation (CONISOFT). Piscataway: IEEE, 2021. 152–160.

SP7. Xu, Chi et al. "A Systematic Mapping Study on Machine Learning Methodologies for Requirements Management." IET software 17.4 (2023): 405–423.

SP9. Ahmad, Arshad et al. "A Systematic Literature Review on Using Machine Learning Algorithms for Software Requirements Identification on Stack Overflow." Security and communication networks 2020 (2020): 1–19.

SP10. Zamani, Kareshna, Didar Zowghi, and Chetan Arora. "Machine Learning in Requirements Engineering: A Mapping Study." 2021 IEEE 29th International Requirements Engineering Conference Workshops (REW). Vol. 2021-. Piscataway: IEEE, 2021. 116–125.

SP11. Aberkane, Abdel-Jaouad, Geert Poels, and Seppe Vanden Broucke. "Exploring Automated GDPR-Compliance in Requirements Engineering: A Systematic Mapping Study." IEEE access 9 (2021): 1–1.

SP12. Zhao, Liping et al. "Natural Language Processing for Requirements Engineering: A Systematic Mapping Study." ACM computing surveys 54.3 (2022): 1–41.

SP14. Cheligeer, Cheligeer et al. "Machine Learning in Requirements Elicitation: A Literature Review." Artificial intelligence for engineering design, analysis and manufacturing 36 (2022)

SP15. Raharjana, Indra Kharisma, Daniel Siahaan, and Chastine Fatichah. "User Stories and Natural Language Processing: A Systematic Literature Review." IEEE access 9 (2021): 53811–53826.

SP16. Lyu, Yijing et al. "A Systematic Literature Review of Issue-Based Requirement Traceability." IEEE access 11 (2023): 13334–13348.

SP17. Perkusich, Mirko et al. "Intelligent Software Engineering in the Context of Agile Software Development: A Systematic Literature Review." Information and software technology 119 (2020): 106241-.

SP19. Sofian, Hazrina, Nur Arzilawati Md Yunus, and Rodina Ahmad. "Systematic Mapping: Artificial Intelligence Techniques in Software Engineering." IEEE access 10 (2022): 51021–51040.

SP20. Nazir, Farhana et al. "The Applications of Natural Language Processing (NLP) for Software Requirement Engineering - A Systematic Literature Review." Information Science and Applications 2017. Vol. 424. Singapore: Springer Singapore, 2017. 485–493.

SP21. Alsalemi, Ahmed Mubark, and Eng-Thiam Yeoh. "A Systematic Literature Review of Requirements Volatility Prediction." 2017 International Conference on Current Trends in Computer, Electrical, Electronics and Communication (CTCEEC). IEEE, 2017. 55–64.

SP23. Magableh, Aws A. "Towards Leveraging Explainable Artificial Intelligent (XAI) in Requirements Engineering (RE) to Identify Aspect (Crosscutting Concern): A Systematic Literature Review (SLR) and Bibliometric Analysis." 2023 International Conference on Information Technology (ICIT). IEEE, 2023. 319–326.

SP24. Corral, Alexandra, Luis E. Sanchez, and Leandro Antonelli. "Building an Integrated Requirements Engineering Process Based on Intelligent Systems and Semantic Reasoning on the Basis of a Systematic Analysis of Existing Proposals." JUCS - Journal of Universal Computer Science 28.11 (2022): 1136–1168.

SP25. Dabrowski, Jacek et al. "Analysing App Reviews for Software Engineering: A Systematic Literature Review." Empirical software engineering: an international journal 27.2 (2022)

SP26. Quintana, Manuel A. et al. "Agile Development Methodologies and Natural Language Processing: A Mapping Review." Computers (Basel) 11.12 (2022): 179-.

SP27. Sonbol, Riad, Ghaida Rebdawi, and Nada Ghneim. "The Use of NLP-Based Text Representation Techniques to Support Requirement Engineering Tasks: A Systematic Mapping Review." IEEE access 10 (2022): 62811–62830.

SP28. Genc-Nayebi, Necmiye, and Alain Abran. "A Systematic Literature Review: Opinion Mining Studies from Mobile App Store User Reviews." The Journal of systems and software 125 (2017): 207–219

SP29. Aguilar, Alfonso Robles et al. "A Systematic Mapping Study of Artificial Intelligence in Software Requirements." Res. Comput. Sci. 149 (2020): 179–188.

SP30. Ahmad Haji Mohammadkhani et al. "A Systematic Literature Review of Explainable AI for Software Engineering." arXiv.org (2023)

SP31. Wang, Simin et al. "Synergy between Machine/Deep Learning and Software Engineering: How Far Are We?" arXiv.org (2020)

SP32. Hou, Xinyi et al. "Large Language Models for Software Engineering: A Systematic Literature Review." arXiv.org (2023)

References

1. Bano, M., Zowghi, D., Ikram, N.: Systematic reviews in requirements engineering: a tertiary study. In: 2014 IEEE 4th International Workshop on Empirical Requirements Engineering (EmpiRE), pp. 9–16. IEEE (2014)
2. Börstler, J., bin Ali, N., Petersen, K.: Double-counting in software engineering tertiary studies-an overlooked threat to validity. Inf. Softw. Technol. **158**, 107174 (2023)
3. Cico, O., Cico, B., Cico, A.: Ai-assisted software engineering: a tertiary study. In: 2023 12th Mediterranean Conference on Embedded Computing (MECO), pp. 1–6. IEEE (2023)
4. Cleland-Huang, J., Settimi, R., Zou, X., Solc, P.: Automated classification of non-functional requirements. Requirements Eng. **12**, 103–120 (2007)
5. Fernández, D.M., et al.: Naming the pain in requirements engineering: contemporary problems, causes, and effects in practice. Empir. Softw. Eng. **22**, 2298–2338 (2017)
6. Ferrari, A., Spagnolo, G.O., Gnesi, S.: Pure: a dataset of public requirements documents. In: 2017 IEEE 25th International Requirements Engineering Conference (RE), pp. 502–505. IEEE (2017)
7. Kitchenham, B.: Procedures for performing systematic reviews. Keele, UK, Keele University **33**(2004), 1–26 (2004)
8. Kitchenham, B., Madeyski, L., Budgen, D.: Segress: Software engineering guidelines for reporting secondary studies. IEEE Trans. Software Eng. **49**(3), 1273–1298 (2022)
9. Kitchenham, B., et al.: Systematic literature reviews in software engineering-a tertiary study. Inf. Softw. Technol. **52**(8), 792–805 (2010)
10. Knauss, E., Houmb, S., Schneider, K., Islam, S., Jürjens, J.: Supporting requirements engineers in recognising security issues. In: Berry, D., Franch, X. (eds.) REFSQ 2011. LNCS, vol. 6606, pp. 4–18. Springer, Heidelberg (2011). https://doi.org/10.1007/978-3-642-19858-8_2
11. Kotti, Z., Galanopoulou, R., Spinellis, D.: Machine learning for software engineering: a tertiary study. ACM Comput. Surv. **55**(12), 1–39 (2023)

12. Kudo, T.N., Bulcão-Neto, R.F., Vincenzi, A.M.: Requirement patterns: a tertiary study and a research agenda. IET Software **14**(1), 18–26 (2020)
13. Pieper, D., Antoine, S.L., Mathes, T., Neugebauer, E.A., Eikermann, M.: Systematic review finds overlapping reviews were not mentioned in every other overview. J. Clin. Epidemiol. **67**(4), 368–375 (2014)
14. Wiegers, K.E., Beatty, J.: Software Requirements. Pearson Education, London (2013)
15. Wohlin, C., Runeson, P., Höst, M., Ohlsson, M.C., Regnell, B., Wesslén, A.: Experimentation in software engineering. Springer, Berlin (2012). https://doi.org/10.1007/978-3-642-29044-2

Exploring LLMs' Ability to Detect Variability in Requirements

Alessandro Fantechi[2,3], Stefania Gnesi[3], and Laura Semini[1,3]

[1] Dipartimento di Informatica, Università di Pisa, Pisa, Italy
laura.semini@unipi.it
[2] Dipartimento di Ingegneria dell'Informazione, Università di Firenze, Florence, Italy
alessandro.fantechi@unifi.it
[3] Istituto di Scienza e Tecnologie dell'Informazione "A. Faedo", ISTI-CNR, Pisa, Italy
stefania.gnesi@isti.cnr.it

Abstract. In this paper, we address the question of whether general-purpose LLM-based tools may be useful for detecting requirements variability in Natural Language (NL) requirements documents. For this purpose, we conduct a preliminary exploratory study considering OpenAI chatGPT-3.5 and Microsoft Bing. Using two exemplar NL requirements documents, we compare the variability detection capability of the chatbots with that of experts and that of a rule-based NLP tool.

1 Introduction

Software Product Line Engineering (SPLE) is a paradigm that has been proposed to support the development of a set (*family*) of different, but similar, software products [4, 11]. In a software product line (SPL), a product family description is composed of a common part and a variable part. Product *variants* can be derived from a product family specification and dealing with *variability* is the key issue that has caused SPLE to mature into a specific software engineering discipline. The concept of *features* is used to model an observable or distinctive aspect of a product, and software product lines (SPL)s are usually modelled in terms of the composition of features, specifying which of them are common to all products (mandatory features) and which are to be included only in some products of the family (optional features). The variability depends on which features are included and which are not in each variant (product). The decision between inclusion or not is located at specific places in design artifacts, named *variation points*. A well-known goal in the SPL community is to create a feature model by extracting features and variability from *natural language (NL) requirement documents*. Some studies [2, 5, 7] employ methods grounded in NL processing tools that apply deterministic rules. Other studies rely on AI models trained for the task [8, 10, 12].

These face a challenge attributed to the well-known lack of a corpus of annotated requirements documents available for model training.

D. Mendez and A. Moreira (Eds.): REFSQ 2024, LNCS 14588, pp. 178–188, 2024.
https://doi.org/10.1007/978-3-031-57327-9_11

There are also tools that combine the power of machine learning for the linguistic analysis of natural language with developer-defined rules. A notable example is FeatureX [13], which first identifies candidate features with lexical analysis and machine learning techniques, then searches for points of variability using rules, e.g., looking for conjunctions between words that represent identified features, or weak and modal verbs that are indicators of optionality.

To address the matter of detecting variability in NL requirements documents using general-purpose AI-based tools, we opted to exploit two of the better known chatbots based on Large Language Models (LLMs) available, specifically ChatGPT and Bing. The first is created by OpenAI and, at the time of writing this paper, utilizes the GPT-3.5 model, the latter is the new Microsoft's Bing chat, powered by GPT-4, able to answer user's questions, search the web for information, and generate creative content.

The design and implementation details of GPT-4 are still not published, while those of GPT-3 are, in [3].

Specifically, in this paper, we discuss the results of some experiences aimed at assessing LLMs' ability to extract features and variability points from requirements in text form. We measure chatbots' performance and compare it with expert judgment and with a more traditional rule-based NL processing tool. To perform the comparison, we use documents and data presented in [5]. The aim of the paper is to answer, although in a preliminary way, the research questions:

RQ1 How does the LLMs performance for feature detection compare to expert judgment?[1]

RQ2 How does the LLMs performance for variation points detection compare to expert judgment? and to a rule-based NLP tool?

RQ3 Is there a difference in performance between chatGPT and Bing?

2   Data Preparation

We have used two requirements documents that satisfy the length constraints of LLMs: **Coffee machine** (Table 1) containing few requirements of a coffee vending machine, and **Eshop** (Table 2) that describes an online shopping system. We chose these documents also because they were annotated in previous work using VIBE [5].

To collect the data, we proceeded according to the following steps: i) design of an appropriate prompt; ii) construction of a Google form to collect responses to the query from LLMs; iii) establishment of a ground truth to be used in the next phase of data review and data analysis.

2.1   Prompt Design

The first step of the experiments concerned the design of a suitable prompt to submit to the LLMs. Prompt design is becoming a new challenge and a topic

---

[1] We cannot compare with the referenced rule-based tool because it only detects variation points and it is not able to identify features.

Table 1. Coffee-machine requirements

C1	After inserting a suitable coin, the user shall choose a beverage and select the amount of sugar
C2	The machine shall offer, as beverages, Coffee and Cappuccino or Tea
C3	The machine shall always offer coffee
C4	A ringtone possibly has to be played after beverage delivery
C5	After the beverage is taken, the machine returns idle
C6	The British market requires tea and excludes any ringtone

Table 2. Eshop requirements

Re1	The system shall enable the search text to be entered on the screen
Re2	The system shall display all the matching products based on the search
Re3	The system possibly notifies with a pop-up the user when no matching product is found on the search
Re4	The system shall allow a user to create his profile and set his credentials
Re5	The system shall authenticate user credentials to enter the profile
Re6	The system shall display the list of active and/or the list of completed orders in the customer profile
Re7	The system shall maintain customer email information as a required part of customer profile
Re8	The system shall send an order confirmation to the user through email
Re9	The system shall allow a user to add and remove products in the shopping cart
Re10	The system shall display various shipping methods
Re11	The order shall be shipped to the client address or to an associated store, if the click&collect service is available
Re12	The system shall enable the user to select the shipping method
Re13	The system may display the current tracking information about the order
Re14	The system shall display the available payment methods
Re15	The system shall allow the user to select the payment method for an order
Re16	After delivery, the system may enable the users to enter their reviews or ratings
Re17	In order to publish the feedback on the purchases, the system needs to collect both reviews and ratings
Re18	The click & collect service excludes the tracking information service

that will surely be developed shortly. Prompt engineering refers to the process of defining specific rules and guidelines to interact with an LLM to guide the model response toward an optimized output. In the literature, some patterns have already been defined and cataloged [6,14]. In this study, we have considered some patterns, and used the following three, Persona, Instruction Prompting, and Template patterns, that seemed the most adequate for the specific case. Each of the selected patterns has a distinct function in guiding the LLM to generate responses in a specific way. In addition, they can be combined to produce a unique prompt: indeed, the students have little experience with LLMs and we could not base the experiment on their ability to have a complex conversation with the chatBots. In the future, other experiments can be done, for example, using query refinement patterns or interaction patterns.

The Persona pattern suggests the users to ask the LLM to act as a specific person, e.g., Napoleon or William Shakespeare, or more generally as a person with a specific role, e.g., a software engineer or a Thai cook. The pattern helps users to communicate some context to the LLM in order to guide it in its response. In this case, we tell the chatbot to pretend to be an analyst in charge of analysing software requirements to define a software product line.

Instruction Prompting Pattern tells to structure a prompt with a list of instructions that define the steps to be taken to accomplish the task and ask the LLM to return partial outputs that altogether define the complete response. Here, we have asked to first return the list of features, then the variation points.

The Template Pattern allows users to instruct an LLM to produce its output in a precise and structured format. In our case, we have provided the set of tags to be used to label the identified variation points.

By applying these patterns, we have defined the prompt that we have used in all tests, with both requirements documents:

You are an analyst charged with analyzing software requirements to define a software product line. Begin by listing all the features in the following requirements document. Then, identify the mandatory features and all points of variation within the document. To this end, apply the following tags to each feature or set of features: "mandatory" for individual mandatory features, "optional" for individual optional features, "or" for disjoint pairs or tuples of features, in which at least one must be present but two or more can coexist, "mutually exclusive" for pairs or tuples of features in which at least one must be present but are alternatives and cannot coexist, "excluded" for pairs of features that are not explicitly allowed together, and "requires" for pairs of features that depend on each other. < requirements list >.

2.2   Google Forms and Assignment to Students.

We have created a Google form[2] with instructions for the students: copy the text that contains the prompt and the list of requirements, paste it in the chatbot, copy the answer and paste it into the form[3]. The forms were filled out by student volunteers: each student queried either ChatGPT3.5 or Bing. To interact with the Bing chat at the time of the experiment, the web interface offered by the Edge browser had to be used. So, Bing was used by the students with Microsoft Edge installed, ChatGPT by the others. We conducted the experiment in parallel with various students to ensure a more reliable outcome, recognizing that obtaining responses by different accounts enhances the soundness of our findings compared to repeating the same query multiple times from a single profile.

2.3   Ground Truth Construction

We defined, in a joint meeting among the authors, the benchmark features and the variation points to be used as ground truth.

---

[2] https://docs.google.com/forms/d/e/1FAIpQLSfscV_uiATaWSJngBH9ruXJBMAnJ Dvtw6TVLOMXgToFZQ1n8Q/viewform.

[3] In the form we have also asked students for an analysis exercise of the results, but it was just a classroom assignment, not used for the purpose of this paper.

Table 3. Coffee Machine Ground truth

features	variation
Coin insertion	mandatory
Beverage selection	mandatory
Sugar selection	mandatory
Coffee	mandatory
Cappuccino, Tea	or
Machine returns idle	mandatory
Suitable coin	alternative
British market, tea	requires
British market, ring tone	excludes
Ringtone	optional

Table 4. Eshop Ground truth

Features	Variation
Search functionality	mandatory
Displaying matching products	mandatory
Pop-up notification	optional
User profile creation	mandatory
Set user credentials	mandatory
User authentication	mandatory
Displaying orders	mandatory
Displaying active orders; Displaying complete orders	or
Order confirmation	mandatory
Add to shopping cart	mandatory
Remove from shopping cart	mandatory
Displaying shipping methods	mandatory
Selection of shipping method	mandatory
Shipping option: client address; to store	or
To store; Click&collect	requires
Enter feedback	optional
Enter reviews; enter ratings	or
Publish feedback	optional
Publish feedback; enter review	requires
Publish feedback; enter ratings	requires
Click&collect service	optional
Click&collect service; tracking	excludes
Tracking information	optional
Displaying payment methods	mandatory
Selection of payment method	mandatory
Customer email information	mandatory

For the coffee machine example, we judged the features and points of variation in Table 3 to be the ones to be identified. Similarly, in Table 4 we list the features and points of variation that we judged should be identified for eshop.

3 Data Review and Data Analysis

The data review and analysis has been divided into two steps: first, in the review phase, we have made a check of the responses uploaded by the students, which contained the output of the LLMs at the prompt defined in Sect. 2.1; then, in the analysis phase, we have calculated the performance of the chatbots.

3.1 Data Review

We received 13 responses from students using chatGPT and 8 responses from students using Bing. In some cases (three for chatGPT and two for Bing), the LLMs identified the features with the requirements, simply returning the list of requirements with the variability tags. We have decided to discard these "lazy" responses which are not reported in the table.

For each relevant response, we have filled a column in the tables in Figs. 1 and 2, putting an "x" if the feature or variation point in the row was found. In the figures, we have put the chatGPT responses region (c1-c10) on the left and coloured in yellow, and the Bing responses region on the right (b1-b6) and

	c1	c2	c3	c4	c5	c6	c7	c8	c9	c10	b1	b2	b3	b4	b5	b6	c sum	b sum	metric
																	c sum	b sum	c aver
																			0,750
Coin insertion		x			x						x	x	x	x	x	x	2	6	b aver
Beverage selection	x	x	x		x	x	x		x	x	x	x	x	x	x	x	8	6	0,958
Sugar selection	x	x	x	x	x	x	x			x	x	x	x	x	x	x	8	6	c tp
Coffee	x	x	x		x	x	x	x	x		x	x	x	x	x	x	8	6	60
Cappuccino	x	x	x		x	x	x	x	x		x	x		x	x	x	8	5	c fp
Tea	x	x	x		x	x	x	x	x	x	x	x		x	x	x	9	5	1
Ringtone	x	x	x		x	x	x	x	x	x	x	x	x	x	x	x	9	6	b tp
Machine returns idle	x	x	x		x	x	x		x	x	x	x	x	x	x	x	8	6	46
																	c fn	b fn	b fp
False Positive	0	0	0	0	0	0	0	0	0	1	0	2	2	0	0	0	20	2	4

	c1	c2	c3	c4	c5	c6	c7	c8	c9	c10	b1	b2	b3	b4	b5	b6	c sum	b sum	metric
																	c sum	b sum	c aver
Coin insertion (mandatory)		x			x						x	x	x	x	x	x	2	6	0,560
Beverage selection (mandatory)	x	x	x		x	x	x			x	x	x	x	x	x	x	7	6	b aver
Sugar selection (mandatory)	x	x	x	x		x	x			x	x			x		x	7	3	0,80
Coffee (mandatory)	x	x	x		x	x	x		x		x	x	x	x	x	x	7	6	c tp
Cappuccino, Tea (or)		x		x	x	x	x	x	x		x	x		x	x	x	7	5	56
Machine returns idle (mandatory)	x	x	x		x	x	x		x	x	x	x	x	x	x	x	8	6	c fp
Suitable coin (alternative coins)																	0	0	12
British market (requires) tea	x		x	x			x				x		x	x	x	x	5	5	b tp
British market (excludes) ring tone	x		x	x							x	x	x	x	x	x	4	6	48
Ringtone (optional)	x	x	x		x	x	x	x	x	x	x	x	x	x	x		9	5	b fp
																	c fn	b fn	6
False Positive	1	0	0	1	1	0	2	3	3	2	0	2	2	1	1	0	44	12	

Fig. 1. Feature and variation points detection on Coffee Machine

	c1	c2	c3	c4	c5	c6	c7	c8	c9	c10	b1	b2	b3	b4	b5	b6	c sum	b sum	c aver
																			0,875
Search functionality	x	x	x	x	x	x	x	x	x	x	x	x	x	x	x	x	10	6	b aver
Displaying matching products	x	x	x	x	x	x	x	x	x	x	x	x	x	x	x	x	10	6	0,868
Pop-up notification	x	x	x	x	x	x	x	x	x		x	x	x		x	x	9	5	c tp
User profile creation	x	x	x	x	x	x	x	x	x	x	x	x	x	x	x	x	10	6	210
Set user credentials	x	x	x	x	x			x	x			x	x		x	x	7	4	c fp
User authentication	x	x	x	x	x	x	x	x	x	x	x	x	x	x	x	x	10	6	4
Displaying active orders	x	x	x	x	x	x		x	x		x	x	x		x	x	8	5	c fn
Displaying complete orders	x	x	x	x	x	x		x	x		x	x	x		x	x	8	5	30
Order confirmation	x	x	x	x	x	x	x	x	x	x	x	x	x	x	x	x	10	6	b tp
Add to shopping cart	x	x	x	x	x	x	x	x		x	x	x	x	x	x	x	9	6	125
Remove from shopping cart	x	x	x	x	x	x	x	x		x	x	x	x	x	x	x	9	6	b fp
Display shipping methods	x	x	x	x	x	x	x	x	x	x	x	x	x	x	x	x	10	6	3
Selection of shipping method	x	x	x	x	x	x	x	x	x	x	x		x	x	x	x	10	5	b fn
Shipping option : client address	x	x		x	x		x	x	x		x		x	x		x	7	4	19
Shipping option : store	x	x		x	x		x	x	x		x			x		x	7	3	
Click&collect service	x	x		x	x	x	x	x	x		x		x	x		x	8	4	
Tracking information	x	x	x	x	x	x	x	x	x		x	x	x	x	x	x	9	6	
Display payment methods	x	x	x	x	x	x	x	x	x	x	x	x	x	x	x	x	10	6	
Selection of payment method	x	x	x	x	x	x	x	x	x	x	x	x	x	x	x	x	10	6	
Publishing feedback		x		x	x	x	x	x	x		x	x	x			x	6	4	
Enter feedback			x	x	x	x	x	x			x		x			x	6	3	
Enter ratings	x	x	x	x	x	x	x	x	x	x	x	x	x	x	x	x	10	6	
Enter reviews	x	x	x	x	x	x	x	x	x	x	x	x	x	x	x	x	10	6	
Customer email information			x	x	x	x	x	x	x		x	x	x		x	x	7	5	
false positives	0	0	0	0	0	0	0	0	4	0	0	0	1	0	2	0			

	c1	c2	c3	c4	c5	c6	c7	c8	c9	c10	b1	b2	b3	b4	b5	b6	c sum	b sum	c aver
																			0,812
Search functionality (mandatory)		x	x	x	x	x	x	x	x			x	x		x	x	8	4	b aver
Display match. products (mandatory)	x	x	x	x	x	x	x	x	x	x		x	x		x	x	10	4	0,577
Pop-up notification (optional)	x	x	x	x	x	x	x	x	x	x	x	x			x	x	10	4	c tp
User profile creation (mandatory)	x	x	x	x	x	x	x	x	x	x		x	x		x	x	10	4	211
Set user credentials (mandatory)	x	x	x	x	x			x	x			x	x		x	x	7	4	c fp
User authentication (mandatory)	x	x	x	x	x	x	x	x	x	x		x	x		x	x	10	4	5
Displaying orders (mandatory)	x	x	x	x	x	x			x			x		x		x	8	3	c fn
Displ. active ord. (or) complete ord.	x	x	x	x	x	x		x	x	x	x	x			x	x	9	4	49
Order confirmation (mandatory)	x	x	x	x	x	x	x	x	x	x			x		x	x	10	3	b tp
Add to shopping cart (mandatory)	x	x	x	x	x	x	x		x			x	x		x	x	9	4	90
Remove from shop. cart (mandatory)	x	x	x	x	x	x	x		x			x	x		x	x	9	4	b fp
Display shipp. methods (mandatory)	x	x	x	x	x	x	x	x	x	x		x	x		x	x	10	4	10
Selection of shipp. meth. (mandatory)	x	x	x	x	x	x	x	x	x	x					x	x	10	2	b fn
Shipp. opt.: client address (or) store	x	x	x	x	x		x	x	x		x			x		x	8	3	66
To store (requires) Click&collect	x					x	x	x			x		x			x	4	3	
Enter feedback (optional)	x	x	x	x	x	x	x	x	x	x	x	x				x	10	3	
Enter reviews (or) enter ratings	x	x				x	x				x	x		x			4	3	
Publish feedback (optional)				x			x	x			x	x					3	2	
Pub. feedback (requires) enter review	x		x	x	x	x	x				x	x	x	x	x	x	7	6	
Pub. feedback (requires) enter ratings	x			x	x	x	x					x	x	x	x	x	6	5	
Click&collect service (optional)	x					x	x				x						3	1	
Click&collect serv. (excludes) tracking	x	x	x	x	x	x	x	x			x	x	x	x		x	9	5	
Tracking information (optional)	x	x	x	x	x	x	x	x			x	x			x	x	9	4	
Displ. payment methods (mandatory)	x	x	x	x	x	x	x	x	x			x			x	x	10	3	
Select payment method (mandatory)	x	x	x	x	x	x	x	x	x						x	x	10	2	
Customer email info. (mandatory)			x	x	x	x	x	x	x						x	x	8	2	
false positives	0	2	0	0	0	0	2	0	0	1	3	0	4	2	1	0			

Fig. 2. Feature and variation points detection on Eshop

coloured in light blue. Moreover, in both figures, the upper part deals with feature identification and the lower part with variation points.

Coffee Machine Review. We have first considered the coffee machine case study, and collected all data in Fig. 1. As for feature identification, it can be seen that Bing, which found an average of 95.8% (46/48) of the features in the ground truth ("b aver"), performed better than chatGPT, that found a 75% (60/80) of the features ("c aver"). Also in variability identification, Bing outperformed chatGPT with an average of 80% (48/60) versus 56% (56/100) of variation points identified.

Both chatbots have returned a very small number of false positives: in features identification, chatGPT returned only 1 false positive in one of the answers, while Bing in 2 answers returned 2 false positives. Few more false positives were returned in the search for variation points: chatGPT returned an overall 13 false positives, and Bing returned an overall 6 false positives.

Eshop Review. The data of eshop case study are collected in Fig. 2. As far as feature identification is concerned, we can observe that both tools found a good number of features with an average by chatGPT of the 87,5% (210/240) ("c aver") vs an average of 86,8% (125/144) found by Bing ("b aver;;) and a small number of false positives: chatGPT returned 4 false positives in 1 of the answers, while Bing in 2 answers returned 1 and 2 false positives, respectively.

In variability identification, chatGPT performed better than Bing: the chatbots identified an average of 81,2% (211/260) and 57,7% (90/156) of variation points identified, respectively. We finally consider false positives: in the search for variation points chatGPT returned an overall 5 false positives and Bing returned an overall 10 false positives.

3.2 Data Analysis

We have performed a quantitative analysis of the data, both for features and variation point detection. The analysis is based on the following performance metrics, where tp are true positives, fp are false positives and fn are false negatives.

$$precision = \frac{tp}{tp + fp} \quad recall = \frac{tp}{tp + fn}$$

Performance measures are in Fig. 3. In this Figure, besides reporting LLMs' performances, we also report precision and recall of an analysis performed with VIBE, a rule based tool presented in a previous study [5]. It is important to mention that VIBE is only able to detect variation points (with the exclusion of mandatory ones), and cannot directly identify features, hence the "n. a." cells.

	coffee machine						eshop					
	rule-based		chatGPT		Bing		rule-based		chatGPT		Bing	
	precision	recall	precision	recall	precision	recall	precision	recall	precision	recall	precision	recall
features	n.a.	n.a.	0.983	0.75	0.92	0.958	n.a.	n.a.	0.981	0.875	0.976	0.868
var. points	1	0.61	0.811	0.56	0.888	0.8	1	0.414	0.976	0.811	0.9	0.576

Fig. 3. Performance measures.

4 Discussion and Threats to Validity

With the data collected from the experiences, we are able to answer to the RQs:

RQ1: The LLMs' performances in features detection are quite good, with precision in the range 0.92–0.981 and recall in the range 0.75–0.958.

RQ2: Also in variation points detection the LLMs' performances are good, with precision in the range 0.811–0.976 and recall in the range 0.56–0.811. Moreover, where comparable, the rule based tool outperforms LLMs when precision is considered, while with respect to recall, with the only exception of Bing on the coffee machine example, the LLMs perform better than the deterministic tool.

RQ3: Overall, Bing performs better than chatGPT, this is especially true for precision, while as far as recall is concerned, there is one case, the detection of eshops variation points, where chatGPT beats Bing. In our opinion, for the type of task required which is to analyze some syntactic and semantic characteristics of a text, Bing's better performance could be ascribed to the underlying GPT-4, which is a larger language model with a wider training dataset, but this is only a conjecture, as the architecture of the two tools is still unknown.

As a lesson learned about the use of prompt templates, we observed that in applying the Template Pattern, the prompt designer should avoid the use of concrete examples of output, as these can influence the LLM's response.

Construct Validity: We have used precision and recall as parameters to evaluate the performance of chatGPT. These metrics have shown effective for the two requirement documents. *Internal validity:* A threat could be the involvement of the authors in the definition of the ground truth, but this threat is mitigated by the unanimous agreement during the ground truth definition meeting. With regard to *external validity*, we presented an initial study, limited to two case studies and a limited number of experiments. We consider that the fact that making the same query from independent profiles yields different responses only partially mitigates this threat.

5 Conclusions and Future Work

This paper has addressed the question of whether general-purpose AI-based tools may be employed to detect variability in Natural Language (NL) requirements documents. The preliminary experiments conducted applying chatGPT (with GPT-3.5) and Bing have shown their effectiveness on the two analyzed requirements documents, but should be extended and confirmed by a larger experimental basis.

A major direction of future work would be feature model extraction from NL requirement documents, on the basis of the feature and variability identification addressed for the preliminary case studies in this paper. We made an initial attempt by asking chatGPT to construct a feature diagram for the coffee machine, with poor results. This topic needs to be addressed in a more systematic study.

To this end, and in general to improve performance, we plan to experiment with other prompt patterns [1,9] than those used in this work, in which we used the models that we found most suited, given the collaboration with students.

Acknowledgements. The research has been partially supported by the MIUR, Italy project PRIN 2022 STENDHAL. We gratefully thank the reviewers for their thoughtful comments and the students of the software engineering course in Pisa for their help.

References

1. Bach, S.H., et al. PromptSource: an integrated development environment and repository for natural language prompts. In: Proceedings of the 60th Annual Meeting of the Association for Computational Linguistics, ACL 2022 - System Demonstrations, Dublin, Ireland, pp. 93–104. Association for Computational Linguistics, 22–27 May (2022)
2. Bakar, N.H., Kasirun, Z.M., Salleh, N.: Feature extraction approaches from natural language requirements for reuse in software product lines: a systematic literature review. J. Syst. Softw. **106**, 132–149 (2015)
3. Brown, T.B., Mann, B., et al.: Language models are few-shot learners. In: 33rd Annual Conference on Neural Information Processing Systems, 6–12 Dec (2020)
4. Clements, P., Northrop, L.M.: Software product lines-practices and patterns. SEI series in Software Engineering. Addison-Wesley (2002)
5. Fantechi, A., Gnesi, S., Semini, L.: VIBE: looking for variability in ambiguous requirements. J. Syst. Softw. **195** (2023)
6. Galindo, J.A., Dominguez, A.J., White, J., Benavides, D.: Large language models to generate meaningful feature model instances. In: Proceedings of the 27th ACM International Systems and Software Product Line Conference - Volume A, SPLC 2023, pp. 15–26. Association for Computing Machinery, New York (2023)
7. Itzik, N., Reinhartz-Berger, I., Wand, Y.: Variability analysis of requirements: considering behavioral differences and reflecting stakeholders' perspectives. TSE **42**, 687–706 (2016)
8. Li, Y., Schulze, S., Saake, G.: Reverse engineering variability from natural language documents: a systematic literature review. In: Proceedings of the 21st International Systems and Software Product Line Conference, SPLC, pp. 133–142. ACM (2017)

9. Liu, P., Yuan, W., Fu, J., Jiang, Z., Hayashi, H., Neubig, G.: Pre-train, prompt, and predict: a systematic survey of prompting methods in natural language processing. ACM Comput. Surv. **55**(9), 195:1–195:35 (2023)
10. Mefteh, H.B.-A.M., Bouassida, N.: Mining feature models from functional requirements. Comput. J. **59**, 1784–1804 (2016)
11. Pohl, K., Böckle, G., van der Linden, F.: Software Product Line Engineering - Foundations, Principles, and Techniques. Springer (2005). https://doi.org/10.1007/3-540-28901-1
12. Schulze, S., Li, Y.: Feature and variability extraction from natural language requirements. In: Lopez-Herrejon, R.E., Martinez, J., Assunção, W.K.G., Ziadi, T., Acher, M., Vergilio, S.R. (eds.) Handbook of Re-Engineering Software Intensive Systems into Software Product Lines, pp. 31–52. Springer International Publishing (2023). https://doi.org/10.1007/978-3-031-11686-5_2
13. Sree-Kumar, A., Planas, E., Clarisó, R.: Extracting software product line feature models from natural language specifications. In: Proceeedings of the 22nd International Systems and Software Product Line Conference - Volume 1, SPLC 2018, Gothenburg, Sweden, 10–14 September, pp. 43–53. ACM (2018)
14. White, J., et al.: A prompt pattern catalog to enhance prompt engineering with chatgpt. CoRR, abs/ arXiv: 2302.11382 (2023)

**Natural Language Processing
for Requirements Engineering**

Designing NLP-Based Solutions for Requirements Variability Management: Experiences from a Design Science Study at Visma

Parisa Elahidoost[1,2]([✉]) [iD], Michael Unterkalmsteiner[1] [iD], Davide Fucci[1] [iD], Peter Liljenberg[3], and Jannik Fischbach[2,4] [iD]

[1] Software Engineering Research Lab SERL, Blekinge Institute of Technology, Valhallavägen 1, 37179 Karlskrona, Sweden
[2] fortiss GmbH, Guerickestraße 25, 80805 Munich, Germany
elahidoost@fortiss.org
[3] Visma, Sambandsvägen 5, 35236 Växjö, Sweden
[4] Netlight Consulting GmbH, Prannerstraße 4, 80333 Munich, Germany

Abstract. **Context and motivation**: In this industry-academia collaborative project, a team of researchers, supported by a software architect, business analyst, and test engineer explored the challenges of requirement variability in a large business software development company. **Question/problem**: Following the design science paradigm, we studied the problem of requirements analysis and tracing in the context of contractual documents, with a specific focus on managing requirements variability. This paper reports on the lessons learned from that experience, highlighting the strategies and insights gained in the realm of requirements variability management. **Principal ideas/results**: This experience report outlines the insights gained from applying design science in requirements engineering research in industry. We show and evaluate various strategies to tackle the issue of requirement variability. **Contribution**: We report on the iterations and how the solution development evolved in parallel with problem understanding. From this process, we derive five key lessons learned to highlight the effectiveness of design science in exploring solutions for requirement variability in contract-based environments.

Keywords: Industry-academia collaboration · Requirements variability management · Lessons learned

1 Introduction

Variability management is a key factor when considering the frequent necessity to modify and update software products. Variability signifies the capacity for alterations in both software products and models [16]. Variability management

---

P. Liljenberg affiliated with Visma, Sambandsvägen 5, 35236 Växjö Sweden at the time of the research.

includes activities such as representing variability in software artifacts explicitly throughout their lifecycle, managing interdependencies, and facilitating implementation. This management process is intricate and challenging, necessitating methodologies, techniques, and tools to be effectively executed [15].

In this study, Blekinge Institute of Technology (BTH) collaborated with Visma, a software company producing a payroll solution that supports hundreds of collective labour agreements (CLA). CLAs are contractual documents written in natural language that represent the basis for the development, testing, and configuration of the payroll system. CLAs within the same domain and related industries often share aspects relevant to implementation. Similarities across these documents usually represent a shared software configuration. A pain point for the company is identifying and managing such similarities, particularly considering that CLAs are renewed regularly. Currently, business analysts need to *manually* sift through documents to assess similarity and maintain traceability between these shared configurations and CLAs. This task is labor-intensive, time-consuming, and error-prone. The initial aim of the collaboration between BTH and Visma was to investigate the challenges related to the variability and traceability of configurations extracted from CLAs, and their impact on requirements analysis and testing. In this paper, we report our experience applying design science [19] to build, together with Visma, several iterative, Natural Language Processing (NLP) based solutions to address problems related to traceability when using CLAs as a source for requirements specifications. We report five lessons learned from this experience. This paper presents insights for future industry-academia collaborations in the same area. Furthermore, it provides an example of the challenges that can arise when dealing with requirements extracted from regulatory documents, such as contracts.

The rest of the paper is organized as follows. Section 2 presents a brief overview of the existing literature on the application of NLP to traceability problems in software engineering and variability extraction from natural language documents. We detail our research goals and methodology in Sect. 3. In Sect. 4, we report the results from five design iterations with associated lessons learned. Finally, Sect. 5 concludes the paper.

2   Related Work

A recent systematic review of 96 primary studies published between 2013 and 2021 on the application of NLP to traceability problems in software engineering [10] highlighted the different syntax and semantics across artifacts as well as the difficulty to modelling tacit knowledge as the main challenges in this research area. Michelon et al. [7] addressed the challenges of maintaining, evolving, and composing variants of systems that evolve over space and time. They identify four key challenges in this area, focusing on the complexities introduced by the temporal and spatial evolution of system variants. Their work emphasizes the need for effective strategies to manage these evolving variants, aiming to improve the maintenance and evolution processes. Ferrari et al. [2] developed tools, CMT and FDE, to facilitate the translation of requirements expressed in natural language into feature models, specifically feature diagrams. Their work aims to bridge the gap between the informal documentation of requirements and the structured representation needed for software product line engineering. These tools support the automated analysis of natural language documents, extracting relevant features, and structuring them into a formal model, thus streamlining the process of feature diagram creation and enhancing the accuracy and efficiency of software product development. Reinhartz-Berger et al. [13] explored the use of semantic and ontological methods to analyze the variability within Software Product Lines (SPLs). Their work aims to understand and manage the complexity and diversity of SPLs by applying advanced information systems engineering techniques. Through their research, they propose a structured approach to capture and reason about the variability in SPLs, leveraging semantic and ontological considerations to improve the design and customization of software products. In regulatory requirements engineering, which is closely related to our study, NLP techniques have been employed to create traceability links between regulations and specifications. Jain et al. [4] utilized five semantic similarity methods to connect use cases derived from specifications to regulatory documents, achieving a Medium Average Precision (MAP) of 0.72 in their evaluation of 69 insurance domain documents. Sleimi et al. [17] enhanced the systematic traceability of legal requirements by proposing a method to automatically extract metadata using domain-specific semantics. This approach was based on parsing rules developed from constituency and dependency parsing, refined with expert annotations of 200 legal statements in the traffic and vehicle regulation domain. Li et al. [5] provide a comprehensive and detailed systematic literature review of approaches and tools in the area of feature and variability extraction from natural language documents. Besides showing that requirements and product descriptions are most commonly used in existing research, the authors show that several of the proposed approaches are neither accurate nor complete, limiting their practical use. The authors present a pipeline, consisting of several NLP, information retrieval, and machine learning-based techniques. In our research, we draw on NLP techniques related to paraphrase mining and plagiarism detection, as discussed by Pavlick et al. [11]. They adapted a machine translation method to learn paraphrases from bilingual corpora, achieving a 4.2% improvement over existing approaches with a 43.7% Area Under the Curve (AUC) in their best model. Ma et al. [6] introduced an unsupervised technique for extracting paraphrases from small topic-specific datasets. Their method, based on word alignment to form a graph for generating paraphrase candidates, outperformed state-of-the-art approaches, significantly increasing recall by 247 to 460% across various corpora while maintaining precision comparable to existing methods. Foltýnek et al. [3] examined six word embedding models and five classifiers for identifying plagiarism, with their best classifier achieving 83.4% to 99.0% accuracy, surpassing human experts and current plagiarism detection systems Wahle et al. [18] assessed five pre-trained word embedding models and eight neural language models on various texts, including paraphrased research papers and articles, with their top models outperforming human evaluators by an average of 2.5%, achieving a 74.8% to 80.5% micro F1-score.

3   Research Methodology

In 2018, the Software Engineering Research Lab group at BTH initiated an eight-year research program[1] with initially nine companies from the telecom and finance domains. We had no prior collaboration experience with six of these companies. Hence, our focus was to understand their challenges in software engineering and match them with our expertise and interests. Visma, a large international accounting software company, was interested in our previous work on using test cases as requirements specifications [1]. In particular, they were interested in "getting a mapping of requirements from laws and collective agreements and see how well they are covered by our test cases." In Sect. 4.1, we set out to address this research agenda. The collaboration included three champions [20] who acted as our contact points within Visma, a software architect (SA), a business analyst (BA), and a test engineer (TE). A significant challenge for business analysts is the manual identification and management of similarities between CLAs. This situation presents a clear research goal: *to develop an automated solution that can efficiently and accurately identify and manage the similarities among CLAs, thereby reducing the reliance on manual processes, minimizing errors, and improving overall efficiency in the management of software requirements configurations.*

We followed the design science paradigm [14,19] to design, implement, and evaluate a technical solution to fulfill our research goal. This paper reports on five iterations in which we develop, evaluate, and improve the solution design collaboratively with the three champions. The design science methodology encompasses three principal phases: initial *problem conceptualization*, addressed in the first two iterations (Sects. 4.1-4.2) through prototype development and cognitive task analysis; the *solution design*, involving CLA analysis and the collection of feedback, primarily during iterations three to five (Sects. 4.3-4.5); and a final phase of *empirical evaluation*, where we have preliminary indications of the solution effectiveness. Each iteration is characterized by specific inputs, the actions undertaken, the resulting outputs, and the identified lessons learned.

4   Design Iterations and Results

This section presents the design science iterations we conducted to understand the challenges of practitioners working with CLAs, build a set of candidate solutions to these challenges, and evaluate them. We report on the iterations and how the solution development evolved in parallel with problem understanding. Lessons learned synthesize the key takeaways from each iteration.

---

[1] rethought.se.

4.1   Iteration One

We scheduled an online meeting with SA, BA and TE to understand the problem better. We agreed to initiate a feasibility study to explore whether current NLP techniques can support BAs in evaluating to what extent test cases, which in the case of the company represent de-facto the requirements, cover statements in CLAs. SA provided us with CLAs and software feature configurations, which we used to implement a simple keyword extractor that matched terms between the two artifacts. We visualize the terms matching in Fig. 1. Each row in the heatmap represents a feature configuration (43 in total), and each column represents a page (72 in total) from one CLA. The color in each cell represents the frequency of the terms (lower-cased, stemmed), extracted from the feature configuration, encountered on the page from the collective agreement. A cell marked with a T signifies that this page is the source for a feature configuration—i.e., the ground truth. We used this visualization to discuss with Visma the feature configurations in the areas of the heatmap where the approach worked well (upper left) and where it did not (lower right). We also discussed using the heatmap to discover areas in the CLA that contain potentially relevant information (upper right area) or to guide the BA in reading unseen collective agreements. Although we received positive feedback from SA, BA and TE, the heatmap did not solve BA's and TE's specific problems when working with CLAs, which they could not articulate then. This misalignment between problem and candidate solution highlighted other challenges the BA and TE encountered when analyzing the CLAs, which the researchers or practitioners had not fully understood yet.

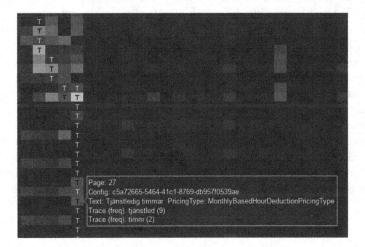

Fig. 1. Visualization of term matching frequencies in feature configurations (rows) and pages in a CLA (columns). Ground truth marked with a "T".

Lesson learned:

Even though the initial meeting with the SA, BA, and TE produced an actionable problem description that could be solved quickly through a proof-of-concept prototype, we did not yet have an in-depth problem understanding. We realized that articulating these challenges in a vacuum, without illustrating the tasks these stakeholders struggle with, is ineffective. However, the heatmap helped researchers and practitioners realize that our understanding of the problem was not aligned. Concrete artifacts created early in a research project can catalyze the discovery of such gaps.

4.2 Iteration Two

We used Applied Cognitive Task Analysis (ACTA) [8] to elicit a detailed description of how the BA and TE perform a particular task, described below, involving the CLAs. We designed a simple data collection instrument in a spreadsheet, capturing the performed sub-tasks, the knowledge needed to perform them, and the associated challenging cognitive activities. Our goal was to understand which tasks require cognitive skills that, with the increase of supported CLA, can be difficult to perform. Such tasks are candidates for designing and evaluating a technical solution. As a basis for the analysis, we identified together with BA and TE the following tasks: (1) develop, based on a CLA, a product configuration for salary payments and adapt, if necessary, the existing product configuration structure (BA); (2) assemble, based on a collective agreement, a test suite and create, if necessary, new test cases (TE). Based on the tasks they are performing, BA and TE have related, yet slightly different, challenges using the CLAs. The variation points (i.e., where the agreement can vary) and the variants (i.e., the options at a particular variation point) of CLAs regarding salary payment were not yet completely defined in the company. Hence, when the BA analyzes a new agreement, a change in the configuration may be necessary. When a new agreement is analyzed, it is tedious and error-prone to manually scan the agreements and compare them to find new variants. The decision to test a new agreement with an existing test suite or whether to change or add a test case relies on TE ability to recall the contents of CLAs. Wrong assumptions can lead to incomplete or wrong test cases. Furthermore, identifying missing test cases is difficult as there is currently no structured way to analyze and measure test coverage for a CLA.

> **Lesson learned:**
>
> When we dissected the task performed by BA and TE into individual steps, we understood *why* their approach, while effective on a small number of agreements, did not scale. The identified problem is certainly actionable—i.e., we can devise several solutions, built on existing research, that can support them in analysing hundreds of agreements. We can also evaluate the effectiveness of the solutions by comparing them to a baseline, contributing evidence to the research area. Making the thought process of experts explicit and uncovering the atomic actions they perform during analysis and their connections are powerful ways to strip away the uncertainty surrounding tasks that stakeholders perceive as challenging. Accordingly, ACTA can be a useful tool to gain a common problem understanding.

4.3 Iteration Three

We used the results from ACTA to refine the problem statement. Accordingly, our new goal was to identify common variants, allowing BA to analyze one aspect that spans all CLAs. Fulfilling this goal allows the BA to validate the configuration developed based on a few agreements against the remaining CLAs. To devise a technical solution to the problem, we obtained 128 CLAs[2], written in the Swedish language, and a list of 17 keywords that the BA used to search specific passages in the CLAs. Our proposed solution for this challenge employs topic modeling and information filtering to identify key topics and significant themes in each document. Topic modeling is particularly well-suited for this problem as it excels in uncovering hidden variants and themes in large text collections by detecting groups of frequently co-occurring words. In our analysis, we employed the Latent Dirichlet Allocation (LDA) technique, which views documents as compositions of various topics identified by the LDA algorithm. This aligns closely with our aim to efficiently filter and categorize document content.

We use Fig. 2 to illustrate the variations in input, pre-processing, and analysis we introduced in the remaining iterations of solution development. In iteration three, we started by pre-processing the text that had been extracted from 128 CLAs. We extracted paragraphs and their associated document name and page number. We removed stop words, lemmatized the paragraph text, and applied part-of-speech (POS) tagging.

To identify the number of topics in all documents, we calculated the coherence score—i.e., a metric for measuring the semantic interpretability of the top terms [9]. We represented topics with the top-N tokens with the highest likelihood of belonging to a specific topic. The coherence score quantifies the extent to which these words exhibit similarity to one another. In our analysis, 23 was the best number of coherent topics. We used the BA's input keywords to validate to

---

[2] Contracts are available upon request https://zenodo.org/records/10640865.

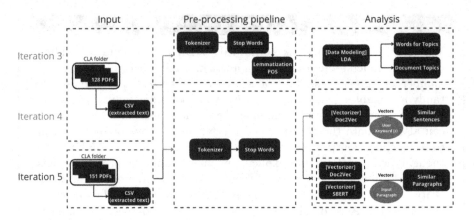

Fig. 2. Graphical depiction of the proposed method - Iterations three, four and five

ascertain whether the identified topics effectively captured these keywords. This approach aids BA in determining which documents encompass the specified keywords by locating them within the topics. We shared our results in a meeting with the three champions.

While BA and TE found the results interesting, there were reservations about their practicality and effectiveness in simplifying their tasks. The feedback from our discussion indicated that topic modeling might not adequately address the problem. Instead, a more detailed analysis that pinpoints the exact page and location of keywords in the documents was deemed more beneficial. Therefore, the next iteration focuses on providing these specific, granular details to better meet the champions' needs.

> **Lesson learned:**
>
> To ensure that identifying variability in CLAs produces practical and actionable insights, the variants and keywords we identify must be easily traceable. This involves pinpointing the exact location of these variants, down to the specific sentence on a designated page in a CLA. Such precision bridges the gap between general observations and detailed, applicable data, which is essential for effectively managing variability in CLAs. Additionally, conducting meetings that include all the beneficiary roles (BA, TE, and SA) simultaneously has proven beneficial. These interdisciplinary discussions enhanced the understanding of each role's responsibilities and improved the collective workflow.

4.4   Iteration Four

This iteration aims to pinpoint the exact page and sentence associated with each keyword. This precision ensures useful traceability and enriches the context, empowering BA to determine the pertinence of every finding. To this end, we explored the possibility of utilizing text representation approaches like word embeddings. In our analysis, we used Doc2Vec as we wanted to identify similar sentences. Doc2Vec generates distributed representations of variable-length pieces of text, such as sentences, paragraphs, or entire documents. During this iteration, conducted in 2021, the availability of models for less common languages like Swedish was limited. Therefore, we trained a Doc2Vec model based on the Swedish Wikipedia. The outline of the proposed method for iteration four can be seen in Fig. 2. After pre-processing, we used the pre-trained Doc2Vec for vectorizing all the input documents and keywords and calculated the cosine similarity between them. We distributed a spreadsheet to BA and TE featuring a list of keywords, each accompanied by relevant paragraphs from the CLAs and page numbers of semantically similar paragraphs, to assess our findings. We recognized some difficulty in understanding how a spreadsheet could be beneficial for daily tasks. We created a prototype user interface to display and filter the contents, tailoring it to the BAs' intended use cases. This prototype served as a practical demonstration of the tool potential benefits, making it easier for BA to see its applicability and usefulness in their daily work.

Figure 3 shows the user task flow diagram for the CLA analysis interface mockup.

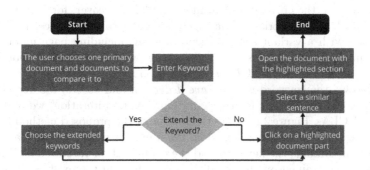

Fig. 3. User task flow diagram for CLA analysis

We showcased the prototype during a workshop. The prototype displayed the key elements of the final application design and aided the analysts in better articulating their requirements. The proposed direction for the forthcoming iteration is to gather additional insights from the analysts to further refine the prototype. The desired functionality is to retrieve all semantically similar paragraphs from various CLAs given an input paragraph. We aim to enrich the context by extracting paragraphs before and after the target text, highlighting differences from the input to streamline the analysis.

> **Lesson learned:**
>
> Prototyping emerged as a crucial factor in the latter stages, significantly influencing the integration of user-specific preferences and requirements into the core structure of our final research artifact. This strategy markedly augmented the engagement and interaction levels among our champions, showcasing how prototyping serves as an essential tool in understanding the research outcome. It emphasizes the transition from preliminary problem understanding to a more refined, user-centered research phase, ensuring that we not only addresses the identified issues but also aligns closely with the users' practical needs and expectations.

4.5 Iteration Five

In this iteration, the analysis goal is to identify all the paragraphs of all the CLAs that are similar to the given paragraph and, therefore, likely associated with the same keyword. This allows the BA to know which CLAs are not covered by the current manually created workflow for each keyword. The BA can also analyze the newly identified paragraphs to identify variations that are not yet considered in the current configuration model. Moreover, for each keyword, we identify all shared variants relating to that keyword. We received 14 keywords, each with a variant sentence from a CLA, including page numbers, to identify similar variants across all CLAs for software configuration. For example, in the sentence *"The sick leave deduction must be 20% of the average sick pay per week."* the keyword is *sick leave deduction* and *20% of the average sick pay per week* is the variant for configuration. For this iteration, we received a total of 151 CLAs. Figure 2 presents the outline of the proposed method for iteration five. Unlike the previous iteration, the input for similarity calculation is a sentence that contains the keyword and the configuration variant. The Doc2Vec model has been trained on the Swedish Wikipedia and the existing CLAs. We determine the similarity between paragraphs by computing their cosine similarity and then ranking them in descending order. We add context—i.e., previous and next paragraphs—to the selected ones. We also experimented with **SBERT Semantic Search** which generates embeddings that have multiple vector representations of the same word based on its context. Thus, BERT embeddings are context-dependent. There exist several pre-trained models for different purposes like Semantic Textual Similarity, Clustering, and Paraphrase mining [12]. We used SBERT for vectorizing sentences and paragraphs and then used the community detection function (which calculates cosine similarity) to list the similar paragraphs to the input paragraph. Additionally, we highlighted differences between each input paragraph and its similar counterparts for a clearer presentation. Figure 4 presents a sample output from the Doc2Vec model corresponding to the keyword input paragraph "Holiday deduction"[3] and the sentence

---

[3] Swedish: *Semesteravdrag.*

"For each unpaid vacation day taken, a deduction of 4.6 % of the monthly salary is made from the employee's current monthly salary."[4]. We processed 10 paragraphs, each matched to a keyword, using both the Doc2Vec and SBERT models, and presented the top-20 results from each to the SA for review. They evaluated the results marking relevant and non-relevant matches.

Document	Page	Paragraph	Similarity	Relevance
		För varje uttagen obetald semesterdag görs avdrag från tjänstemannens aktuella månadslön med 4,6 % av månadslönen.		
avtal_mellan_svensk_scenko nst_och_ledarna_fr_2011.pdf	7	semesterersättning beräknas som 4 6 % av den aktuella månadslönen per out- tagen betald semesterdag jämte semestertillägg. semesterersättning för sparad semesterdag beräknas som om den sparade dagen tagits ut det semesterår anställningen upphörde. för varje uttagen obetald semesterdag görs avdrag från den anställdes aktuella månadslön med 4 6 % av månadslönen. mom 4 semesterersättning annan anställd i fråga om övriga anställningar än sådan med tidlön enligt ovan utgår enbart semesterersättning med 12 % beräknad på utbetalda löner gager inställelse- honorar och tillägg enligt ovan.	0.860371	1
installationsavtalet-2017-2020.pdf	41	§ 4 avdrag för obetald semesterdag för varje uttagen obetald semesterdag görs avdrag från arbetstagarens aktuella månadslön med 4 6 procent av månadslönen. § 5 sysselsättningsgrad	0.852512	1
Svemek 2013-2016.pdf	49	semestertillägg ska utbetalas vid det avlöningstillfälle som infaller före huvudsemestern. för varje uttagen obetald semesterdag görs avdrag med 4 6 % av arbetstagarens aktuella månadslön. semesterlön och semesterersättning beräknas på samma sätt för sparade semesterdagar som övriga betalda se- mesterdagar. hänsyn tas dock till arbetstidsmåttet under det intjänandeår under vilket semesterdagen tjänats in.	0.852239	1

Fig. 4. Snippet of the result

To assess both models, we computed their average precision (AvP)—i.e., a metric condensing the precision-recall relationship into a single value [21].

Table 1 presents a comparative summary of the outcomes. Doc2Vec achieved marginally superior results compared to SBERT. Notably, both models yielded better performance with inputs that included formulas and numbers, as opposed to the more brief, text-only inputs like paragraph 4.

Table 1. Average precision results for Doc2Vec and SBERT

Input paragraphs	Doc2Vec AveP	SBERT AveP
paragraph 1	1	0.8987
paragraph 2	0.6044	0.56
paragraph 3	0	0
paragraph 4	0	0
paragraph 5	1	1
paragraph 6	0.8571	0.7656
paragraph 7	0.9601	0.8807
paragraph 8	1	0.8
paragraph 9	1	1
paragraph 10	1	0.887
Mean Average Precision	0.74216	0.679

---

[4] Swedish: *För varje uttagen obetald semesterdag görs avdrag fråntjänstemannens aktuella manadslon med 4,6 % av månadslönen.*

Based on the insights of iteration four, we developed a new user interface prototype that embodied our latest analysis, offering a clearer visualization of the use case. During the concluding meeting, upon reviewing the prototype, BA acknowledged that the analytical findings are be beneficial for their routine work.

> **Lesson learned:**
>
> We recognized that automating the identification of shared variants cannot be fully achieved. When it comes to legal or contractual analysis, expert input from those with domain knowledge remains essential, especially when considering the context of the input. The iterations helped the champions refine their needs and objectives. Design Science enabled continuous improvement in design, enhancing both the user interface and user experience through iterative refinements. Each phase is carefully enhanced, drawing on user feedback and suggestions. Implementing this iterative methodology proves beneficial in industry-academia partnerships, particularly when there is ambiguity about the problem context for both collaborators.

5   Conclusion

The collaboration between the BTH and Visma provided a context for investigating the complexities of requirement traceability in the presence of high variability, typical of contractual documents. In this work, we showed the efficacy of design science methodology in addressing the industry-relevant challenge of managing variability and maintaining traceability among CLAs. The lessons learned from this effort underscore the potential benefits of applying design science to gain problem understanding. This is further illustrated through the iterative nature of our solution development, which evolved with our understanding of the problem. Throughout this experience, one of the significant findings was the realization that the goal of completely automating the identification of shared configurations is not entirely feasible. Our results underscore the need for expert intervention, particularly in tasks involving legal or contractual analysis. Nevertheless, the stakeholders involved in our study acknowledged that even a modest degree of automation can be beneficial. These findings, while specific to this context, hint at broader implications for similar challenges in other industries, suggesting a potential for generalizing the approach to address variability and traceability in different domains. Our insights contribute to both academic research and industry practices, offering tangible examples of the hurdles faced when managing contract-based requirements.

Acknowledgment. We would like to thank all employees at Visma who supported our study. This work was further supported by the KKS foundation through the S.E.R.T. Research Profile project at Blekinge Institute of Technology.

References

1. Bjarnason, E., Unterkalmsteiner, M., Borg, M., Engström, E.: A multi-case study of agile requirements engineering and the use of test cases as requirements. Inf. Softw. Technol. **77**, 61–79 (2016)
2. Ferrari, A., Spagnolo, G.O., Gnesi, S., Dell'Orletta, F.: CMT and FDE: tools to bridge the gap between natural language documents and feature diagrams. In: Proceedings of the 19th International Conference on Software Product Line. SPLC '15, New York, NY, USA, pp. 402–410. Association for Computing Machinery (2015)
3. Foltýnek, T., et al.: Detecting machine-obfuscated plagiarism. In: Sundqvist, A., Berget, G., Nolin, J., Skjerdingstad, K.I. (eds.) iConference 2020. LNCS, vol. 12051, pp. 816–827. Springer, Cham (2020). https://doi.org/10.1007/978-3-030-43687-2_68
4. Jain, R., Ghaisas, S., Sureka, A.: Sanayojan: a framework for traceability link recovery between use-cases in software requirement specification and regulatory documents. In: Proceedings 3rd International Workshop on Realizing Artificial Intelligence Synergies in Software Engineering, pp. 12–18 (2014)
5. Li, Y., Schulze, S., Saake, G.: Reverse engineering variability from natural language documents: A systematic literature review. In: Proceedings 21st International Systems and Software Product Line Conference - Volume A, pp. 133–142. ACM (2017)
6. Ma, D., Chen, C., Golshan, B., Tan, W.C.: Essentia: mining domain-specific paraphrases with word-alignment graphs. In: EMNLP (2019)
7. Michelon, G.K., Obermann, D., Assunção, W.K.G., Linsbauer, L., Grünbacher, P., Egyed, A.: Managing systems evolving in space and time: four challenges for maintenance, evolution and composition of variants. In: Proceedings of the 25th ACM International Systems and Software Product Line Conference - Volume A. SPLC '21, New York, NY, USA, pp. 75–80. Association for Computing Machinery (2021)
8. Militello, L.G., Hutton, R.J.: Applied cognitive task analysis (acta): a practitioner's toolkit for understanding cognitive task demands. Ergonomics **41**(11), 1618–1641 (1998)
9. O'Callaghan, D., Greene, D., Carthy, J., Cunningham, P.: An analysis of the coherence of descriptors in topic modeling. Exp. Syst. Appl. **42**, 5645–5657 (2015)
10. Pauzi, Z., Capiluppi, A.: Applications of natural language processing in software traceability: a systematic mapping study. J. Syst. Softw. **198**, 111616 (2023)
11. Pavlick, E., Ganitkevitch, J., Chan, T.P., Yao, X., Durme, B.V., Callison-Burch, C.: Domain-specific paraphrase extraction. In: Proceedings of the 53rd Annual Meeting of the Association for Computational Linguistics and the 7th International Joint Conference on Natural Language Processing, pp. 57–62 (2015)
12. Reimers, N., Gurevych, I.: Sentence-Bert: sentence embeddings using Siamese Bert-networks. In: Proceedings of the 2019 Conference on Empirical Methods in Natural Language Processing. Association for Computational Linguistics (2019)
13. Reinhartz-Berger, I., Itzik, N., Wand, Y.: Analyzing variability of software product lines using semantic and ontological considerations. In: Jarke, M., et al. (eds.) CAiSE 2014. LNCS, vol. 8484, pp. 150–164. Springer, Cham (2014). https://doi.org/10.1007/978-3-319-07881-6_11
14. Runeson, P., Engström, E., Storey, M.-A.: The design science paradigm as a frame for empirical software engineering. In: Felderer, M., Travassos, G. (eds.) Contemporary Empirical Methods in Software Engineering, pp. 127–147. Springer, Cham (2020). https://doi.org/10.1007/978-3-030-32489-6_5

15. Schmid, K., John, I.: A customizable approach to full lifecycle variability management. Sci. Comput. Program. **53**(3), 259–284 (2004)
16. Sinnema, M., Deelstra, S., Hoekstra, P.: The COVAMOF derivation process. In: Morisio, M. (ed.) ICSR 2006. LNCS, vol. 4039, pp. 101–114. Springer, Heidelberg (2006). https://doi.org/10.1007/11763864_8
17. Sleimi, A., Sannier, N., Sabetzadeh, M., Briand, L., Dann, J.: Automated extraction of semantic legal metadata using natural language processing. In: 26th International Requirements Engineering Conference (RE). IEEE (2018)
18. Wahle, J.P., Ruas, T., Foltýnek, T., Meuschke, N., Gipp, B.: Identifying machine-paraphrased plagiarism. In: Smits, M., et al. (eds.) Information for a Better World: Shaping the Global Future. LNCS, vol. 13192, pp. 393–413. Springer, Cham (2022). https://doi.org/10.1007/978-3-030-96957-8_34
19. Wieringa, R.: Design Science Methodology for Information Systems and Software Engineering. Springer, Heidelberg (2014). https://doi.org/10.1007/978-3-662-43839-8
20. Wohlin, C., Aurum, A., Angelis, L., Phillips, L., Dittrich, Y., Gorschek, T., Grahn, H., Henningsson, K., Kagstrom, S., Low, G., et al.: The success factors powering industry-academia collaboration. IEEE Softw. **29**(2), 67–73 (2011)
21. Zhang, P., Su, W.: Statistical inference on recall, precision and average precision under random selection. In: 9th International Conference on Fuzzy Systems and Knowledge Discovery, pp. 1348–1352. IEEE (2012)

Natural2CTL: A Dataset for Natural Language Requirements and Their CTL Formal Equivalents

Rim Zrelli[1]($\boxtimes$) (iD), Henrique Amaral Misson[1](iD), Maroua Ben Attia[2](iD),
Felipe Gohring de Magalhães[1](iD), Abdo Shabah[2](iD), and Gabriela Nicolescu[1](iD)

[1] Polytechnique Montréal, Montreal, QC H3T 1J4, Canada
rim.zrelli@polymtl.ca
[2] HumanITas Solutions, Montreal, QC H2Y 1N3, Canada

Abstract. The design of contemporary critical systems involves numerous requirements that must be clearly and coherently articulated, posing significant challenges for system designers. This paper addresses the challenge of translating ambiguous Natural Language (NL) requirements into unambiguous Computation Tree Logic (CTL) specifications, an essential task for maintaining consistency and precision in system design. We introduce Natural2CTL, a novel dataset comprising 2,095 pairs of NL requirements and their CTL specifications. A key aspect of this research includes a detailed methodology for data collection and annotation. The robustness of Natural2CTL is established through rigorous validation processes, including evaluations by academic and industry experts, inter-rater reliability assessments, and practical verification using UPPAAL case studies. These validation efforts underscore the dataset's reliability and its potential applicability in both research and educational domains within Requirements Engineering (RE) and formal methods.

Keywords: Natural Language Requirements · Ambiguity · CTL · Specifications · Dataset Development · Validation

1 Introduction

The rapid development and proliferation of critical systems in modern times have culminated in an era marked by heightened complexities in system design. Being pivotal across sectors from aerospace to healthcare, these systems necessitate precision and reliability at every design step. Software requirements specification is unquestionably important throughout the entire software development life-cycle. Many of the issues faced in software systems can be attributed to gaps in collecting, specifying and managing requirements. High-quality requirements are crucial for catching errors early, thereby reducing project risks and costs. The continuous evolution of Requirements Engineering (RE) underscores its significance in contemporary software engineering research and practice.

Requirements, predominantly written in natural language (NL), are susceptible to ambiguity and vagueness, leading to different interpretations by stakeholders and potentially insufficient precision for critical software safety. Moreover, these ambiguities could introduce inconsistencies within a system's requirements, presenting further challenges in system design and validation. To overcome this challenge, several approaches can be found in the literature to deal with ensuring precise NL-based software requirements specifications. A systematic literature review [1] identifies strategies to tackle ambiguities at lexical, syntactic, semantic, and pragmatic levels, ranging from the establishment of word glossaries to the application of NLP techniques for automatic translation into models. The paper [1] also highlights the absence of holistic solution approaches that consider ambiguity at multiple linguistic levels.

In addition to these linguistic strategies, formal specifications emerge as a tool for addressing the nuances of NL. Formal methods provide structured representations and techniques for precise specification and systems validation. Such formalized techniques navigate the inherent ambiguities of NL, imparting specifications with clarity and an unambiguous representation. Through a precise translation mechanism, there's a potential to detect and resolve inconsistencies early in the design phase, further fortifying the role of formal methods.

Transitioning from NL-based requirements to their structured formalized specification necessitates a precise approach. Computation Tree Logic (CTL) [2] [3] is widely acknowledged as an effective specification language, noted for its proficiency in articulating temporal aspects of reactive systems. CTL, in its fundamental nature, enables a methodical representation of system behaviours across temporal dimensions, rendering it indispensable for critical systems wherein temporal dynamics are paramount. It is, however, crucial to recognize the limitations of CTL, particularly its limited capacity to capture complex relational dynamics between entities. Despite these constraints, CTL was adopted for its robust toolset support, its aptness for extant models, and its potential to bring robustness to RE.

Significant efforts have been made to translate NL requirements into formal specifications. For instance, the study in [4] demonstrates methods to automatically generate CTL properties from English comments, reflecting the ongoing efforts to enhance verification processes. Similarly, authors in [5] discuss the transformation of informal requirements to formal ones, albeit with limitations in capturing temporal constraints. The work in [6] further underlines the potential of language models in this domain, notably in the field of first-order logic and linear-time temporal logic. Additionally, [7,8] illustrate the challenges and state-of-the-art in extracting formal models and LTL (Linear Temporal Logic) formulas from NL texts. The research in [9] provides a comprehensive view of the various approaches and the need for advancements in formalizing NL requirements into temporal logics like LTL and CTL. These studies collectively highlight significant progress in the field, yet also underline persisting challenges. There remains a distinct lack of capturing the intricacies of temporal logic and complex relational dynamics.

The challenge is intensified by the absence of systematically organized datasets linking NL requirements with CTL equivalents. The deployment of translation techniques in real-world scenarios depends heavily on the availability of representative datasets for training and evaluation.

To address this gap, we introduce Natural2CTL, a dataset designed to seamlessly integrate the nuances of NL with the precision of formal logic. Natural2CTL encompasses a structured compilation of NL requirements, each paired with its pertinent CTL specification. This dataset not only facilitates rigorous validation in scenarios demanding precision and structured representation but also serves as a resource for research and development in automated translation processes within the domains of formal methods and RE.

The Natural2CTL dataset has undergone a rigorous validation process, ensuring its reliability and effectiveness in accurately translating NL requirements into CTL. The validation included phases such as an initial evaluation by two Ph.D. candidates with expertise in formal methods and CTL, an inter-rater reliability assessment to ensure consistency in translations, and UPPAAL model-checking case studies for practical verification. Additionally, an independent industry expert's evaluation of the dataset further strengthened its credibility, aligning closely with our original CTL annotations.

The remainder of the paper is structured as follows. Section 2 describes the data creation process and outlines its characteristics. Section 3 is dedicated to a detailed evaluation of the dataset. Finally, Sect. 4 concludes the paper and addresses future research directions.

2 Dataset Development and Overview

2.1 Data Collection

To construct a comprehensive dataset, we embarked on an extensive process of data collection and annotation, aiming to provide a repository that connects software requirements with their corresponding CTLs. The approach outlined below is fusing academic rigour and practicality that is useful to the industry.

Our primary sources for software requirements included public datasets ([11,12]), academic papers ([13–18]), and Kaggle [19], a popular platform for datasets. These sources provided a rich variety of functional and non-functional requirements essential for our study.

Our initial effort involved a literature search to identify papers containing software requirements paired with their CTL counterparts ([15–18]). This was an essential step to ensure that a portion of our CTL translations aligned with existing research standards and ensuring that they were externally validated, complementing our manual annotations.

We adopted a targeted approach in selecting requirements. Emphasis was placed on labels such as functional, availability, fault tolerance, performance, and security. These were chosen for their clarity and directness in depicting software behaviour, making them suitable for effective CTL representation.

Our decision to concentrate on these specific requirements stemmed from their relevance and capability to yield precise and clear CTL specifications.

Labels such as maintainability, portability or scalability give fundamental insights into software design. However, their qualitative and subjective nature may not be a perfect fit for the CTL structure. We gave precedence to requirements demonstrating distinct temporal behaviour. This focus stemmed from CTL's ability to capture system dynamics over time. By selecting requirements that emphasized time-anchored system behaviours, we ensured they were amenable to CTL formulations, thus maintaining fidelity within the CTL framework.

In selecting requirements for translation into CTL, it was crucial to distinguish between those that were suitable and those that were not. For instance, consider the requirement from the Kaggle dataset [19]: 'The average number of recycled parts records per day shall be 50,000.' This requirement, representative of the scalability label, illustrates a challenge for CTL translation. CTL focused on specifying system behaviour over time, is less suited for capturing quantitative performance metrics like this one. Such requirements, while important, do not align with the temporal and behavioural focus of CTL and were therefore excluded from our dataset. This decision reflects our prioritization of requirements that could be effectively represented in CTL, focusing on clear temporal and behavioural aspects. This example highlights the rationale behind our selective approach and underscores the importance of aligning the dataset's content with the strengths and limitations of CTL as a formal specification language.

The selection process entailed an initial review of available requirements, followed by an analysis of their suitability for CTL translation. This approach guaranteed the inclusion of only the most relevant and clear requirements, aligning with the foundational principles of temporal logic in system specification. This systematic and focused approach to data collection was fundamental in assembling a dataset that not only meets academic standards but is also practically relevant to the industry. The Natural2CTL dataset thus stands as a comprehensive repository, bridging the gap between NL and formal logic specifications.

2.2   Annotation Process

The annotation of the Natural2CTL dataset was a meticulous process that aimed to transform NL requirements into precise CTL specifications. This journey began with a methodical approach to manual annotation, carefully considering the variability and ambiguity inherent in NL.

Understanding the formal semantics of CTL was pivotal [20]. CTL's systematic representation incorporates temporal operators to capture system behaviours over time. The Key path quantifiers-"A" (for all paths) and "E" (there exists a path)-alongside temporal operators like "G" (globally), "F" (Eventually, or in the future), "X" (next), and "U" (until), provide a flexible language for specifying system properties. For instance, "$AG(p)$" denotes 'p' holding globally on all paths, while "$EF(q)$" indicates the existence of a path where 'q' eventually becomes true.

The "Property Pattern Mappings for CTL" [21] provided a structured approach for translation. This method helped bridge NL requirements to their formalized representations, covering patterns such as Absence, Existence, Bounded Existence, Universality, Precedence, and Response. These patterns were instrumental in reducing the variability in interpreting NL requirements and ensuring consistent CTL translations. For a more in-depth exploration of the patterns and their respective sub-variants, more details can be found at http://patterns. projects.cis.ksu.edu/.

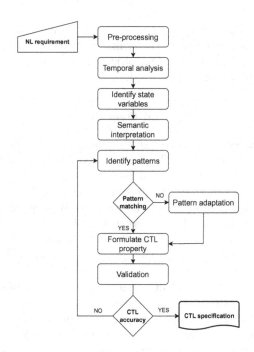

Fig. 1. Flowchart demonstrating the translation process from NL to CTL

Our process (Fig. 1) began with pre-processing to clarify ambiguities in the NL requirements, followed by temporal analysis to identify time-anchored behaviours. State variables and actions were then defined, leading to a deep semantic interpretation to contextualize the requirement. We identified CTL patterns that matched the interpreted semantics, adapting or combining patterns as necessary to accurately reflect the requirement's intent. To embody the inherent variability of CTL and ensure the robustness of our translations, we implemented a validation step to assess the accuracy of the CTL properties. Any deviations led to a re-evaluation of the pattern identification or adaptation process. This approach ensured that the final CTL specification was a true representation of the original NL requirement, even for atypical cases that did not conform to standard patterns.

The annotation process was iterative, with a focus on addressing requirements that presented ambiguous temporal dynamics. To ensure the precision of CTL translations, we engaged in a thorough analysis of the formal aspects of CTL. Our collective expertise facilitated a comprehensive internal review process, where each translation was scrutinized and refined to maintain accuracy and adherence to the semantic depth of the requirements.

We adopted a modular translation approach, breaking down complex requirements into manageable components for individual translation and subsequent integration. This practice significantly reduced errors and ambiguities, enabling a more granular and precise translation process.

The annotation phase was more than just a simple translation process. It unfolded an effort to balance and synchronize the inherent flexibility of NL and the rigorous determinism of CTL. Our objective was to ensure that the fidelity of each original requirement is preserved.

2.3 Dataset Characteristics and Overview

Our dataset comprises a comprehensive compilation of 2,095 entries, sourced from various application domains. It encapsulated different software systems like traffic management systems, which address smart traffic control and vehicular networks, and Cloud Computing. It also reflects the synchronized collaboration inherent in distributed systems in our dataset. Database Systems especially concerning transactions and consistency, as well as E-commerce Systems, further enrich our data with insights into online shopping dynamics. Moreover, telecommunication nuances like handovers between base stations form an essential part of the collection.

The dataset's richness is evident not just in the variety of systems it covers but also in the range of requirements it addresses. The dataset represents both functional and non-functional aspects. Requirements range from performance-related criteria and safety precautions to security measures and usability features. These requirements explore the intricacies of the system, addressing aspects like startup and shutdown behaviours, communication protocols, mechanisms for error handling, and adaptive responses to unexpected inputs.

A significant portion of our dataset is dominated by liveness properties as evident from the response pattern's widespread usage. This emphasizes the focus on system behaviours that guarantee certain actions or events following previous conditions or triggers. Our dataset's dimensions, however, extend beyond just these liveness-related characteristics. It robustly encompasses safety properties, ensuring that the system avoids undesirable or potentially hazardous states. By covering both these aspects, the dataset provides a well-rounded perspective on system expectations. Additionally, the presence of fairness and stability properties augments the dataset's comprehensive nature, ensuring coverage of potential system behaviours. This ensures not just the absence or presence of certain actions but also their equitable distribution and consistency over time.

For a more granular insight into the dataset's structure and composition, we present below a representative Table 1 illustrating the transformation of each requirement from an NL stipulation to its formal CTL specification.

Table 1. Excerpt from the dataset illustrating the structure of our CSV file

ID	requirement_text	CTL
01	The Disputes System shall prevent the creation of duplicate dispute requests 100% of the time	$AG(\neg\ duplicateDisputeRequest)$
02	The system shall filter data by: Venues and Key Events.	$AG(filterByVenues) \wedge$ $AG(filterByKeyEvents)$
03	When given a takeoff command, the _Internal Simulator_ shall move the UAV to the takeoff altitude corresponding to UAV's current longitude and latitude.	$AG(takeoffCommand \rightarrow$ $AF(moveToTakeoffAltitude))$
04	The TCS shall provide an interface between the TCS and an external hard copy printer.	$AF(interfaceWithExternalPrinter)$

3   Dataset Evaluation

3.1   Initial Evaluation

To establish the dataset's credibility, we initiated a first validation phase involving two Ph.D. candidates, each bringing distinct expertise in formal methods and CTL. The first, with a background in machine learning and formal verification, complemented the second candidate's practical experience in the industry, particularly in formal analysis and verification of airborne software.

Throughout a focused four-week period, both validators independently selected dataset entries at a ratio of 1 in every 5 for evaluation. This approach resulted in each validator independently reviewing 400 entries. Due to the random nature of the selection process, there were instances where both validators independently chose and reviewed the same entry. Each entry was rigorously assessed and annotated as 'correct,' 'partially correct,' or 'incorrect,' with validators providing potential corrections where necessary. For guidance in their evaluations, they used the Property Pattern Mappings for CTL [15], which formed the foundation of our annotations.

The feedback from this initial validation phase revealed diverse interpretations of requirements, reflecting variations in CTL translation patterns. Particularly, the use of time operators showed significant variability. Approximately 20% of the entries reviewed were flagged for rectification, indicative of the complex nature of translating NL requirements into CTL.

3.2   Inter-Rater Reliability and Review Process

In our commitment to ensure the robustness of our dataset, we focused on assessing inter-rater reliability. This crucial step involved having both validators re-annotate the entries initially evaluated by their counterpart, applying a structured ordinal rating system. The translations were categorized into three levels: 'correct' indicated a CTL translation that accurately and fully represented the original NL requirement; 'partially correct' denoted translations that were generally on the right track but required minor adjustments or refinements; and 'incorrect' pointed to translations that deviated from accurately capturing the requirement's essence. To quantify the consistency in ratings, we calculated Krippendorff's Alpha [22]. A score of 0.7659 was achieved, which falls within the range of substantial agreement. This score is crucial as it implies a significant level of consistency and reliability in the ratings provided by our validators.

After the reliability assessment, we engaged in a detailed review process to address instances where the validators' annotations differed. In such cases, we held joint discussions with both validators to reach a consensus on the most accurate CTL representation for each entry. These discussions were particularly crucial in instances where discrepancies might have indicated potential misinterpretations or ambiguities in the original translations. As a result, we carefully revisited and, where necessary, revised our translations to better align with the validated interpretations, ensuring a higher degree of accuracy and consistency across the dataset.

3.3   Validation and Interpretative Variability: UPPAAL Case Study

To rigorously validate the CTL translations in our dataset, we conducted a series of case studies using UPPAAL [10], a model-checking tool renowned for verifying real-time systems. UPPAAL was chosen for its advanced capabilities in simulating and verifying real-time systems, making it a more suitable choice for our complex dataset scenarios. This decision was driven by UPPAAL's ability to handle a broader range of temporal logic expressions and its more intuitive model-building environment. These studies were crucial in testing various CTL interpretations and highlighting the inherent variability in formalization.

We selected the requirement: "*When given a takeoff command, the Internal Simulator shall move the UAV to the takeoff altitude corresponding to UAV's current longitude and latitude.*" Two CTL translations were considered:

1. *AG(takeoffCommand → AF(moveToTakeoffAltitude))*, represents a response pattern, suggesting a direct causation.
2. *AF(moveToTakeoffAltitude)*, embodies an existence pattern, indicating eventual achievement of the state without a direct causal link.

Both CTL expressions were verified in UPPAAL (Fig. 2). The analysis aimed to determine which CTL more accurately represented the requirement's intended meaning. The first CTL was more contextually aligned with the requirement's intended meaning. This choice was reinforced by the fact that the second CTL, despite being valid in terms of CTL syntax, did not capture the causative relationship implied in the requirement. This distinction emphasizes the need for careful consideration of the context and semantics in CTL translations.

This example is an illustration of a broader validation process applied to multiple entries in the dataset, ensuring the contextual accuracy of our CTL

(a) UPPAAL model representation for the UAV takeoff requirement

takeoffCommand --> moveToTakeoffAltitude	A<> (takeoffCommand)
Verification/kernel/elapsed time used: 0s / 0.015s / 0.013s.	Verification/kernel/elapsed time used: 0s / 0s / 0.007s.
Resident/virtual memory usage peaks: 16,620KB / 61,844KB.	Resident/virtual memory usage peaks: 16,852KB / 62,172KB.
Property is satisfied.	Property is satisfied.

(b) Verification results for the two CTL translations.

Fig. 2. UPPAAL Model and Verification Results for UAV Takeoff Requirement

annotations. The insights gained from this case study, and others like it, have been instrumental in refining our dataset, ensuring each entry's CTL translation is contextually sound and logically consistent. The selection criteria for the requirement included its ability to demonstrate the diverse range of CTL interpretations possible from a single NL requirement.

3.4 Expert Industry Evaluation

Following the UPPAAL case studies, we engaged an industry expert with extensive experience in model checking and formal methods. The expert independently translated 100 randomly selected entries from our dataset into CTL. The translations were then analyzed by the annotator and the two initial validators. This analysis aimed to compare the expert's translations with our original annotations, focusing on aspects like exact match, minimal corrections needed, and complete deviations. This step was crucial in identifying areas where our initial translations could be further refined.

The analysis of the expert's independent translations yielded promising outcomes, with 84% of the translations aligning closely with our original CTL annotations, thereby affirming the accuracy and reliability of our initial work. The need for adjustments in approximately 16% of the entries primarily stemmed from challenges in interpreting nuanced requirements and resolving ambiguities inherent in complex sentence structures within the original requirements. In cases where significant disparities existed between the expert's translations and our own, we undertook an in-depth, case-by-case analysis. This involved meticulously scrutinizing each CTL expression to dissect the differences in interpretations and determine which version most faithfully captured the essence of the NL requirement. These findings were instrumental in guiding our subsequent refinement efforts.

3.5   Future Validation Plans

Building on the validation work already undertaken, our plans focus on establishing a more comprehensive and continuous evaluation process for the Natural2CTL dataset. To achieve this, we intend to form a dedicated validation committee, consisting of experienced academics and industry professionals with expertise in formal methods, CTL, and RE. This committee, including a seasoned professor in formal methods, will collaboratively review and refine the dataset.

Beyond the expert committee, we are committed to fostering community engagement around the Natural2CTL dataset. The dataset is publicly accessible at https://github.com/RimZrelli/PublicDataset/, and we will actively seek feedback from the broader research community and potential users. This feedback will be invaluable in identifying real-world challenges, usage scenarios, and opportunities for improvement. We envision this as a dynamic process where the dataset is not only a static resource but evolves and adapts in response to the valuable input from its diverse user base.

4   Conclusion

The creation and development of the Natural2CTL dataset represent a pivotal step forward in bridging the gap between the nuances of NL requirements and the structured precision of CTL specifications. This dataset, encompassing 2,095 pairs of NL requirements and their CTL equivalents, stands as a testament to the potential synergy between RE and formal methods. It underscores the importance of integrated approaches in enhancing system design and refining the software development process.

Throughout its development, the Natural2CTL dataset underwent validation processes, including rigorous reviews by academic and industry experts, inter-rater reliability assessments, UPPAAL case studies, and comprehensive evaluation by an independent industry professional. These steps were instrumental in ensuring that the dataset not only meets academic rigour but also holds practical relevance for the industry.

The Natural2CTL dataset is anticipated to contribute to the advancement of research and development in the area of automated translation of NL requirements into formal specifications. Its potential applications are vast, ranging from aiding in the development of tools for automated requirements analysis to serving as a benchmark for new methodologies. Furthermore, its utility in educational settings cannot be overstated, as it offers a practical resource for teaching and learning the intricacies of formal methods and their application in RE.

Acknowledgment. We extend our sincere thanks to P.S. Nouwou Mindom and L. Elfatimi for their vital role in the dataset validation, greatly enriching its integrity and validity. This research was funded by Mitacs under grant IT19246 and grant IT30530.

References

1. Amna, A.R., Poels, G.: Ambiguity in user stories: a systematic literature review. Inform. Softw. Technol. **145**, 106824 (2022)
2. Clarke, E.M., Emerson, E.A.: Design and synthesis of synchronization skeletons using branching time temporal logic. In: Grumberg, O., Veith, H. (eds.) 25 Years of Model Checking. LNCS, vol. 5000, pp. 196–215. Springer, Heidelberg (2008). https://doi.org/10.1007/978-3-540-69850-0_12
3. Emerson, E.A., Halpern, J.Y.: "Sometimes" and "not never" revisited: on branching versus linear time temporal logic. J. ACM (JACM) **33**(1), 151–178 (1986)
4. Harris, C.B., Harris, I.G.: Generating formal hardware verification properties from natural language documentation. In: Proceedings of the 2015 IEEE 9th International Conference on Semantic Computing (IEEE ICSC 2015), pp. 49–56. IEEE (2015)
5. Li, F.-L., Horkoff, J., Borgida, A., Guizzardi, G., Liu, L., Mylopoulos, J.: From stakeholder requirements to formal specifications through refinement. In: Fricker, S.A., Schneider, K. (eds.) REFSQ 2015. LNCS, vol. 9013, pp. 164–180. Springer, Cham (2015). https://doi.org/10.1007/978-3-319-16101-3_11
6. Hahn, C., Schmitt, F., Tillman, J.J., et al.: Formal specifications from natural language. arXiv preprint arXiv:2206.01962 (2022)
7. Ghosh, S., Singh, A., Merenstein, A., et al.: SpecNFS: a challenge dataset towards extracting formal models from natural language specifications. In: Proceedings of the Thirteenth Language Resources and Evaluation Conference, pp. 2166–2176 (2022)
8. Brunello, A., Montanari, A., Reynolds, M.: Synthesis of LTL formulas from natural language texts: State of the art and research directions. In: 26th International symposium on temporal representation and reasoning (TIME 2019). Schloss Dagstuhl-Leibniz-Zentrum fuer Informatik (2019)
9. Buzhinsky, I.: Formalization of natural language requirements into temporal logics: a survey. In: IEEE 17th International Conference on Industrial Informatics (INDIN), vol 2019, pp. 400–406 (2019). IEEE
10. Uppsala University, Sweden, Aalborg University in Denmark, "UPPAAL 5" (2023). https://uppaal.org/
11. Ferrari, A., Spagnolo, G.O., Gnesi, S.: PURE: a Dataset of Public Requirements Documents", National Research Council of Italy (2018). [dataset]. https://nlreqdataset.isti.cnr.it/
12. Hayes, J.: "CM1/Requirements Tracing", University of Ottawa (2015). [dataset]. https://promise.site.uottawa.ca/SERepository/datasets-page.html
13. Tjong, S.F.: Avoiding ambiguity in requirements specifications. Faculty Eng. Comput. Sci. (2008)
14. Masuoka, E., Fleig, A., Ardanuy, P., et al. MODIS. Volume 1: MODIS level 1A software baseline requirements (1994)
15. Aditi, F., Hsiao, M.S.: Hybrid rule-based and machine learning system for assertion generation from natural language specifications. In: 2022 IEEE 31st Asian Test Symposium (ATS), pp. 126–131. IEEE (2022)
16. Cosler, M., Hahn, C., Mendoza, D., et al.: nl2spec: Interactively Translating Unstructured Natural Language to Temporal Logics with Large Language Models. arXiv preprint arXiv:2303.04864 (2023)
17. Harris, C.B.: Generating formal verification properties from natural language hardware specifications. University of California, Irvine (2015)

18. Diamantopoulos, T., Roth, M., Symeonidis, A., et al.: Software requirements as an application domain for natural language processing. Lang. Resources Evaluat. **51**, 495–524 (2017)
19. Souvik, "Software Requirements Datasett", Kaagle (2020). [dataset]. www.kaggle. com/datasets/iamsouvik/software-requirements-dataset?datasetId=560206& sortBy=dateRun&tab=collaboration
20. Baier, C., Katoen, J.-p.: Principles of model checking. MIT press (2008)
21. Dwyer, M.B., Avrunin, G.S., Corbett, J.C.: Patterns in property specifications for finite-state verification. In: Proceedings of the 21st International Conference on Software Engineering, pp. 411–420 (1999)
22. Krippendorff, K.: Computing Krippendorff's alpha-reliability (2011)

Requirements Engineering for Artificial Intelligence

Towards a Comprehensive Ontology for Requirements Engineering for AI-Powered Systems

Eran Sadovski[1]([envelope]), Itzhak Aviv[2], and Irit Hadar[1] [ORCID]

[1] University of Haifa, City Campus, 3303220 Haifa, Israel
`sdv.eran@gmail.com`
[2] The Academic College of Tel-Aviv Yaffo (MTA), 6818211 Yaffo, Israel

Abstract. Context and motivation: Artificial intelligence (AI) provides computer systems problem-solving and decision-making features mimicking human behavior. As AI becomes widely adopted, AI-powered systems become increasingly ubiquitous. Requirements engineering (RE) is fundamental to system development, including AI-powered systems, which provide novel RE challenges. **Question/problem:** Developing means for addressing these challenges, which include increased need and importance of specifying and addressing social requirements, (e.g., responsibility, ethics, and trustworthiness); achieving a comprehensive understanding of all RE aspects, given the substantial growth in the diversity and complexity of requirements and the emergence of new and often contradictory ones; and, employing relevant methods and techniques that are suited for addressing these challenges. **Principal ideas/results:** We propose an RE4AI ontology as a first step toward addressing the above challenges. The development of the ontology was based on a meta-synthesis of relevant publications for identifying recurring themes and patterns, resulting in a set of themes categorized into RE stages, topics, stakeholders' roles, and constraints that formed the developed ontology. **Contribution:** The ontology provides a systematic and unambiguous representation of the accumulated RE knowledge about the system, including requirement themes, relationships between requirements, constraints, and stakeholders needed in the RE process. This ontology provides the basis for a complete AI RE methodology (AI-REM) framework that will incorporate methods to develop and manage AI-powered system requirements.

Keywords: Requirement Engineering · Ontology · Artificial Intelligence · Machine Learning · FR · NFR · RE4AI

1 Introduction

The emergence of artificial intelligence (AI) has introduced a new dimension to enhancing the capabilities of information systems. AI-powered systems are characterized by their ability to automate and improve various operations, for example, spanning image classification, natural language processing, fraud detection, content generation, and

D. Mendez and A. Moreira (Eds.): REFSQ 2024, LNCS 14588, pp. 219–230, 2024.
https://doi.org/10.1007/978-3-031-57327-9_14

human-machine interactions [1]. The computational power of modern processors, coupled with the availability of vast datasets, has been instrumental in driving the disruption of AI-powered systems [2]. Nevertheless, the integration of AI capabilities into information systems poses significant challenges, particularly in the areas of design, validation, and deployment. These challenges are exacerbated by the unique concerns introduced by AI components, such as performance, data quality, ethics, explainability, safety, trustworthiness, and accountability [3].

The literature highlights many challenges of RE4AI associated with functional and non-functional requirements (FRs and NFRs). FR are hindered by the complexity and unpredictability of AI system performance [2, 4], while NFRs have gained prominence, encompassing aspects such as ethics, responsibility, transparency, explainability, fairness, privacy, safety, and more [5, 6]. Managing the trade-offs between FRs and the extensive range of unique NFRs and among various NFRs has emerged as a significant difficulty in the context of AI-powered systems. While extensive research has been conducted in the field of requirements engineering for AI (RE4AI), there remains a lack of a comprehensive and validated RE4AI ontology tailored to AI-powered systems. In software development, ontologies are increasingly being used to represent knowledge about the world. Ontologies are explicit descriptions of how we conceptualize the world, and they can be used to reason about that knowledge and share it with others [7].

In RE, ontologies are pivotal, tackling both internal complexities and the quest for precise, reliable requirements [5, 8, 9]. They provide a framework to address incomplete or inconsistent requirements, improve domain understanding for quick adjustments, enable automated validations, and support the RE process with software guidance and checklists [10]. Ontologies also clarify requirements, lower the risk of miscommunication, ensure detailed specifications to reduce developer and client dissatisfaction, support comprehensive requirement definitions for improved project management, and allow for updates to cater to client needs. Addressing both internal and external challenges, ontologies enhance the precision, efficiency, and success of software projects [9]. They reduce requirement ambiguity, facilitate clearer understanding, improve stakeholder communication and collaboration, enable better management and traceability of requirements, and encourage requirement reuse [7, 8]. Furthermore, ontologies support quality by providing structured requirement representation and automated validation, offering guidance to avoid inconsistencies, streamlining validation to lessen manual work, and enhancing overall project quality [8, 9]. They also improve domain knowledge, helping to identify and adjust to changing requirements.

This research proposes an ontology for RE4AI challenges and adaptations to the RE process when designing AI-powered systems. The research contribution is the holistic view of the RE stages, topics, and stakeholders' roles involved in developing and managing requirements for AI-powered systems. It highlights stages and topics that may need modification or addition to accommodate the unique aspects of AI-powered systems. The RE4AI ontology provides insights for practitioners, including (1) Reduced ambiguity: The ontology introduces a shared vocabulary for structured and unambiguous AI-powered requirement representation. We clarify the vocabulary's significance using first-order logic, ensuring each term is clearly and distinctly understood, minimizing misinterpretations among engineers. (2) Improved understanding: It provides a holistic view of AI-powered system RE, guiding stakeholders and requirements engineers to address both FRs and NFRs throughout the RE process. It highlights interconnections between requirements that are often treated independently. (3) Enhanced communication and collaboration: It establishes connections between requirements and stakeholders, facilitating their timely involvement and conflict resolution. (4) Requirement traceability: The ontology breaks down requirements into tasks across each RE stage, enabling engineers to track their development. It emphasizes the impact of NFRs on FRs and the evolving nature of requirements in AI-powered systems.

The rest of the paper is organized as follows: Sect. 2 describes the research method, Sect. 3 details the proposed RE4AI ontology, and Sect. 4 concludes with further discussions and suggestions for future work.

2   Research Method

To define the RE4AI ontology, we followed a multi-step research process. The first research step was a systematic literature review (SLR), using the Prisma methodology to identify relevant research contributing to the RE4AI ontology components [11]. We used the Dimension.ai[1] search engine to find articles published in the last five years. The main goal in selecting articles was to identify those contributing to the development of our ontology, with a focus on their alignment with specific dimensions including RE stages, requirement topics, and stakeholder roles (to be elaborated in Sect. 3).

Our search began with keywords relevant to the RE4AI domain including requirement engineering, elicitation, analysis, specification, validation, management, functional, non-functional, artificial intelligence, machine-learning, ethics, responsibility, trust, privacy. The keywords were combined in several separate strings; 'and' relationships were defined between AI\ML keyword, each RE stages and each requirements topic (e.g., AI & elicitation & ethics), 'or' relationships was established between different RE stages and different requirements topics (e.g. Elicitation ‖ Analysis ‖ Validation) & (Ethics ‖ Human-AI interaction ‖ Privacy).

This step yielded 834 articles for potential inclusion. In the first round of screening, articles were assessed for their relevance to our study by considering whether they (1) discussed RE activities, (2) covered topics pertinent to AI systems, and (3) were broadly applicable to AI systems rather than focused on a specific domain. Titles and abstracts were screened first, to ascertain relevance, leading to the removal of articles with a specific domain focus (e.g., healthcare, computer vision) and articles that focused on the use of AI for RE, rather than on RE for AI, narrowing the pool to 128 articles. Next, a deeper analysis involved further inclusion/exclusion criteria to pinpoint articles that provided actionable insights for the ontology and to eliminate those that did not offer new information on key RE topics. For example, articles that merely outlined challenges without offering solutions were excluded. This process ensured the selection of articles that made significant, innovative contributions in the context of our research.

---

[1] The engine search included the following databases: ACM Digital Library, Elsevier Scopus, IEEE Xplore, SpringerLink, Web of Science, ScienceDirect, Directory of Open Access Journals (DOAJ), Dimensions Datasets and arXiv.

We outlined a final set of 85 relevant articles[2], which we than analyzed step using a meta-synthesis methodology [12] to develop the RE4AI ontology. The data extraction procedures focused on study context (e.g., participants, research methods, setting), findings (e.g., themes, patterns, categories) and interpretations (e.g., authors' explanations of their findings). Next, we coded the extracted data using a thematic analysis approach [13], identifying recurring themes and patterns, and grouping similar data together. Then, we synthesized the coded data to identify common and divergent findings across the studies and explored the relationships between different themes and patterns. Finally, we interpreted the synthesized findings to better understand the requirements for AI-powered systems.

We performed deductive thematic data analysis, a top-down approach in which the investigator employs their pre-existing knowledge or a relevant theory as a foundational structure to scrutinize the data and identify themes, which are iteratively utilized to organize the data analysis [14]. In this research, the pre-existing knowledge contained the basic categories which construct the RE methodology, including the standard stages of RE, the requirement topics, and the stakeholders participating in the process. These categories also construct the suggested dimension of the ontology (as will be demonstrated in the next paragraph). The codes generated through the thematic analysis were associated with the categories[3]. The subsequent inquiries established the associations: 1) What are the primary topics pertaining to the requirements of AI-powered systems? 2) Which attributes are associated with each requirement? 3) What are the specified relationships between requirements? 4) Which stakeholders are involved in the RE process for different topics? The first author coded the data and the second author reviewed it, resulting in Cohen's Kappa[4] inter-rated agreement of 94.4%. Two disputed codes were agreed to be deferred to in future research: maintainability and scalability.

The RE4AI ontology was constructed in accordance with the methodology proposed in "Ontology 101: A Guide to Creating Your First Ontology" [15]. This paper guided other ontological frameworks that were developed in the RE domain, for example RE for cyber security systems [8] which demonstrated its effectiveness in the development of ontologies. The foundational principles of "ontology 101" [15] lay out a hierarchical organization for the elements within the domain meta-model on which the ontology is developed. This hierarchical organization encompasses classes and their subclasses, descending to instances, the attributes of each instance, and the facets pertaining to each attribute. Initiating the construction of the ontology necessitates the identification of the domain meta-model, which is then structured hierarchically following the specified ontology methodology. The meta-model for the RE4AI domain, including its classes, instances, and relationships, was clarified by the SLR. The primary dimensions of the ontology—stages of RE, topics of requirements, and roles of stakeholders—were informed by previous publications on RE ontology. The first dimension, requirement topics, forms the backbone of the RE ontology, drawing from earlier works that presented a comprehensive ontology focusing on functional and non-functional requirements, thereby underscoring the centrality of this dimension in the overarching RE ontology

---

[2] Link to included articles.
[3] Link to demonstrative thematic analysis excerpts.
[4] https://builtin.com/data-science/cohens-kappa.

[9]. The second dimension incorporates the stages of RE by mapping its various components, activities, and outputs as concepts within the ontology. This organization aids in the formal structuring and categorization of the RE process, illustrating connections, dependencies, and limitations among different elements of the process. For example, stages such as *Elicitation, Analysis, Specification,* and *Validation* can be represented as classes within the ontology, detailed with properties that outline their interrelations and dependencies [8]. The stakeholder's involvement constitutes the third dimension, highlighting their essential role in the requirements articulation within the Requirements Ontology. Their engagement is critical for accurately, capturing the needs and expectations for the software system being developed. Incorporating stakeholders into the Requirements Ontology enables the systematic documentation and management of their input throughout the RE process, leading to a more effective and stakeholder-responsive software development approach [8].

The validity of this work is compromised by threats to external, internal, and construction validity. *External validity* depends on our capacity to comprehensively include all existing knowledge, which might be hindered by inadvertent exclusion of valuable publications during the search or screening process. The majority of the excluded articles were context-specific; a small number of these might have offered valuable additional knowledge that can possibly be generalized beyond the specific context of the paper. In order to address this concern, we made the decision to adopt a more lenient approach throughout the screening process of the papers, even if some of the publications were repetitious or too specific to a particular domain, with the aim of strengthening our conclusions. *Internal validity* may be threatened by misinterpretations of the data retrieved from the selected papers. To mitigate this threat, two researchers were involved in the theme coding process, resulting in a high inter-rater agreement level (as reported above). In the next steps of the research, we intend to further validate our interpretations and resulting ontology via interviews with experts. *Construction validity* may be threatened by our decision to construct the ontology at this initial stage based solely on a literature review. Employing a deductive thematic analysis based on existing theoretical frameworks mitigates this risk to some extent, by utilizing predetermined categories. Still, the ontology development process might have suffered from the exclusion of other aspects of the ontology that have not been previously documented in the literature, leading to a lack of stability in its structure. In order to address this potential threat, our strategy involves advancing and enhancing the ontology through the application of grounded theory and real-life use cases in our next research.

3 RE for AI Ontology

This section delves into the details of the RE4AI ontology, elucidating its importance, structure, and the rationale behind its construction. Figure 1 shows the results for the 85 included articles contributing to this ontology, indicating the number of articles contributing to each topic. It should be noted that due to the hierarchy of requirements, there exist certain requirements that are not explicitly discussed in specific articles, but can be derived from many of them. This is particularly evident in the case of data, model and human-AI interaction which are derived from other requirements and consequently mentioned in multiple articles. For instance, while we found only one article that directly and explicitly discusses Human-AI interaction, more than 10 other publications that focus on explainability, responsibility, and trustworthiness indirectly provide insights into this topic. "General" refers to articles covering the entire development process or addressing NFRs without focusing on specific requirements. Some ontology topics lack dedicated articles and are not displayed in the figure.

Figure 2 shows the RE4AI ontology, which was created by coding the data according to the proposed dimensions. The ontology scheme included the following dimensions: (1) RE stage; (2) Requirement topic; (3) Stakeholders' roles. It further describes high-level relationships between requirements which include contradictions and trade-offs.

Total Articles (85)			
RE Topics (59)		RE Stages (26)	
NFRs (57)	FRs (2)		
Explainability\ Transparency (16)	Human-system Interaction (1)	General Development (16)	Elicitation (2)
Responsibility (9)	System behavior (1)	Analysis (1)	Specification (1)
Ethics (7)		Validation (4)	Management (2)
Trust (5)			
Fairness (5)			
General NFRs (5)			
Data & Model (4)			
Safety (3)			
Privacy & Security (3)			

Fig. 1. Included articles – No. of articles contributing to each topic

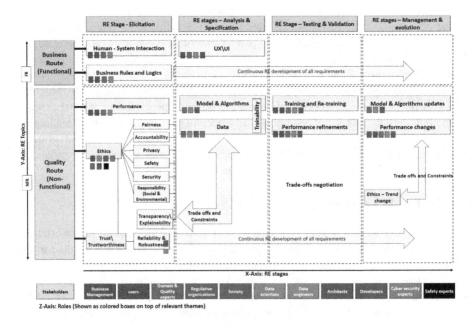

Fig. 2. RE4AI Ontology

A comprehensive explanation of each dimension of the ontology, together with its constituent elements, is provided in the online annex accessible through the following link. The subsections below provide concise definitions of each dimension represented as an axis in Fig. 2.

3.1 X-Axis: RE4AI Stages

The traditional RE stages include elicitation, analysis, specification, validation, and management. The RE process begins with two parallel routes: The functional route, which aligns with traditional systems, focusing on business goals and what the system should do to achieve them. The non-functional route considers system qualities, including conventional aspects like availability, reliability, and performance, with some of them becoming more important and complex in AI-powered systems, like ethics and responsibility [5].

Building AI systems demands a unique **requirement elicitation** approach. AI's dynamic nature and complex dependencies require iterative, agile processes. The RE4AI ontology guides stakeholders in defining requirements sustainable in the face of this dynamic nature [2, 16]. **Requirement analysis** tackles challenges like conflicting needs and trade-offs, impacting data models and ethical considerations. Adaptability is key, as models evolve and data change. Key outputs include chosen AI models, data sources, performance criteria, and ethical considerations [17–20]. **Requirement specification** translates stakeholder input into concrete ML algorithms, training methods, and data requirements. Metrics for efficiency, scalability, and availability are also defined [2, 21]. **Requirement validation** ensures the system meets goals, even with new and untrained data. This involves diverse testing and continuous monitoring. Validating non-functional requirements, such as safety, demands formal methods for assessing accuracy, reliability, and limitations [3, 4, 22–24]. **Requirement management** for AI is essential due to their evolving nature. Systems must be designed with iterative models to adapt to changing trends and user needs. RE for AI is about embracing uncertainty and continuous adaptation, ensuring successful development and deployment of these dynamic systems [25, 26].

3.2 Y-Axis: Requirements

Functional Requirements (FRs). FRs describe the behavior of a system, specify what a system should or should not do, and what the system must perform to meet the needs of the stakeholders. FRs include the system business goals and objectives, and the human-system interaction [25, 27]. FRs for AI-powered systems address the same types of FRs of any other system but need some refinements pertaining to their implementation on the interactions between models, algorithms, and data [28]. The SLR revealed two main topics related to the needed refinement of FRs for AI-powered system:

Business Goals and Objectives. For AI-powered systems, the RE process entails the identifications of tasks and procedures that may be automated, or cases in which AI can improve decision-making or productivity [26].

Human-AI Interaction. Human-AI interaction refers to the method in which users interact with AI-powered system. This method contributes significantly to the system's trustworthiness. The RE process of AI-powered systems should incorporate requirements to prioritize the amplification, augmentation, and enhancement of human performance, rather than seeking to replace it [29, 30].

Non-functional Requirements (NFRs). NFRs dictate system quality attributes and constraints, specifying how a system should behave rather than what it should do [27]. Adapting NFRs for AI-powered systems and analyzing the trade-offs they introduce to the system requires new RE methods and solutions for AI [3, 6].

Traditional NFRs encompass reliability, availability, maintainability, usability, security, and more. However, the advent of AI introduces supplementary NFRs that center around a set of requirements designed to collectively attain an ethical and trustworthy system. Within this set, a multitude of articles have exposed over thirty distinct requirements, including, but not limited to, transparency, explainability, fairness, safety, privacy, accountability, and responsibility [31]. The online annex provides a comprehensive explanation of each NFR as derived from the SLR.

The proposed ontology encapsulates the primary needs that necessitate the attention of requirements engineers. It further uncovers the hierarchical structure of the extensive collection of NFRs. This hierarchy commences with the fundamental needs for ethical and trustworthy systems, progresses to the subdivision of these requirements into extensive sets encompassing fairness, accountability, privacy, and more, and concludes with the transparency and explainability requirements that are indispensable for fulfilling the higher tier AI-oriented NFRs.

Both FRs and NFRs, as described above, are initially defined during the elicitation phase, and continue to evolve throughout the RE process. The requirements for models, data, and trainability are derived from these FRs and NFRs, and therefore the evolution of these requirements typically begins only at the analytics and specification phases.

Model Requirements. AI models span various domains and tasks, with each algorithm and model designed for specific requirements, including the type and size of the dataset, the complexity of the problem, the available computing resources, and the desired level of accuracy [18].

Data Requirements. Data requirements include the specification of data quantity, quality, structure, testing, and selection for training [20]. Notable data characteristics for ML development encompass accuracy, completeness, consistency, provenance, cleanliness, lineage, bias, relevance, and validity [32].

Testability and Trainability. RE for AI-powered systems must address the need of continuous revalidation and retraining of the model. Within the requirements for testability and trainability, it is essential to incorporate tools into the system that endow it with the capabilities required to perform these tasks, whether offline or online [4, 22].

Requirements Relationships. In the proposed ontology, we also address relationships that exist between requirements and the presence of contradictions, constraints, and trade-offs. Many of these relationships have been discussed earlier, with a focus on several significant connections, as described below.

Model and Data Requirements. The specification of model and data requirements is dependent on the system's objectives, which are fundamental components of FRs, as well as on ethical and trustworthiness requirements, which are NFRs. As a result, it is only possible to elicit model and data requirements following the analysis phase. For instance, the determination of the most suitable XAI (eXplainable AI) model is only possible following the analysis of the explainability requirements [33].

NFRs' Hierarchy. The ontology manifests the hierarchical structure of NFRs. It commences with ethics and trustworthiness requirements, which exhibit substantial overlaps and subsequently break down into an extensive array of sub-requirements. Notably, transparency and explainability requirements emerge as essential components of several higher-level NFRs, such as fairness, privacy, safety, security, and accountability.

Validation. It is vital to emphasize that in AI-powered systems, the process of validation is not confined to a discrete, standalone stage. It is rather an ongoing and integral part of the system's life cycle. The additional motivation for continuous validation is the system's learning nature and the evolving trends that influence the system's requirements. Furthermore, it exerts a profound and continuous influence on the requirement management stage.

We note that due to space limitations, not all properties and relationships can be described in this research preview paper; these will be reported in the future upon the completion of the ontology development.

3.3 Z-Axis: Roles

References related to the engagement of stakeholders with various roles in the RE4AI process are scarce and are only hinted at in the literature while discussing other RE topics. The ontology presented herein elucidates the specific roles that are deemed essential for active participation in the RE process, drawing insights from prior research and applying methodologies from traditional software engineering. In the context of RE for AI-powered systems, significant emphasis is placed on the critical and indispensable contributions of three principal roles: *data professionals,* who are required to be consistently engaged throughout all phases of the RE process [2]; *quality experts,* who are concerned with social and ethical aspects, including regulatory bodies, societal organizations, experts in safety and security, and the broader community [34, 35]; and, *domain experts*, whose significance, while evident for the development of any system, necessitates in our context a deeper understanding of the impact of AI technology [33, 36]. Further description of these roles is given in the online annex.

4 Discussion and Conclusion

In recent years, the field of RE, and particularly in the context of AI-powered systems, has received significant research attention. Notably, a substantial amount of attention has been given to NFRs, with a strong emphasis on ethics and trustworthiness for AI-powered systems [6, 23]. These NFRs have been extensively explored, with a particular focus on their influence on the system model and data [26, 37]. However, while challenges associated with NFRs have been extensively addressed in research publications, practical solutions have remained somewhat scarce.

FRs, in contrast, have been subject to a limited amount of academic research. Topics pertaining to FRs are customarily examined in articles that concentrate on particular stages of RE, domains, or applications, such as computer vision in healthcare or autonomous vehicles [38], with an emphasis on the utilization of AI to promote their capabilities. A full understanding of the challenges and best practices for defining objectives and optimizing user experience in AI-powered systems is yet to be achieved. The prevailing perspective often emphasizes how AI can assist in achieving existing system objectives, rather than how AI capabilities can introduce new and innovative ones.

In this paper we propose a RE for AI ontology that provides requirements engineers and system stakeholders a compass based on the topics engaged in the RE process as well as the challenges and practices gathered from the SLR. This research builds on current available knowledge and emphasizes the need for additional research to complete a comprehensive framework and fill existing gaps.

The foundational phase of developing the RE4AI ontology began with the performance of the SLR, laying the groundwork for subsequent enhancements aimed at validating and refining the ontology. Key areas for improvement include: 1) Transitioning from the traditional waterfall methodology to an agile approach, which better facilitates iterative cycles suitable for AI projects; 2) Involving stakeholders throughout the process to address diverse requirements; 3) Analyzing the interplay between various AI-specific requirements and their features; and 4) Customizing the ontology to accommodate a range of AI technologies, particularly focusing on generative AI and Large Language Models (LLMs). Due to space limitations, we were able to present here only the main findings of the study, excluding some of the details of the full complexity of the structure of the ontology, (e.g., attributes of each requirement topic and the relationships between requirements). Our next objective is to continue our efforts in this research direction, resulting in a comprehensive and stable ontology.

The forthcoming research phase will focus on developing actionable strategies that organizations can readily incorporate into RE processes. This includes exploring the integration of LLMs, which have demonstrated significant promise in task performance, albeit with certain limitations and challenges [39]. By employing a grounded theory approach to gather best practices for RE procedures, and through the implementation of these strategies in real-world industry scenarios, the research aims to attach practical methodologies to the proposed ontology framework. This would provide a robust tool for organizations transitioning to AI-powered operations. To achieve these goals, the next steps will involve expert interviews and surveys to deepen the understanding about the usefulness of the proposed ontology, specifically for RE4AI, validate or question the existing ontology, introduce new requirements and attributes, and uncover connections within the ontological structure.

References

1. Daewon, L., Park, J. Hyuk.: Future trends of AI-based smart systems and services: challenges, opportunities, and solutions. J. Inf. Process. Syst. **15**(4), 717–723 (2019)
2. Vogelsang, A., Borg, M.: Requirements engineering for machine learning: perspectives from data scientists. In: Proceedings of the IEEE 27th International Requirements Engineering Conference Workshops (REW) (2019)

3. Ahmad, K., Bano, M., Abdelrazek, M., Arora, C., Grundy, J.: What's up with requirements engineering for artificial intelligence systems? In: Proceedings of the IEEE 29th International Requirements Engineering Conference (RE), pp. 1–12 (2021)
4. Hand, D.J., Khan, S.: Validating and verifying AI systems. Patterns 1(3) (2020)
5. Horkoff, J.: Non-functional requirements for machine learning: challenges and new directions. In: Proceedings of the IEEE 27th International Requirements Engineering Conference (RE) (2019)
6. Habibullah, K.M., Gregory, G., Horkoff, J.: Non-functional requirements for machine learning: an exploration of system scope and interest. In: Proceedings of the IEEE/ACM 1st International Workshop on Software Engineering for Responsible Artificial Intelligence (2022)
7. Confalonieri, R., Weyde, T., Besold, T.R., del Prado Martín, F.M.: Using ontologies to enhance human understandability of global post-hoc explanations of black-box models. Artif. Intell. **296**, 103471 (2021)
8. Siegemund, K.: Contributions to ontology-driven requirements engineering. Doctoral dissertation, Dresden, Technische Universität Dresden, Diss. (2014)
9. Castañeda, V., Ballejos, L., Caliusco, M.L., Galli, M.R.: The use of ontologies in requirements engineering. Glob. J. Res. Eng. **10**(6), 2–8 (2010)
10. Blanco, C., Rosado, D.G., Varela-Vaca, Á.J., Gómez-López, M.T., Fernández-Medina, E.: Onto-CARMEN: ontology-driven approach for cyber–physical system security requirements meta-modelling and reasoning. Internet Things **24** (2023)
11. Page, M.J., et al.: The PRISMA 2020 statement: an updated guideline for reporting systematic reviews. Int. J. Surg. **88**, 105906 (2021)
12. Leary, H., Walker, A.: Meta-analysis and meta-synthesis methodologies: rigorously piecing together research. TechTrends **62**(5), 525–534 (2018)
13. Alhojailan, M.I., Ibrahim, M.: Thematic analysis: a critical review of its process and evaluation. West East J. Soc. Sci. 1(1), 39–47 (2012)
14. Proudfoot, K.: Inductive/deductive hybrid thematic analysis in mixed methods research. J. Mixed Methods Res. **17**(3), 308–326 (2023)
15. Noy, N.F., McGuinness, D.: Ontology 101: A Guide to Creating Your First Ontology. Standford University, Viitattu (2012)
16. De Sousa Silva, A.F., Silva, G.R.S., Canedo, E.D.: Requirements elicitation techniques and tools in the context of artificial intelligence. In: Xavier-Junior, J.C., Rios, R.A. (eds.) BRACIS 2022. LNCS, vol. 13653, pp. 15–29. Springer, Cham (2022). https://doi.org/10.1007/978-3-031-21686-2_2
17. Levy, M., Hadar, I., Aviv, I.: A requirements engineering methodology for knowledge management solutions: integrating technical and social aspects. Requirements Eng. **24**, 503–521 (2019)
18. Rahman, M.S., Khomh, F., Hamidi, A., Cheng, J., Antoniol, G., Washizaki, H.: Machine learning application development: practitioners' insights. Softw. Qual. J. 1–55 (2023)
19. Lavalle, A., Maté, A., Trujillo, J., García-Carrasco, J.: Law modeling for fairness requirements elicitation in artificial intelligence systems. In: International Conference on Conceptual Modeling, pp. 423–432 (2022)
20. Weber, M., Engert, M., Schaffer, N., Weking, J., Krcmar, H.: Organizational capabilities for AI implementation—coping with inscrutability and data dependency in AI. AI Inf. Syst. Front. (2022)
21. Berry, D.M.: Requirements engineering for artificial intelligence: what is a requirements specification for an artificial intelligence? In: Proceedings of the International Working Conference on Requirements Engineering: Foundation for Software Quality, pp. 19–25 (2022)
22. Tao, C., Gao, J., Wang, T.: Testing and quality validation for ai software–perspectives, issues, and practices. IEEE Access **7**, 120164–120175 (2019)

23. Mylrea, M., Robinson, N.: AI trust framework and maturity model: improving security, ethics, and trust in AI. Cybersecur. Innov. Technol. J. **1**(1), 1–15 (2023)
24. Ghallab, M.: Responsible AI: requirements and challenges. AI Perspect. **1**(1), 1–7 (2019). https://doi.org/10.1186/s42467-019-0003-z
25. Ahmad, K., Abdelrazek, M., Arora, C., Bano, M., Grundy, J.: Requirements engineering for artificial intelligence systems: a systematic mapping study. Inf. Softw. Technol. **158** (2023)
26. Paleyes, A., Urma, R.G., Lawrence, N.D.: Challenges in deploying machine learning: a survey of case studies. ACM Comput. Surv. **55**(6), 1–29 (2022)
27. Paech, B., Dutoit, A.H., Kerkow, D., Knethen, A.V.: Functional requirements, non-functional requirements, and architecture should not be separated - a position paper. In: Proceedings of the International Working Conference on Requirements Engineering: Foundation for Software Quality (REFSQ) (2002)
28. Belani, H., Vukovic, M., Car, Ž.: Requirements engineering challenges in building AI-based complex systems. In: Proceedings of the IEEE 27th International Requirements Engineering Conference Workshops (REW) (2019)
29. Kutz, J., Neuhüttler, J., Spilski, J., Lachmann, T.: AI-based services-design principles to meet the requirements of a trustworthy AI. In: International Conference on the Human Side of Service Engineering (2023)
30. O'Grady, K.L., Harbour, S.D., Abballe, A.R., Cohen, K.: Trust, ethics, consciousness, and artificial intelligence. In: 2022 IEEE/AIAA 41st Digital Avionics Systems Conference (DASC), pp. 1–9 (2022)
31. Habibullah, K.M., Gay, G., Horkoff, J.: Non-functional requirements for machine learning: understanding current use and challenges among practitioners. Requirements Eng. **28**(2), 283–316 (2023)
32. Priestley, M., O'Donnell, F., Simperl, E.: A survey of data quality requirements that matter in ML development pipelines. ACM J. Data Inf. Qual. (2023)
33. Akbarighatar, P., Pappas, I., Vassilakopoulou, P.: A sociotechnical perspective for responsible AI maturity models: findings from a mixed-method literature review. Int. J. Inf. Manag. Data Insights **3**(2) (2023)
34. Preece, A., Harborne, D., Braines, D., Tomsett, R., Chakraborty, S.: Stakeholders in explainable AI. arXiv preprint arXiv:1810.00184 (2018)
35. Deshpande, A., Sharp, H.: Responsible AI systems: who are the stakeholders?. In: Proceedings of the 2022 AAAI/ACM Conference on AI, Ethics, and Society, pp. 227–236 (2022)
36. Belle, V., Papantonis, I.: Principles and practice of explainable machine learning. Front. Big Data **39** (2021)
37. Arrieta, A.B., et al.: Explainable artificial intelligence (XAI): concepts, taxonomies, opportunities and challenges toward responsible AI. Inf. Fusion **58**, 82–115 (2020)
38. Ma, Y., Wang, Z., Yang, H., Yang, L.: Artificial intelligence applications in the development of autonomous vehicles: a survey. IEEE/CAA J. Automatica Sinica **7**(2), 315–329 (2020)
39. Arvidsson, S., Axell, J.: Prompt engineering guidelines for LLMs in Requirements Engineering (2023)

Operationalizing Machine Learning Using Requirements-Grounded MLOps

Milos Bastajic, Jonatan Boman Karinen, and Jennifer Horkoff[(✉)] [iD]

Chalmers | University of Gothenburg, Gothenburg, Sweden
`jennifer@horkoff.gu.se`

Abstract. Machine learning (ML) use has increased significantly, [Question/Problem] however, organizations still struggle with operationalizing ML. [Principle results] In this paper, we explore the intersection between machine learning operations (MLOps) and Requirements engineering (RE) by investigating the current problems and best practices associated with developing an MLOps process. The goal is to create an artifact that would guide MLOps implementation from an RE perspective, aiming for a more systematic approach to managing ML models in production by identifying and documenting the goals and objectives. The study adopted a Design Science Research methodology, examining the difficulties currently faced in creating an MLOps process, identified potential solutions to these difficulties, and assessed the effectiveness of one particular solution, an artifact containing guiding Requirements Questions sorted by ML stages and practitioner roles. [Contribution] By establishing a more thorough understanding of how the two domains interact and by offering practical guidance for implementing MLOps processes from an RE perspective, this study advances both the MLOps and RE fields.

Keywords: Machine Learning · Requirements Engineering · ML · RE · MLOps · Design Science

1 Introduction

The use of machine learning (ML) has increased significantly in recent years and organizations utilizing ML simultaneously increase earnings while decreasing spending [3]. In order to fully leverage the benefits of ML in industry, it is critical to have a well-organized and efficient approach for how to operationalize ML models [3]. In the context of this study, ML operations (MLOps) include the end-to-end conceptualization, implementation, monitoring, deployment, and scalability of ML products [7]. Although studies show the economic benefits seen when operationalizing ML models, research indicates that the industry struggles with taking the models to production [6, 7, 15].

Since MLOps is a relatively new topic, studies have been conducted to address the ambiguity of the term [7, 15], and to provide practitioners with suggestions of

© The Author(s), under exclusive license to Springer Nature Switzerland AG 2024
D. Mendez and A. Moreira (Eds.): REFSQ 2024, LNCS 14588, pp. 231–248, 2024.
https://doi.org/10.1007/978-3-031-57327-9_15

tools and architectures for implementing an MLOps process [6–8]. In traditional software development, RE is considered to be a crucial part that ensures software meets the needs of its users and stakeholders [12], and involves identifying, documenting, and managing system or product requirements [11]. The relationship between requirements engineering (RE) and MLOps is that while RE focuses on identifying and documenting stakeholders' needs, MLOps facilitates the efficient and reliable development and delivery of ML systems. We believe that taking an RE perspective in MLOps would be beneficial to a wider, needs-oriented view. However, there exists a gap in research regarding MLOps grounded in RE.

To fill this gap in research, this study undertakes a Design Science Research (DSR) approach to investigate a set of research questions concerning RE for MLOps. An artifact is developed to help organizations curate requirements for their MLOps. The artifact is a list of ML-stage sorted specific Requirement Questions, designated roles to ask these questions, and common responses to these Requirement Questions. The artifact was developed as part of three Design Science Research iterations, starting with a review of relevant literature, followed by rounds of interviews and a workshop with ML experts to iteratively understand the problems in practice and evaluate and improve the artifact. The methodology included 14 interviews with 11 different ML practitioners across four companies, including a workshop with three of the interview participants.

This work was conducted in collaboration with the charging and energy division of the automotive company Polestar, a leading player in the electric vehicle industry, as part of an (unpublished) master's thesis [2].

The paper is structured as follows: Sect. 2 outlines the research questions and design science methodology. Results for each research question are presented in a different section: Sect. 3 presents the related work and problems and best practices, Sect. 4 presents the MLOps requirement form artifact, and Sect. 5 presents the evaluation results of the artifact. Section 6 discusses results, future work, and validity threats, while Sect. 7 concludes the paper.

2   Methodology

This section presents the methodology used in this work: the research approach, collaborating company, context, data gathering methods, and lastly, the data analysis approach.

Research Questions. The study aims to answer three research questions (RQs), each with its own individual primary area of focus, the problem statement, the solution to the problem, and the evaluation of the solution:

RQ1: What are current challenges in designing an MLOps process and how do they relate to requirements knowledge?

RQ2: Which potential solution exists to mitigate the challenges of developing an MLOps process grounded in requirements engineering?

RQ3: How well does the potential solution mitigate the requirements-related problems with developing an MLOps process?

Design Science Research. This study follows a DSR approach. Design science is an iterative research method that involves the development and evaluation of artifacts (such as models, theories, and prototypes) that have the potential to solve practical problems and advance scientific understanding [5]. It is often used to address complex real-world problems that cannot be fully understood or resolved through pure traditional research methods. The design artifact is intended to be used in real-world scenarios to address an identified problem and therefore it is usually developed with a specific audience or user group in mind. Figure 1 presents an overview of all the methods used in each of our DSR iterations.

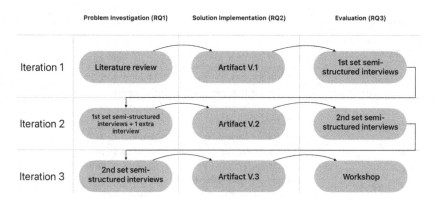

Fig. 1. An overview of the methodology utilized throughout the three DSR iterations.

Collaborating Company Context. Polestar's charging and energy division was in the process of implementing an MLOps architecture to manage the end-to-end ML life cycle. Thus, this division was interested in exploring the potential benefits of applying RE to the design of its MLOps architecture, to ensure that the system meets their needs and aligns with industry regulations and standards.

Data Collection. Multiple methods for data collection were used: two rounds of semi-structured interviews, a literature analysis, and a workshop.

Literature. A literature analysis was conducted at the start of the study to acquire an understanding of existing research and knowledge related to RQ1. The purpose was to identify gaps in knowledge, build on existing research, situate the work within the broader context of the field, and ensure that the initial proposed solution was based on previous findings and theories. The results of the literature analysis guided the development of the first artifact. This approach was considered necessary due to the low number of available people in the industry with adequate knowledge of MLOps and its related challenges.

The literature analysis was performed using literature database search engines Google Scholar and Web of Science. To begin the search, terms such as: "RE for

MLOps", "RE for ML", "MLOps challenges", "DevOps for ML", and "MLOps best practices" were searched. After a few relevant papers had been found, a snowballing approach was utilized to find further relevant studies. Grey literature was selected with great precaution and was only used when peer-reviewed resources were lacking, making up a minority of cited sources. We cannot claim that this process was fully systematic, producing complete results, but we believe we were able to find the greater part of the key literature at the time of our search (Spring 2023). A full systematic literature review is left for future work.

Table 1. Interviewees with ID (ID.round, i.e., ID1.1 and ID1.2 are the same person), current role, and current company.

Interviewee ID	Current Role	Company	Experience (Years)
1st set semi-structured interviews			
ID1.1	Data Analyst	Company 1	8
ID2	Software Engineer	Company 1	3+
ID3	Software Engineer	Company 1	2
ID4	Software Engineer	Company 1	1
ID5	Software Developer	Company 1	2+
ID6.1	Product Owner	Company 1	15
ID7.1	Sr Manager, ML Engineering and Research	Company 2	20+
ID8	Software Engineering Manager	Company 1	8
2st set semi-structured interviews			
ID1.2	Data Analyst	Company 1	8
ID6.2	Product Owner	Company 1	15
ID7.2	Sr. Manager, ML Engineering and Research	Company 2	20+
ID9	Sr. Data scientist	Company 3	15
ID10	ML Researcher	Company 4	5
ID11	Software Engineer	Company 1	0.5

Participant Selection. Participants were practitioners in the field of MLOps, mostly within Polestar. However, since there was a lack of people with experience in MLOps, additional interviews were held with experts outside of Polestar, found via recommendations from experts within Polestar. The sampling was purposive: first, requesting possible individuals with relevant knowledge and expertise from the main company contact, then snowballing by asking the interviewees for recommendations on relevant interview prospects. Specifically, we looked for people with experience with MLOps in practice, without focusing on specific roles. When inviting for the final workshop, care was taken to involve three participants with different roles representing roles in the final artifact.

Interviews. Interviews were held during all iterations of the study and followed a semi-structured format, see Table 1 for a full list of interviewees. Additionally, a pilot interview was held in each iteration to test and improve the interview proto-col before conducting the remaining interviews. These interviews were conducted via video conference, and they were recorded and automatically transcribed using Microsoft's Teams application for later analysis. See [2] for interview questions for both iterations.

In order to evaluate the first version of the artifact developed from the results of the literature analysis and simultaneously continue the problem investigation, the semi-structured interviews were divided into two parts. First, the intervie-wees were asked questions related to the problem investigation for the next cycle. Second, at the end of the interviews, they were asked questions targeting the evaluation of the artifact created from the current cycle's problem investiga-tion. The interview structure of combining problem investigation together with evaluation of the previous artifact was used for all interviews. The full list of interview questions can be found in [2].

Workshop. In the third iteration, we held a workshop to evaluate the final arti-fact. The workshop included three participants who had previously partaken in the interviews: ID1, ID6, and ID7. These individuals were purposely selected as they could cover three of the roles in the artifact. The workshop was arranged in three steps: 1) an imaginary business case involving ML and MLOps was presented to the participants, who role-played as the team who would imple-ment the solution, 2) the group used the artifact to come up with a project plan, which consisted of a multitude of MLOps requirements. While the group worked with the case and the artifact, the hosts noted any apparent struggles and discussions with time stamps, which were later analyzed. 3) the workshop participants answered a questionnaire individually, to capture the participant's thoughts on working with the artifact on a staged real-life business case. The evaluating questions and case description can be found in [2].

Data Analysis. Data collected in this was qualitative, which was analyzed by the first two authors using coding and theme identification methods presented by Saldaña [13]. While reading through the transcriptions of the interviews, open coding, also referred to as initial coding, was used to assign codes to each piece of data that related to the research questions. This was followed by axial coding, where the relationships between the codes were analyzed to identify higher-level themes. Finally, we found overarching themes from the collected data corpus.

3   RQ1 Problem Investigation: Challenges in MLOps Relating to RE

This section presents the results from our problem investigation. Note that three cycles iteratively collected information for each RQ, but for readability, we orga-nize these findings by RQ, merging findings from multiple iterations. The data for this RQ comes from two sources, a literature review and data collected from the companies through interviews.

RQ1: Literature Analysis. The initial challenges or problems (P) and best practices (BP) related to requirements for MLOps were extracted during the literature analysis. We give an overview of related literature then provide a summary of extracted problems and best practices in Table 2.

Machine Learning Operations (MLOps). In their paper, Subramanya et al. discuss how the emergence of the DevOps approach was motivated by the need to improve the efficiency of cross-functional teams in releasing software [14]. DevOps aims to address these challenges by implementing processes that enable faster, more reliable, and repeatable software builds. Inspired by DevOps, there has been a growing interest in the practice of MLOps, which aims to bring data scientists and operations teams together in order to achieve similar benefits [4].

Table 2. RE-related Problems and Best Practices from the Literature

ID	Problem	Description	Sources
P1	Data Drift	The distribution of input data changes from the data used to train the model, but the desired prediction output remains the same	[6,8]
P2	Concept Drift	The functional relationship between a model's inputs and outputs changes, resulting in a modification of the output definition as the input changes.	[1,6,8]
P3	Inter-team Communication	Communication challenges in MLOps can arise due to the different roles and knowledge levels of professionals involved.	[7,10]
P4	Performance During Serving	Performance-related challenges after deployment are commonly related to two different categories: 1) traffic management concerns, such as network latency, ML system throughput, and access points, and 2) ML model performance.	[1,6,10]
P5	Disorganized Data	The data gathered for a model might originate from different sources and is usually disorganized.	[7,9,10]
P6	Sustainable MLOps	Three critical components: Explainability, fairness, and accountability.	[15]
ID	**Best Practice**	**Description**	**Source**
BP1	Versioning	Implementing versioning for data, models, experimentation logs, and code increases a system's reproducibility and traceability.	[7,8]
BP2	Model Deployment and Serving	Defining what model prediction serving pattern to use: Model-as-Service where your model is exposed as an endpoint on the web, Precompute where your model expects batches of input data to do predictions that are saved for later, or Model-as-Dependency where your model is loaded in real-time	[8]
BP3	Data Quality and Labeling	Labeling processes that are coordinated and transparent.	[17]
BP4	Feasibility	Exploring business problems that could be solved with ML, rather than looking for ML problems, should be the first step in the development process.	[10,17]

MLOps has become a trending topic in industry and research. Tamburri [15] identifies that while MLOps involves the orchestration of various software components to support the end-to-end lifecycle of ML models, the complexity of these operations can make them unsustainable. John et al. acknowledge that MLOps still is in its infancy and therefore conduct a systematic literature review and a grey literature review [4]. The authors derive and validate a framework that identifies the activities involved in the adoption of MLOps, together with a maturity model for the implementation of MLOps processes. Kreuzberger et al. explore the concept of MLOps and discuss how it can address the challenges of automating and operationalizing ML products [7]. They provide an overview of the necessary principles, components, roles, architecture, and workflows associated with MLOps, as well as open challenges related to ML.

As MLOps is a relatively new topic area, some information was sourced from specialty courses from reputable experts, papers under review, and leading industry blogs. For example, in an MLOps specialization course given by Andrew Ng and DeepLearningAI, the complete end-to-end MLOps lifecycle is covered and discussed [10]. The course brings forth multiple challenges found within each of the MLOps stages: Scoping, Data, Modeling, and Deployment. A blog post by Microsoft [9] discusses a specific case where an MLOps process was implemented, including tools and technologies, some challenges encountered, as well as requirements for the MLOps infrastructure.

Requirements Engineering (RE) for ML. The publication landscape of RE for ML-based systems was examined in a systematic mapping study by Villamizar et al. [16]. The authors found that the challenges of RE are exacerbated by the unique characteristics of ML. The authors believe that to address these challenges and ensure the quality of ML-based systems, it is necessary to identify best practices and approaches for RE in the context of ML-based systems. In our opinion, this should include an RE-based consideration of MLOps.

We summarize key challenges and best practices from the literature in Table 2. These findings were used to craft the initial version of our artifact, refined through subsequent evaluation, with the final version presented in Sect. 4. A mapping between these findings and artifact elements is available in [2].

RQ1: Iteration 1 Interviews. This subsection presents the results of the thematic analysis from the first part of the first set of semi-structured interviews, further addressing RQ1. The codes and themes that emerged from the analysis, along with their definitions and examples, are summarized in this subsection and Fig. 2. These results were used to improve the artifact.

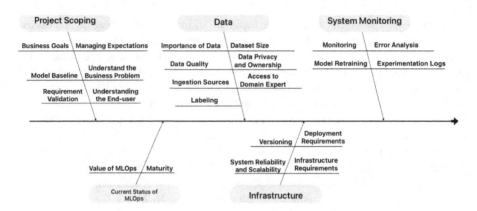

Fig. 2. RE-related Best Practices and Challenges for MLOps According to Interviews.

Project Scoping (7/8 interviewees). The Project Scoping theme encompasses a set of codes related to the scoping stage of an MLOps project, including: Business Goals, Understand the Business Problem, Model Baseline, Manage Expectations, and Requirement Validation. Multiple interviewees discussed the importance of incorporating various perspectives and requirements into the development process to create ML products that are deemed successful. Additionally, several interviewees described the challenge of capturing business goals and visions, incorporating them into ML research.

To ensure effective implementation of business objectives, it is crucial to thoroughly understand and validate requirements and avoid misinterpretation or insufficient detail. Due to ML and MLOps being complex and resource-demanding, it is important to evaluate whether a business problem requires an ML solution or can be solved using traditional optimization. *"One of the challenges is to understand what's an appropriate problem to be solved with machine learning. And right now there are so many areas where machine learning can be introduced that you can deliver very good results, but there are a lot of areas where you just shouldn't introduce machine learning. ...* - ID7.1

Another important aspect to consider in the scoping stage of an MLOps project is what baseline the final model should have. We see that it is important to consider previous solutions and performance requirements. We also see that it is sometimes important to consider other aspects such as ML versus manual labor and its significance to business value. *" ... if that's a 90% accurate solution during this mechanical way, discrete way, you can maybe get an 85% (with ML) ... fully automated (solution) with no manual labor and maybe even works better with scale. So actually improves with scale.* - ID7.1

Five interviewees named the importance of establishing clear expectations from the end-user of the ML system when implementing ML solutions, as there is often a knowledge gap between what ML can actually accomplish and what people assume it can do. Interviewees emphasized the significance of identifying the end-users and their intended use of the ML solution. This can be a challenging task since it can be difficult to determine the end-user, yet it significantly impacts the nature of the final ML solution.

Current Status of MLOps (7/8 interviewees). The theme Current Status of MLOps consists of two codes: Value of MLOps and Maturity. The theme encapsulates the current status of MLOps, including why MLOps is necessary and why it is not used more in the industry. Interviewees highlighted the benefits of utilizing MLOps in general. Specifically, one interviewee mentioned how DevOps practices are essential for developing and maintaining ML products. The maturity of MLOps was found to be a reoccurring topic. Three out of the eight interviewees stressed that MLOps currently is very immature, with one of them being an ML researcher who stated that many of the available tools in this domain are too ambitious in what they try to achieve. This leads to challenges faced during the MLOps architecture selection phase, which could potentially be solved with predefined requirements, but it could also simply be a case of immaturity.

Data (5/8 interviewees). The data theme includes technical aspects such as ingestion sources, data quality, ownership, and privacy. Additionally, the theme highlights the importance of domain knowledge and expertise in understanding and effectively working with and labelling the data.

Traditional software projects focus on the code as the primary product, while ML projects are highly dependent on the data used to train a model. When the available data is insufficient in terms of size and coverage to produce a reliable model, it can be challenging to generate meaningful insights or predictions. It is essential to carefully consider the availability and quality of data before starting development on an ML solution, as these factors can significantly impact the outcome. Furthermore, privacy and data ownership should be taken into account, especially when working with different entities. It is also important to have a thorough understanding of the source of data and its corresponding metadata to assess the potential presence of biases in the data, as well as to consider the method of data collection, including whether it is automated or manual.

Another interviewee emphasized the importance of having a consistent ingestion source that provides data according to what the model is trained on. In cases where time series data or data in windows of time are being collected, there may be instances where the selected window does not capture a natural cycle, leading to incomplete or inaccurate results. Furthermore, when working with supervised learning, proper labeling of data is critical, as it serves as the foundation for the model's understanding of the task at hand.

Several interviewees stated that access to domain experts is vital in ensuring accurate and effective data analysis. While data analysts may possess a versatile skill set, they may not always have the necessary domain expertise to fully understand the data being analyzed.

Infrastructure (8/8 interviewees). The theme Infrastructure includes the codes: Versioning, System Reliability and Scalability, Infrastructure Requirements, and Deployment Requirements. These codes relate to the importance of developing and maintaining a robust infrastructure to support data management. This includes effective versioning and storage practices, as well as the need for automation and testing to ensure the reliability and scalability of infrastructure. Additionally, the theme includes codes related to the challenges and complexities of implementing and migrating infrastructure, particularly due to the high level of technical competence necessary in order to implement it.

System Monitoring (6/8 interviewees). The System Monitoring theme covered the importance of monitoring a deployed system to maintain and improve the performance of an ML system within an MLOps process, including Monitoring, Model Retraining, Experimentation Logs, and Error Analysis. These codes primarily discuss the importance of monitoring ML systems, retraining models to maintain accuracy, tracking experiments to improve reproducibility, and analyzing errors to enhance system reliability.

RQ1: Iteration 2 Interviews. During the problem investigation in iteration 2, several previous themes were confirmed by multiple interviewees. We summarize confirmed and new themes from this second round of interviews.

Multiple interviewees from the second set of interviews confirmed the Project Scoping theme by discussing many similar or identical codes such as Understanding the Business Problem and Business Goals. In addition to what was reinforced by the interviewees regarding this theme, a new code emerged, the importance of Understanding ML. Given that MLOps requires collaboration from a multidisciplinary team, ensuring that everyone has a comprehensive understanding of how ML works can evidently be a challenge.

As part of this iteration, the new theme Developing a Model emerged (2/6 interviewees), consisting of challenges and best practices linked to model development. This theme delved into aspects such as Data leakage, Versioning, Development environments, and CI/CD. In addition, 3/6 interviewees talked about the new theme Requirement Management, which discusses some important nonfunctional requirements to have in mind while developing production-ready ML models, the difference between pure ML projects and ML projects with MLOps introduced, and the importance of continuously working with requirements.

4   RQ2: RE for MLOps Artifact

The following section describes the final version of the artifact that was designed through three iterations to address the challenges identified as results of the RQ1, acting as the solution and results of RQ2. The artifact is structured to align with the end-to-end stages of an MLOps process: Scoping, Data, Modeling, and Deployment. Figure 3 gives an overview of the stages and how information from one stage feeds into another.

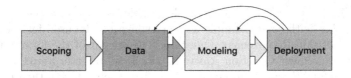

Fig. 3. Overview of the Stages of MLOps and its Iterative Nature.

The MLOps Requirements Form consists of two parts: The first part serves as an introduction that guides users on how to use the form effectively, and is available in our online resource [2]. The second part is the form itself, which is the result of cycle three's problem investigation and the evaluation from cycle two, see Figs. 4, 5, 6, and 7. The order of the questions in the different stages was also considered in order for them to mirror how these questions might come up naturally in an ML project which needs to be operationalized.

The form was developed in order to assist teams or individuals in eliciting MLOps requirements when implementing MLOps. Each question in the form has been designed to serve as an adaptation or mitigation to one or more of the best practices and challenges identified during the literature analysis and interviews. For example, in Fig. 4, scoping questions "What are the business problems?", "What are metrics for success?", and "What are the resources needed?" are derived from BP4 Feasibility in Table 2. A full mapping from Requirement Questions to problems, best practices, and interview findings can be found in [2].

Roles to ask	Requirement Question	Requirement Question Answer	Examples
Scoping:			
Business stakeholder	What are the business problems?		Battery optimization, Fraud detection, Demand forecasting
Data Scientist	Can the business problems be solved with ML, how?		Has it been done before, research proves it possible, still unclear
Product owner	What are the metrics for success?		ROI, customer wishes
Product owner	What are the resources needed?		Data, time, people
Product owner, Business stakeholder, Data scientist	What is the budget limit for the computation necessary to train the model?		If on premise: 100h allowed, 50h, Unlimited If on cloud: Budget is $1,000, $5,000, $500
Business stakeholder	Who is the end user?		Demographical information, Internal company users, Customers
Business stakeholder	How will the users interact with the model, what interface will they need?		App, Voice activated feature, Web page, API
Business stakeholder, Product owner, Data scientist, Data engineer	Who is the domain expert and can we access them?		Doctors, Lawyers, Domain specific researcher

Fig. 4. Part one of the Form: Requirement Questions Regarding Scoping.

Each Requirement Question in the form is intended to be answered by specified roles within a team and sometimes roles outside of the implementation team. Additionally, each question includes a field for documenting the answer, paired together with some example answers which can be used for reference or clarification of said Requirement Question. The output of the form is the documented Requirement Question Answers, which can be interpreted as informal requirements for MLOps requirements. These requirements can then be used as they are, or as a foundation for creating more formal requirements.

5   RQ3: Design Science Artifact Evaluation Results

We present the artifact evaluation from two sets of interviews and the workshop.

RQ3: Iteration 1 Interviews. In the first round of interviews with eight people from two companies, the artifact as derived from the literature was presented. We asked the interviewees about the redundancy and clearness of that artifact. Regarding redundancy, there were no complaints; however, regarding clearness,

Roles to ask	Requirement Question	Requirement Question Answer	Examples
Data:			
Business stakeholder, Product owner, Data scientist, Data engineer	Where does the data come from?		Owned data, crowdsourced, purchase data, purchase labels
Data scientist, Data engineer	What data format will be used?		Structured, unstructured
Data scientist, Data engineer	How should the data be preprocessed?		Remove data, remove duplicates
Data scientist, Data engineer, Domain expert	What are the data labeling guidelines?		On images: Label each scratch independently on the screen, label each animal separately in the field
Product owner, Business stakeholder	Who will label the data?		In-house resources, Crowdsourced, Outsourced, Mixture of resources
Data scientist, Data engineer, Product owner	What meta-data should be collected?		Time, system model, factory, device type
Data engineer, Legal team, Business stakeholder, Product owner	Are there any privacy concerns regarding the data?		Names, Emails, Addresses, Phone numbers, general GDPR concerns
Data engineer, Legal team, Business stakeholder, Product owner	Are there any necessary data ownership considerations?		Data is owned by us, it's open source, another party owns all data
Product owner, Data scientist, Data engineer	How much data is expected to be stored?		~10TB
Product owner, Data scientist, Data engineer, Domain expert	When does the data become irrelevant?		Never, new product version release, annually
Data engineer, Domain expert	Are there any cyclic behaviours to the data?		Seasonal sales cycle, full day cycle
Data scientist	What is the minimum amount of data that is necessary to train the model?		10k images, 100 gb worth of 1080p mp3 video recordings
Data scientist	For streaming data, what is the minimum frequency of data points necessary to meet the business goals?		Every 5ms, Every 1s, Every data point
Product owner, Data scientist, Data Engineer	How will the data be acquired?		Automated tool, manually collected, purchased

Fig. 5. Part two of the Form: Requirement Questions Regarding Data.

Roles to ask	Requirement Question	Requirement Question Answer	Examples
Modeling:			
Product owner, Data scientist	What is the model baseline?		Human-level performance, A previous system's performance, Dummy model
Product owner, Legal	Is it necessary to audit the model? Who should audit the model? What is the audit focus?		Yes/No. Business stakeholder, Third party, Data scientists. Transparency, Equality, Fairness, Accountability, etc.
Data scientist, Data engineer	Which potential risks for bias exists?		Gender bias, Brand bias, Ethnicity bias
Product owner, Data scientist	How is the input data served to the model?		Batch data, Real time data
Data scientist, IT Architect	Where should the experimental data result be stored?		Database, Excel document, JSON-file
Product owner	What are important business goal metrics the ML model should consider?		Business required classifications performance, different from general ML model performance
Data scientist	What experimental data should be tracked?		Dataset used, Hyperparameters, Results, Results with metric summary/analysis, Training resources, Training time),
Data scientist, Software engineer, DevOps engineer, MLOps engineer	What deployment constraints exist?		None, Edge device's hardware capabilities

Fig. 6. Part three of the Form: Requirement Questions Regarding Modeling.

one participant directly asked how the artifact should be used. This highlighted the need for a description of how the artifact could be used in an industry setting.

When asked if the artifact would be useful and why, 7/8 participants said that they think the artifact would be useful while implementing an MLOps process. Several of the interviewees stated that it could manage expectations and be a helpful way to communicate what is needed in order to create an MLOps process. Another reason repeated by several interviewees was that it is a good way to get an overview of what is needed before starting the implementation of the MLOps process. A way to eliminate any creeping issues that might become costly if not spotted early on. Further suggestions for specific changes were given and incorporated into the next version of the artifact. The initial version and more detail on suggestions and changes are available in [2].

RQ3: Iteration 2 Interviews. In the second set of interviews with six people from four companies, the revised artifact was presented. Table 1 provides an overview of the participants involved in this cycle, where three interviewees were new to the study, two from new participating companies, and three participated in previous interviews. The feedback collected focuses on what the interviewees appreciated about the artifact, suggested improvements, concerns, how well the answers can translate to architecture design, how they would use it in practice, if it can lead to requirements, and how generalizable the artifact is. Feedback was incorporated into the final artifact version as presented in Sect. 4.

All of the six interviewees interviewed for this iteration expressed their appreciation for the artifact, all six participants agreed that it would be highly beneficial. They echoed the sentiment that the artifact could manage expectations and

Roles to ask	Requirement Question	Requirement Question Answer	Examples
Deployment:			
Product owner, MLOps engineer, DevOps engineer	How should the deployment process be handled?		Canary releases, A/B releases, Shadow releases
Product owner, MLOps engineer, DevOps engineer	Where should the prediction device be located?		Cloud or edge device
DevOps engineer, MLOps engineer, Software engineer	Which software metrics are important to monitor?		Memory, computing power, latency, throughput, server load
Data scientist, Data engineer, MLOps engineer	Which input metrics are important to monitor?		feature types (INT or String), feature range, Data schema validation
Data scientist, Software engineer, MLOps engineer	Which output metrics are important to monitor?		# times users redo search, avg. prediction accuracy
Product owner, MLOps engineer, Data scientist	How often should the model be retrained on the data gathered from deployment?		Every Monday, once a month, based on deployed input/output metric triggers
Product owner, DevOps engineer, Data scientist	Are there any specific performance requirements?		Latency requirements, Query per seconds requirements

Fig. 7. Part four of the Form: Requirement Questions Regarding Deployment.

serve as a useful communication tool to clarify the requirements and expected outcomes of the MLOps process. Additionally, it was praised for its concrete and clear structure and content.

All six interviewees said that they would use the artifact. When asked about their intended usage, one interviewee stated they would employ it as a checklist for important tasks in the initial stages, and in later stages it would be used as an onboarding tool for new team members. Additionally, two interviewees suggested using it as a planning tool to identify crucial project roles and more accurately estimate project expenses.

Even though interviewees appreciated the artifact, we saw concerns regarding the extensiveness of the artifact. Those participants suggested that it might be unnecessary in situations where one is not aiming for the models to reach production, where the ML project simply is not large enough to make it worth implementing MLOps, or where one would only like to create a high-level overview for pitching the MLOps idea. *"One pitfall could be that you're over-engineering something that should be just a simpler experiment. "* - ID10

During the interviews, suggestions for general improvements were given. One suggestion was to incorporate dependencies between the Requirements Questions, allowing for an indication of the downstream impact of changes made. Another suggestion was to enable the filtering of Requirements Questions based on specific roles. A further interviewee suggested allowing adding and removing questions in order for organizations to be able to personalize the form. It was also suggested that information be included in the artifact regarding the potential impact on a system if any of the questions were left unanswered.

In addition, two interviewees recommended adding questions concerning data leakage, which is when test data leaks into the training data; including a question about the frequency of streaming data; and asking about the location of preprocessing. Two interviewees suggested asking about who should be responsible for labeling the data, whether it should be internal or external.

As most interviewees in the first iteration were from Polestar, generalizability was a risk. Thus, we asked whether the interviewees believed the artifact was generalizable, and all six interviewees, including three who were not from Polestar, agreed that it could be applied beyond Polestar and the automotive sector.

RQ3: Workshop. The workshop with three previous interviewees was deemed a success in the sense that all of the participants could successfully discuss the scoping of the imaginary project case using the artifact. According to the questionnaire answers, the participants agreed unanimously that the artifact helped them discuss all of the necessary parts for the imaginary case. The artifact helped guide their discussion through all of the ML stages, avoiding overlooking factors they might have missed without the artifact. Finally, participants recommended adding the Requirement Question "Can the business problems be solved with ML, how?" to also ask if ML is the right approach for the business problems. Participants suggested that it is not necessary to answer the Requirements Questions sequentially. This could potentially be improved or resolved by digitizing the artifact, also suggested in the interviews. Regarding the time necessary to use the artifact, the participants managed to discuss and fill out the full artifact in relative detail in two hours.

6 Discussion

In the first iteration problem investigation, various challenges and best practices related to requirements knowledge were discovered in the literature and through interviews. The MLOps Requirements Form was designed based on the literature analysis and interviews with relevant industry professionals. The Requirements Questions in the form were designed to work for teams regardless of their specific project or industry, or the type of ML model. While it may seem like the MLOps Requirements Form must be answered in sequential order, this is not necessary. There exists no strict way of using the artifact, leaving space for it to be used in various work settings.

The interviews reveal that there is a gap in the industry for a tool like our artifact, which we believe partly explains the overwhelmingly positive evaluation feedback. However, for those familiar with ML or MLOps in practice, many of the issues, challenges, and Requirements Questions may already be known. Still, we believe our preliminary results show the value of having a consolidated and centralized shared understanding of these questions, particularly, as pointed out by an interviewee, as part of onboarding or training.

In addition to the positive feedback provided, the first two evaluating iterations also disclosed what could be missing, wrong, redundant, or unclear in the first two versions of the artifact. We saw a steep decrease in comments that can be interpreted as negative in the evaluation of the final artifact, compared to the first artifact, indicating artifact completeness and saturation.

However, there are still elements in the form that could be improved. Form digitization is important, particularly allowing users to customize the form and add new, company- or industry-specific questions. At least one interviewee felt the form may be over-engineering in some cases, thus the form should be redesigned to be more dynamic, with certain questions only being shown in certain contexts, allowing for a lightweight version when the application of ML is more uncertain or preliminary.

Previous work has pointed out that data scientists have a different view of ML development, as compared to developers [17]. As the form covers questions from the perspective of a variety of roles, including data scientist, we believe such an artifact can help to bridge the communication gap between developers and data scientists. In addition, we have not extensively explored the process of introducing the form to a company, i.e., what roles would be responsible to ensure the form is used, is there a need for an artifact champion, or will the form be used without enforcement due to obvious benefits. Future studies should focus on evaluating these aspects.

In addition, future work should consider the evaluation of the form over time. This includes both periodic updates of the information in the form for a project in a company and updating the contents of the form itself to consider future issues and challenges. As MLOps is still a relatively new topic, our collective knowledge of challenges and issues will evolve, and thus the artifact should be evolved to address this information in future studies.

While the MLOps Requirements Form is a promising tool to help mitigate the challenges of developing an MLOps process grounded in RE, other potential solutions may also exist. Moreover, the artifact may not be completely suitable for every organization's needs and processes. Nonetheless, the MLOps Requirements Form was thought by participants to be a potentially useful tool that could significantly aid in the process of doing RE for MLOps, which is a previously overlooked step when scoping an MLOps project.

Threats to Validity. In terms of *internal validity*, our literature survey may have missed some previous work; however, snowballing and the use of a popular MLOps course helped to ensure our initial version of the form captured key topics. Our main company contact was included as a data point in the interviews, deemed necessary due to the limited number of individuals with experience in MLOps. However, we refrained from sharing our opinions and findings with this contact prior to the interview. The interviews were conducted by either author and were equally distributed between the two authors; however, a pilot interview was conducted and then reviewed together to improve consistency and the interview protocol. All interviews were held in English, with the option of Swedish to remove any potential language barriers. To ensure the reliability and validity of the data analysis, the first two authors independently analyzed the data, with the results compared to ensure a consistent interpretation of the data. Moreover, where possible, data was triangulated by collecting it from multiple sources, such as interviews, literature, and workshops, to ensure results were not biased by a single data source.

In terms of *external validity*, the majority of interviewees were affiliated with a single company, which could potentially undermine the generalizability of our findings. However, we initially conducted a literature review to inform our problem investigation, and six interviewees, including three outside of Polestar, agreed that the artifact was highly generalizable. Our evaluation has not yet involved application of the Requirements Form to a real project, future work should continue evaluation as part of industrial projects.

7 Conclusions

This study's aim was to advance the knowledge of how MLOps and RE interrelate. This is important since it enables MLOps practitioners to gain similar benefits as those attained by applying RE to traditional software development. Our aim was to produce an artifact that would serve as an itinerary for MLOps adoption from a RE viewpoint. We used a DSR methodology to explore the best practices already in use and the difficulties involved in creating an MLOps process and assess the efficacy of various solutions in addressing the identified challenges. Our work reveals that integrating RE into MLOps processes is thought to be of great value for making sure that ML models are successfully implemented and operationalized. We created the MLOps Requirements Form, an artifact that acts as a tool for practitioners when building MLOps processes from a RE standpoint. We received an positive response from experts in the field, who described the potential effectiveness and value of the MLOps Requirements Form. Future work should explore digitalizing the form, allowing it to be more customizable, and should apply the form to real projects.

Acknowledgements. We are grateful for the support of Emanuella Wallin at Polestar, and for the time and input of all participants. We thank the REFSQ reviewers for helpful feedback which improved the paper

References

1. Baier, L., Kühl, N., Satzger, G.: How to cope with change?-preserving validity of predictive services over time (2019)
2. Bastajic, M., Boman Karinen, J.: Requirements grounded MLOps - a design science study. Master's thesis, Chalmers (2023)
3. Chui, M., Hall, B., Singla, A., Sukharevsky, A.: The state of AI in 2021 (2021). https://www.mckinsey.com/capabilities/quantumblack/our-insights/global-surve y-the-state-of-ai-in-2021
4. John, M.M., Olsson, H.H., Bosch, J.: Towards MLOps: a framework and maturity model. In: 2021 47th Euromicro Conference on Software Engineering and Advanced Applications (SEAA), pp. 1–8. IEEE (2021)

5. Knauss, E.: Constructive master's thesis work in industry: guidelines for applying design science research. In: 2021 IEEE/ACM 43rd International Conference on Software Engineering: Software Engineering Education and Training (ICSE-SEET), pp. 110–121 (2021)

6. Kolltveit, A.B., Li, J.: Operationalizing machine learning models - a systematic literature review. In: 2022 IEEE/ACM 1st International Workshop on Software Engineering for Responsible Artificial Intelligence (SE4RAI), pp. 1–8 (2022)

7. Kreuzberger, D., Kühl, N., Hirschl, S.: Machine learning operations (MLOps): overview, definition, and architecture (2022). https://arxiv.org/abs/2205.02302

8. Kumara, I., Arts, R., Di Nucci, D., Van Den Heuvel, W.J., Tamburri, D.A.: Requirements and reference architecture for MLOps: insights from industry. Authorea Preprints (2023)

9. Microsoft: Machine learning operations (MLOps) framework to upscale machine learning lifecycle with azure machine learning. Microsoft Azure, blog (2024). https://learn.microsoft.com/en-us/azure/architecture/ai-ml/guide/mlops-technical-paper

10. Ng, A.: Machine learning engineering for production (MLOps) specialization. Coursera (2024). https://www.coursera.org/specializations/machine-learning-engineering-for-production-mlops

11. Nuseibeh, B., Easterbrook, S.: Requirements engineering: a roadmap. In: Proceedings of the Conference on the Future of Software Engineering, pp. 35–46 (2000)

12. Pandey, D., Suman, U., Ramani, A.: An effective requirement engineering process model for software development and requirements management. In: 2010 International Conference on Advances in Recent Technologies in Communication and Computing, pp. 287–291 (2010)

13. Saldaña, J.: The Coding Manual for Qualitative Researchers, pp. 1–440 (2021)

14. Subramanya, R., Sierla, S., Vyatkin, V.: From DevOps to MLOps: overview and application to electricity market forecasting. Appl. Sci. **12**(19), 9851 (2022)

15. Tamburri, D.A.: Sustainable MLOps: trends and challenges. In: 2020 22nd International Symposium on Symbolic and Numeric Algorithms for Scientific Computing (SYNASC), pp. 17–23 (2020)

16. Villamizar, H., Escovedo, T., Kalinowski, M.: Requirements engineering for machine learning: a systematic mapping study. In: 2021 47th Euromicro Conference on Software Engineering and Advanced Applications (SEAA), pp. 29–36. IEEE (2021)

17. Vogelsang, A., Borg, M.: Requirements engineering for machine learning: perspectives from data scientists. In: 2019 IEEE 27th International Requirements Engineering Conference Workshops (REW), pp. 245–251. IEEE (2019)

Crowd-Based Requirements Engineering

Unveiling Competition Dynamics in Mobile App Markets Through User Reviews

Quim Motger[1]([✉]) [iD], Xavier Franch[1] [iD], Vincenzo Gervasi[2] [iD],
and Jordi Marco[1] [iD]

[1] Universitat Politécnica de Catalunya, Barcelona, Spain
{joaquim.motger,xavier.franch,jordi.marco}@upc.edu
[2] Universitá di Pisa, Pisa, Italy
vincenzo.gervasi@unipi.it

Abstract. [**Context and motivation**] User reviews published in mobile app repositories are essential for understanding user satisfaction and engagement within a specific market segment. [**Question/problem**] Manual analysis of reviews is impractical due to the large data volume, and automated analysis faces challenges like data synthesis and reporting. This complicates the task for app providers in identifying patterns and significant events, especially in assessing the influence of competitor apps. Furthermore, review-based research is mostly limited to a single app or a single app provider, excluding potential competition analysis. Consequently, there is an open research challenge in leveraging user reviews to support cross-app analysis within a specific market segment. [**Principal ideas/results**] Following a case-study research method in the microblogging app market, we introduce an automatic, novel approach to support mobile app market analysis. Our approach leverages quantitative metrics and event detection techniques based on newly published user reviews. Significant events are proactively identified and summarized by comparing metric deviations with historical baseline indicators within the lifecycle of a mobile app. [**Contribution**] Results from our case study show empirical evidence of the detection of relevant events within the selected market segment, including software- or release-based events, contextual events and the emergence of new competitors.

Keywords: mobile apps · market analysis · competition dynamics · user reviews · event detection · microblogging

1   Introduction

User reviews play a crucial role in providing valuable feedback and shaping the reputation of mobile apps [5]. They offer a wealth of information, including user satisfaction, engagement, and sentiments towards different aspects of the app [4].

© The Author(s), under exclusive license to Springer Nature Switzerland AG 2024
D. Mendez and A. Moreira (Eds.): REFSQ 2024, LNCS 14588, pp. 251–266, 2024.
https://doi.org/10.1007/978-3-031-57327-9_16

Analysis of user reviews can uncover valuable insights into app performance and user preferences, and even identify emerging trends and potential issues [2,6,13]. However, manual analysis of these reviews is time-consuming, subjective, and impractical due to the informal nature and sheer volume of data [9]. As a result, app providers struggle to uncover hidden patterns, identify significant events, and understand the factors driving user satisfaction or dissatisfaction, especially if these emerge from a comparison with other similar and competing apps.

In fact, attention in research has been mostly devoted to internal analysis within the scope of a given mobile app or the catalogue of apps of a given app provider. We hypothesize that by leveraging a combination of quantitative metrics and event detection techniques, app providers can acquire a deeper and more timely understanding of a given app market. This entails any change in users' expectations and in the market in which they are competing, which may constitute a threat or present new opportunities to their business. This knowledge can then potentially support informed decisions that will materialize into new or evolved requirements. To the best of our knowledge, no other proposals exist for leveraging review-based metrics for cross-app and competition analysis within a given app market.

In this paper, we propose an automated, novel approach to unveil competition dynamics and explore mobile app market insights through the continuous analysis of user reviews and the detection of app-related events[1]. We collect user reviews from app repositories and utilize various well-established review-based metrics in the field to quantitatively assess user activity, satisfaction and engagement. To identify significant events, we focus on detecting deviations from baseline metrics. Reported events are used to select a subset of reviews from potentially correlated events. These reviews are summarized and presented to app providers to support explainability of such events. This, in turn, helps inform them about potential threats and opportunities that may inform the requirements elicitation for the future evolution of their app. We conducted a preliminary validation using the microblogging apps market as a case study to explore the potential of the approach (Sect. 4) and assess its novelty (Sect. 6).

2   Research Method

2.1   Goal and Research Question

We perform a field study using a case study methodology [14], in order to facilitate the analysis of user review activity within its natural context (i.e., mobile app repositories) by means of minimal intrusion, limited to data collection. Our goal is **to proactively inform app stakeholders about changes in how users perceive a given app within a specific market segment, ultimately to infer the threats and opportunities stemming from these**

---

[1] Full datasets and complete evaluation results are available in the replication package: replication package https://doi.org/10.5281/zenodo.10125307. Source code is also available at:
code repository https://github.com/quim-motger/app-market-analysis.

changes. To this end, our research employs an exploratory design resulting in insights about the correlation between review-based metrics from a mobile app with respect to potential competitor apps. These insights focus on the automatic, proactive identification of significant events within a specific market segment and aim to monitor user behaviour and detect feedback trends. These events are then used to establish and detect potential user behavioural changes triggered by new market trends. To this end, we define the following research question:

> **RQ.** How can user review-based metrics be leveraged to uncover threats and opportunities from the user review activity of different mobile apps within the same market segment?

2.2 Stakeholder Analysis

We identify three mobile app stakeholders that may benefit from our research through different types of analysis:

- **Providers.** Entities or individuals involved in the lifecycle of mobile apps, such as product owners, developers, marketers, and quality assurance teams. They might benefit by gaining insights into their app's performance and the competitive landscape, including identification of competitors, user trends, market threats and opportunities, aiding in decision-making regarding requirements, release planning, features, and niche identification.
- **Consultants.** Professionals or experts who provide specialized guidance to app providers, helping them optimize their strategies in a given app market. They can conduct in-depth analyses, comparing multiple apps' performance, features, and user feedback within a given market segment. This analysis provides valuable insights to identify market gaps, understand the positioning of different app providers, and propose informed decisions to stay competitive.
- **Users.** Individuals who interact with mobile apps as consumers. Our approach allows users to dynamically compare apps, considering evolving feedback trends and market dynamics. This empowers users to make more informed choices when selecting apps, potentially leading them to discover new and suitable candidates for their needs.

2.3 Case Study: The "Twexit"

As preliminary evaluation for this paper, we selected an instrumental case study offering a contemporary, widely-discussed, and highly polarized event - the acquisition of X (formerly known as Twitter) by Elon Musk [10]. On April 14th, 2022, business magnate Elon Musk made an offer to purchase Twitter, Inc., owner of Twitter, one of the most popular microblogging apps. After several months of uncertainty, the acquisition was completed on October 27th, with opinions deeply divided between supporters and detractors of the operation. While some opinions praised Musk's stance on freedom of speech and no-censorship policies for Twitter, detractors reported great dissatisfaction with the upcoming changes in the Twitter app. This caused disruptions in the microblogging apps market for several months, leading to the migration of a significant amount of both users and advertisers to alternative existing apps, or even to the emergence of new competitors [10]. A popular example is the migration of Twitter users to Mastodon[2], an open-source, decentralized microblogging app, to which Twitter Inc. even reacted by banning Mastodon's accounts on their platform.

We chose to focus on the "Twexit" event for the following reasons:

- **Novelty and relevance:** Being a timely and widely discussed phenomenon within the microblogging app market, which makes it an ideal case to test our approach's ability to capture significant events and their consequences.
- **Real-world market dynamics:** Reflecting common market dynamics where sudden shifts in user perceptions and preferences occur due to high-profile events. Analysing this case provides valuable insights for adapting to evolving mobile app markets.
- **Assessment of generalization:** Serving as a well-known market phenomenon, it also provides a foundation for uncovering less obvious correlations and less popular events in the microblogging app market.

3   Approach

Our approach is composed of four stages (see Fig. 1): review collection, metric extraction, statistical analysis and event summarization. Each stage is described in some detail in the following subsections.

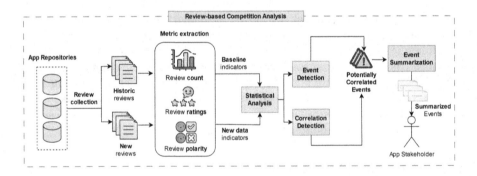

Fig. 1. System design

3.1   Review Collection

Our approach combines web scraping and API consumption techniques developed in previous work [11]. This enables us to gather data from various

---

[2] https://joinmastodon.org/about.

decentralized and heterogeneous sources, including app stores, search engines, and recommendation platforms. We obtain a diverse range of information that includes both historic reviews and a periodic polling of new reviews for multiple apps in parallel in a given market segment. *Historic reviews* serve as a baseline capturing the app's typical activity patterns, enabling us to assess and compare the app's performance over time (see Sect. 3.3). Deviations from these established patterns can be used as robust signals for identifying significant market events. *New reviews* are collected periodically with configurable frequency (e.g., daily, weekly), serving as fresh data points to assess the current state of an app by comparing it with baseline statistical indicators.

3.2 Metric Extraction

To support the event monitoring process, we focus on three established metrics that offer quantitative measures to assess changes in user engagement. This selection is motivated by the analysis of related work (see Sect. 6).

– **Review count (c).** The number of published reviews is an indicator for monitoring disruptive events in the app market. Sudden surges or declines in review numbers can relate to multiple disruptions, such as emerging trends, market saturation, or shifts in user preferences.
– **Review rating (r).** Ratings provide insights into user satisfaction and engagement. A significant increase may signal effective feature adoption, while a notable drop may indicate disruptive events like critical bugs or poorly received updates. Ratings are normalized to an integer value within $[0, 4]$.
– **Review polarity (p).** Polarity measures opinion divergence among users. Extreme reviews indicate intense debate or controversy within an app. Polarity extends review ratings by capturing sentiment-based analysis at the sentence level, addressing limitations and capturing previously overlooked knowledge. Polarity is normalized to an integer value within $[0, 4]$, where 0 and 4 represent extremely negative and positive reviews, respectively.

Metrics c and r are directly collected and measured from the original app repositories through the review polling process. To compute p, we build on the work of a third-party sentiment analysis tool[3] based on a RoBERTa-based model. Notice metrics are computed uniformly using all reviews obtained from each data source.

3.3 Statistical Analysis

Table 1 presents the formalization of the analytic metrics for measuring potentially correlated events. Below, we describe how these metrics are employed.

---

[3] Available at: https://github.com/AgustiGM/sa_filter_tool.

Table 1. Event monitoring analytics

Analytic metric	Description
Time Window $[t_0, t_0 + w)$	Time window within the interval $[t_0, t_0 + w)$, where w is the interval length in number of days
Average Value $\mu(m_{app}, [t_0, t_0 + w_e))$	Average value of metric m_{app} from all reviews published within the time window $[t_0, t_0 + w_e)$
Average Difference $\delta(m_{app}, [t_0, t_0 + w_e))$	Variation between $\mu(m_{app})$ and the average for the preceding time window $\mu(m, [t_0 - w_e, t_0))$
Standard Deviation $\sigma(m_{app}, [t_0, t_0 + w_e), t_\Omega)$	Baseline standard deviation for a given m measured from all $\delta(m_{app})$, where $t_\Omega \leq t_i < t_0$, being t_Ω a configurable date considered as the start date for baseline statistical computation
Correlation $\rho(m_{app_i}, m_{app_j}, [t_0, t_0 + w_c), t_\Phi)$	Pearson correlation between m_{app_i} and m_{app_j} measured from all $\mu(m_{app})$, where $t_\Phi \leq t_i < t_0$, being t_Φ a configurable date considered as the start date for baseline statistical computation

Event Detection. Individual app timelines are used to monitor, detect and report events based on statistically significant deviations for a given metric. We use the complete set of reviews of a mobile app in a given market segment $app \in Apps$ to measure the average difference δ and the standard deviation σ for each time window $[t_0, t_0 + w)$, as depicted in Table 1. Then, events for a given metric $m_{app} \in \{c, r, p\}$ are computed using the following formula:

$$E(m_{app}, t_0, w_e) = \begin{cases} 1 & \text{if } a \geq k \cdot s \\ 0 & \text{if} -k \cdot s < a < k \cdot s \\ -1 & \text{if } a \leq -k \cdot s \end{cases}$$

where $a = \delta((m_{app}, [t_0, t_0 + w_e))$, $s = \sigma((m_{app}, [t_0, t_0 + w_e))$ and k is a positive scalar determining the range of fluctuations that are considered normal: lower values of k lead to higher sensitiveness and more events being reported as significant; conversely, with higher values of k only fewer events of great momentum are reported. A new batch of user reviews collected during a specific time window $[t_0, t_0 + w_e)$ for a catalogue of mobile apps is processed to collect the aforementioned metrics and compute the monitoring analytics listed in Table 1. After this data collection and metric computation stage, we compute for each app in the identified market segment the value of $E(m_{app}, t_0, w_e)$.

Correlation Detection. The event detection analysis is complemented with the statistical assessment of periods of time for which two mobile apps experienced a high (positive or negative) correlation (see Fig. 2). To this end, for each time window $[t_0, t_0 + w_c)$ and each pair of metrics m_{app_i}, m_{app_j} where $app_i, app_j \in Apps$ and $i \neq j$, we measure the Pearson correlation $\rho(m_{app_i}, m_{app_j}, [t_0, t_0 + w_c))$ utilizing the set of reviews of app_i, app_j. Each correlation value is computed using all metric data points in m_{app_i}, m_{app_j} for $[t_i, t_i + w_c) \mid t_\Phi \leq t_i < t_0 + w_c$, where t_Φ is the beginning of the observation period for correlation analysis. Correlated periods between m_{app_i} and m_{app_j} are then computed using the following formula:

$$C(m_{app_i}, m_{app_j}, t_0, w_c) = \begin{cases} 1 & \text{if } r \geq h \\ 0 & \text{if } -h < r < h \\ -1 & \text{if } r \leq -h \end{cases}$$

where $r = \rho(m_{app_i}, m_{app_j}, [t_0, t_0 + w_c))$ and $h \in [-1.0, 1.0]$ is defined as the sensitivity threshold for detecting correlated trends. Similarly to k, lower h values lead to a high sensitiveness for reporting correlated periods, while higher h values lead to a more strict detection. We employ a distinct time window w_c to enhance our ability to detect correlations with a finer granularity. This enables us to identify even subtle or short-lived trends within the data more effectively.

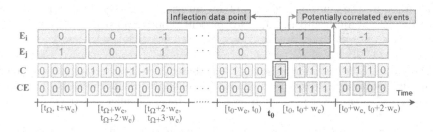

Fig. 2. Illustrative summary of the potentially correlated event detection method.

Potentially Correlated Events Detection. Events and metric-based correlated periods are then used to find intersections of time windows $[t_0, t_0 + w_e)$ and $[t_0, t_0 + w_c)$ for which events were found at the same time in which there was a high correlation period (see Fig. 2). Hence, we define the following formula:

$$CE(m_{app_i}, m_{app_j}, t_0, w_e, w_c) = \begin{cases} 1 & \text{if } E_i = E_j \text{ and } C = 1 \\ -1 & \text{if } E_i \neq E_j \text{ and } C = -1 \\ 0 & \text{otherwise} \end{cases}$$

where we use the abbreviations $E_i = E(m_{app_i}, t_0, w_e)$, $E_j = E(m_{app_j}, t_0, w_e)$, and $C = C(m_{app_i}, m_{app_j}, t_0, w_c)$. Given the longest possible interval $[t_{start}, t_{end})$ from each period for which correlation detection is constantly $C = \pm 1$, we limit our analysis to intersections with respect to the first date interval $[t_{start}, t_{start} + w_e)$. This is motivated by the following reasons: (1) to focus exclusively on inflection data points of C, reflecting a change in the dynamic of the correlation ρ; and (2) to make our approach suitable for proactive detection in real-time, excluding future data from the analysis at each time window $[t_0, t_0 + w_e)$. Furthermore, since potentially correlated events may not always occur simultaneously (especially for lower w_e values), we extend the $E_i = E_j$

assessment when $CE(m_{app_i}, m_{app_j}, t_0, w_e, w_c) = 0$ for the original time window $[t_0, t_0 + w_e)$. This extension includes the immediately preceding time window $[t_0 - w_e, t_0)$, evaluating $E_i = E_j$ for all date interval permutations within $\{[t_0 - w_e, t_0), [t_0, t_0 + w_e)\}$.

3.4 Event Summarization

If $CE(m_{app_i}, m_{app_j}, t_0, w_e, w_c) \neq 0$, then our approach has detected potentially correlated events within $[t_0, t_0 + w_e)$. To support efficient interpretation, we complement our approach with a review summarization stage to make events suitable for human assessment. To this end, we employ ChatGPT (gpt3.5-turbo model) using prompt engineering[4] within a zero-shot learning approach to summarize a random sample of n reviews published within the time window $[t_0, t_0 + w)$. Each reported event E_i, E_j corresponds to a particular summarization of the selected sub set of reviews associated with E_i, E_j. To illustrate this, in Summarization #1 we report the summarized output for an event triggered by the reviews of the Hive Social app (ellipsis added for presentation purposes).

> **Summarization #1 - $E(c_{HiveSocial}, \text{"Nov 17, 2022"}, 7) = 1$**   all
>
> The most significant event raised is the inability of users to create accounts on the Hive Social app. Many users reported encountering issues during the sign-up process, whether it was with email registration, phone number registration, [...]

In addition to generic summarization, our approach leverages the polarity metric p to split the reviews at the sentence level between negative sentences ($p \leq 1$) and positive sentences ($p \geq 3$). Each category is subsequently subjected to the same summarization process described above, enabling stakeholders to receive distinct summaries of diverse user perspectives on a specific event.

The following examples are the summarized output for positive and negative review sentences for the same event reported in Summarization #1. This example demonstrates the value of incorporating polarity-based distinction to extend the insights derived from the event summarization task. The summary of positive reviews (Summarization #2) allows deeper understanding of the underlying reasons for disruptive user activity in Hive Social, which might have been overlooked in the original summarization. Conversely, the summary of negative reviews (Summarization #3) offers a more nuanced analysis of the factors contributing to performance and functionality issues beyond user registration.

> **Summarization #2 - $E(c_{HiveSocial}, \text{"Nov 17, 2022"}, 7)) = 1 \;/\; p \geq 3$**   positive
>
> The most significant event raised is the introduction and positive reception of a new app called "Hive." The reviewers express their initial impressions and enthusiasm for the app, highlighting its potential and features reminiscent of popular platforms like MySpace, [...] and Twitter. [...]

---

[4] Prompt template available in replication package.

Summarization #3 - $E(c_{HiveSocial},$ "Nov 17, 2022", 7) = 1 / $p \leq 1$ negative

The most significant event is the poor performance and functionality of the Hive social app, especially on Android devices. Many users reported issues such as slow loading times, frequent crashes, inability to upload or access photos, difficulties in creating accounts, and problems with basic features [...]

4 Evaluation

4.1 Design

We conducted a preliminary evaluation based on the validation of threats and opportunities as defined in Sect. 2.1 and the case study described in Sect. 2.3.

- **Data collection.** We employ the "Twexit" event as an exemplar to evaluate the effectiveness of our approach in identifying and reporting significant events within the microblogging app environment. To this end, we selected 12 microblogging apps for Android (including Twitter and Mastodon) based on users' recommendations from AlternativeTo[5]. For each app, we collected all reviews available in multiple repositories (Sect. 3.1) published within a time window of 52 weeks from June 9th, 2022 to June 7th, 2023 (included). Two of the apps were then excluded due to insufficient data (i.e., less than 20 reviews per month on average). The 10 remaining apps and corresponding number of reviews (187.639 in total) are listed in Fig. 3.
- **Parameter setup.** For event detection (E), we set $t_\Omega =$ "Jun 09, 2022" as the start date for baseline statistical computation; $w_e = 7$ as a weekly interval length for time window generation; and $k = \pm 2$ as the sensitivity factor. For correlation detection (C), we set $t_\Phi = t_0 - 14$ as start date for using recent data points (i.e., last 14 days); $w_c = 1$ for a fine-grained correlation analysis on a daily level; and $h = \pm 0.5$ as the sensitivity threshold value. Finally, for each potentially correlated event ($CE = \pm 1$), we performed event summarization on $n = 50$ as the average value of reviews published within a time window length of $w = 7$ is ~ 41, excluding Twitter (if the number of reviews in a given time window is < 50, we use all available documents).
- **Result analysis.** Evaluation is focused on: (1) verification of the ground truth case (i.e., "Twexit"); and (2) validation of additional potentially correlation events. To this end, we report statistics on potentially correlated events linked to the "Twexit", as well as complementary events that reflect other non-related competition phenomenon. In addition, event summaries are then categorized using a set of thematic categories of reasons to leave and join an app reported by actual users [1]. These categories include reasons related to usefulness, usability, content, reliability, security, existence of better alternatives, influence of others, popularity and design. We use these results to infer generic types of triggers of user trialling behaviours, supported by the examples and metrics selected in this study.

---

[5] From https://alternativeto.net/software/twitter/?platform=android.

Table 2. Number of potentially correlated events detected for each metric $m = c, r, p$.

	Events (E)			Correlations (C)			Correlated events (CE)		
	+1	−1	±1	+1	−1	±1	+1	−1	±1
count (c)	17	7	24	547	867	1,414	13	13	26
rating (r)	16	14	30	438	772	1,210	2	4	6
polarity (p)	27	23	50	309	717	1,026	1	2	3
Total	60	44	104	647	1,178	1,825	16	19	35

4.2   Results

Table 2 reports the summarized results for events (E), correlations (C) and potentially correlated events (CE) detection. Complementarily, Fig. 3 reports the distribution of events for the data set collected for the case study introduced in Sect. 2.3. From the original set of 104 events with $E = \pm 1$ for some metric $m_{app} \in \{c, r, p\}$, a correlation-based analysis reduces the scope of analysis to 35 potentially correlated event pairs E_i, E_j covering 29 unique events with $CE = \pm 1$. Within the observation period (i.e., one year), from 35 pairs of potentially correlated events E_i and E_j, 25 pairs (71.4%) are scoped within a 4-week span after the Twexit on October 27th, the trigger event of our ground truth. These 25 event pairs encompass 14 out of the 29 (48.3%) unique events E found in the set of potentially correlated events, and it affects 6 out of the 10 apps in our dataset, including Twitter and Mastodon (BlueSky was released a few months later). Among these, potentially correlated events include[6]: the migration of users from Twitter to Mastodon as a preferred alternative to some users (Summarization #4); the popularity of TruthSocial as a consolidated competitor motivated by the influence of public figures like Donald Trump (Summarization #5); and the emergence of Hive Social, a new competitor (Summarization #1).

> **Summarization #4** - $E(c_{Mastodon},$ "Oct 27, 2022", 7) = 1   all
>
> The most significant event [...] is the mixed reception and criticism of a Twitter alternative, likely referring to the Mastodon social media platform. [...]

> **Summarization #5** - $E(c_{TruthSocial},$ "Nov 10, 2022", 7) = 1 / $p \geq 3$ positive
>
> The most significant event [...] is the launch of a new social media app called "Truth Social", which is associated with former President Donald Trump. [...]

To extend the analysis to all potentially correlated events, Fig. 3 reports the results on the thematic-based classification of reasons to leave and join mobile apps by Al Shamaileh et al. [1]. We include the polarity-based distinction to identify how positive and negative reviews might infer different user perspectives in the summarization process. Frequency of each reason is consistent with

---

[6] Complete summaries for all potentially correlated events are available in the replication package.

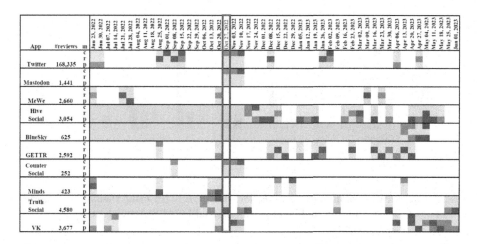

Fig. 3. Distribution of detected events for each app and each metric $m \in \{c, r, p\}$. Green and red cells represent positive and negative deviations for which $E(m_{app}, t_0, w_e) = \pm 1$. Yellow cells represent $E(m_{app}, t_0, w_e) = 0$ assessments for which another $m\prime$ reported $E(m\prime, t_0, w_e) = \pm 1$ for that same app and $[t_0, t_0 + w_e)$. Grey cells represent lack of data. The red box highlights the time window containing the date of the trigger event for the case study in Sect. 2.3 (i.e., acquisition of Twitter by Elon Musk).

results in the original study, being usability (24), usefulness (17) and content (16) the most frequent aspects mentioned in summaries. Filtering on positive reviews showcases a great impact on discussing the app as a potential better alternative to another one (19), being Twitter the reference app to which an alternative is being considered in all scenarios. Furthermore, influence of public figures like Musk or Trump also becomes significantly mentioned, both for negative (6) and positive (9) summaries. From this analysis, and using a sample representation of reported events, generic types of potentially correlated events can be inferred. Below, we highlight the most prominent ones, while also providing some examples for which the used metric was strictly necessary to detect that specific event. To explore the potential of our approach to detect and explain events beyond the ground truth event (i.e., Twexit), these examples refer to additional, complementary events, motivated by the classification in Fig. 4.

- **Software-based events.** Changes in the app's functionality, performance, updates, and user interface. They involve modifications in the codebase, backend infrastructure, or frontend interface that impact the user experience, system stability, or overall performance of the app.

Summarization #6 - $E(r_{VK}, \text{"Apr 06, 2023"}, 7) = -1 \ / \ p \leq 1$ negative

[...] Users have reported various issues, including login problems, account recovery issues, problems with OTP (One-Time Password) verification, excessive advertising, issues with video playback [...]

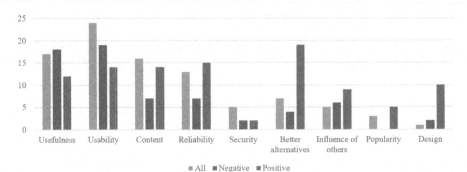

Fig. 4. Classification of unique events E_i for which $CE = \pm 1$ using thematic categories of reasons to leave and join mobile apps as reported by Al Shamaileh et al. [1]

- **Context-based events.** External factors such as market trends, user preferences, regulations, or technological advancements that influence the app's usage, adoption, or user behavior. They highlight the dynamic interplay between the app and its surrounding environment.

> **Summarization #7 - $E(p_{Gettr}, \text{"Oct 20, 2022"}, 7) = 1$**   all
>
> The most significant event is the popularity of a social media app called Gettr [...] for various reasons, such as following and supporting a public figure named Andrew Tate, seeking an alternative to other social media platforms like Facebook and Twitter, [...]

- **New competitors.** Introduction of rival apps or platforms offering similar or alternative solutions in the same market space. It involves the entry of new players, their features, unique selling propositions, and potential impact on the user base, market share, and competitive landscape.

> **Summarization #8 - $E(c_{BlueSky}, \text{"Apr 20, 2023"}, 7) = 1 \; / \; p \geq 3$**   positive
>
> The most significant event [...] is the excitement and interest surrounding a new social media app called "BlueSky." [...] They express positive sentiments about the app, stating that it's better than Twitter and that they prefer it.

5   Discussion

5.1   Findings

Our research question aims at identifying statistical analysis tasks (i.e., methods) that are useful for the automatic detection of significant threats and opportunities in a given market, which can be a source for new or evolved requirements. Evaluation results illustrate that the trigger event of the case study is actually reported by our approach. Furthermore, when examining the event detection results in Fig. 3 independently, it is not immediately evident that a disruptive pattern or event-based phenomenon related to the Twexit event is discernible. It is only when we augment the review-based analysis with a correlation analysis that we observe a significant concentration of events during the Twexit time frame. This observation underscores the enhanced utility of a cross-app statistical analysis. Moreover, beyond Twitter-to-Mastodon user migration, multiple examples of minor events which might not have such a major impact (e.g., new competitors, positive/negative reactions to new releases from competitors) are also proactively detected. Finally, it is essential to emphasize that our approach does not assess the actual cause-and-effect relationships between potentially correlated events. Instead, our focus is on detecting significant events that represent anomalies in a single app's timeline, which occur simultaneously in time. These events might ultimately provide insights into situations where users may be inclined to leave or join a particular app.

Beyond ground truth events, event summarization (as shown in Summarization #1–#8) and categorization (in Fig. 4) showcase the potential of our approach to support market analysis use cases (depicted in Sect. 2.2). An in-depth analysis of a well-known period of observation (i.e., one year) can serve app providers entering a new market or app consultants to conduct SWOT analysis on a given app's market and support informed decisions on strategic alignment, requirements elicitation and release planning. Moreover, metric analysis and methods are designed to support timely, up-to-date detection of events exclusively based on past historical data. Hence, app providers within a given market can use these insights to support continuous development, software maintenance and user feedback analysis. Finally, these results illustrate our approach under a specific set up for the defined configurable parameters (enumerated in Sect. 4.1). Given that the significance of an event is a subjective and contextual notion, we do not seek to define optimal values for these parameters. Moreover, they could be set up differently at different times, according to stakeholder-specific criteria, and to the specific information need that has to be satisfied at any point in time.

5.2   Threats to Validity

We examine our study's constraints by addressing the validity issues outlined by Wohlin et al. [16]. Concerning construct and internal validity, the accuracy of the statistical analysis relies heavily on the quantity of timely available data. Continuous access to a sufficient amount of reviews is essential to avoid missing events. On the other hand, different review metrics and metric computation may lead to varying results. In this research, we have focused on the use of established metrics covered by related work (see Sect. 6). Examples in Sect. 4 illustrate multiple scenarios in which each metric $m_{app} \in \{c, r, p\}$ disjointly reports multiple events. These examples contribute to the hypothesis that the selection of metrics for our approach is appropriate to respond to the research question. Furthermore, we plan to extend the analysis to complementary review metrics as future work. Additionally, the use of specific parameters for the evaluation might introduce bias, as different parameter settings could yield varied results, potentially impacting the study's internal consistency. We do not claim that the selection of parameters answers to a specific validity of an optimal setting of our approach. As mentioned in Sect. 5.1, parameters must be properly selected according to stakeholders interests. Concerning event summarization, using a random sample of n reviews might introduce a bias in the summarization process. To this end, we plan to extend our approach by integrating an internal summarization process (see Sect. 7). Finally, the use of the Twexit event as ground-truth poses a generalization bias, which we attempted to mitigate by extending evaluation with the analysis of all potentially correlated events and the thematic-based classification of reasons to leave and join mobile apps.

Concerning external validity, we rely on gpt3.5-turbo model instance from OpenAI on an unsupervised setting for event summarization. This decision was made at this research stage to focus on the exploratory nature of our research question, based on validating potentially correlated events. We used gpt3.5-turbo based on its demonstrated accuracy on document summarization on unsupervised tasks [3]. We also rely on an external tool for computing the polarity (p) metric. Finally, concerning conclusion validity, main limitations emerge from generalization beyond the microblogging market or even beyond the set of apps and reviews in the dataset. To mitigate these limitations, we utilize a diverse dataset with various sources and reviews in the microblogging market from a large period of observation (i.e., 1 year), increasing the depth and breadth of our insights and minimizing potential generalization issues.

6   Related Work

There is an increasing body of research on the analysis of user reviews in software engineering [5]. Most of this body of research focuses on user reviews for a single app, mainly with the purpose of extracting feature requests, bug reports or more generally, requirements for next releases. Similar to our approach, Gao et al. [8] analyse time slices corresponding to releases of a given app in order to detect emerging issues, namely bugs or unfavourable app features. Our concept of event is more general and includes other possible market events. Strønstad et al. [15] proposed a supervised machine learning-based pipeline for the automatic detection of anomalies in single-app time series, based on metrics related to (1) statistical data, (2) ratings and upvotes, (3) sentiment analysis, and (4) review content. However, being single-app oriented, they do not try to connect reviews from different competitive apps in order to get a holistic view of the marketplace.

In fact, approaches considering user reviews for multiple apps are scarce, and even less aim at assessing the position of an app in the market. Shah et al. [13] compare a reference app with its competitors in terms of sentiment, bug reports and feature requests. Dalpiaz and Parente [6] perform a SWOT analysis for a reference app after comparing with the sentiment of competitor apps as expressed in user reviews. Assi et al. [2] identify high-level features mentioned in user reviews and create a comparative table that summarizes users' opinions for each identified feature across competing apps. The three approaches provide snapshots of the marketplace in a given moment of time, without including the time dimension, as our approach does. While timely evaluation of user reviews has been explored before for single app analysis [7], our approach could be integrated with any of these methods, providing the missing time dimension in the context of whole-market segment analysis. In addition, these methods use classical feature extraction and topic modelling techniques at the feature/topic granularity level. Instead, we focus on the app as a whole and delegate onto ChatGPT3.5 the summarization of the cause of events. Finally, there are a few recent studies assessing the causes for app switching behaviours in the context of mobile app markets [1,12]. However, these are limited to survey and interview-based methods, and do not refer to automated proposals for identifying or reporting these phenomenona.

7   Conclusion and Future Work

In this paper, we have presented a novel approach for automated event monitoring in a mobile app market segment. Evaluation results illustrate several timely and actionable examples regarding market dynamics, including software-level key changes, contextual factors, and the entry of new competitors. A systematic approach to continuously report these types of events could be applied in multiple use cases, including: app providers to enhance tasks such as requirements elicitation, prioritization and release planning; a consulting company to provide market consultancy in software and requirements engineering tasks; a large organization to decide which app to select for internal use to conduct a business activity. Moreover, the selected review metrics and statistical methods provide valuable insights to support the research hypothesis, while also opening the scope for future work. We plan on expanding the original metrics, evaluating their covariance structure, and exploring potential inner correlations. We will also enhance the selection mechanism and employ quality-based filtering techniques to improve the set of reviews used for summarization. Additionally, we aim to extend the reported knowledge and enhance the explainability of event monitoring results by analysing variations in outcomes across different experimentation setups. Last, we plan to combine our approach with some of the methods mentioned in the related work to combine our time dimension with the feature-oriented granularity that these methods provide.

Replication package

The complementary replication package contains the following artefacts:

- **Dataset:** The dataset of reviews used for the evaluation of our approach.
- **Software Components:** Employed for metric extraction, event detection, correlation detection, and identification of potentially correlated events.

- **Metric Values:** The resulting metric values for c, r and p.
- **Statistical Analysis:** Detailed records of the identified events, correlations, and potentially correlated events for all pairwise combinations of c, r and p.
- **Summaries:** Complete summaries for all potentially correlated events.

Acknowledgment. With the support from the Secretariat for Universities and Research of the Ministry of Business and Knowledge of the Government of Catalonia and the European Social Fund. This paper has been funded by the Spanish Ministerio de Ciencia e Innovación under project/funding scheme PID2020-117191RB-I00/AEI/10.13039/501100011033.

References

1. Al-Shamaileh, O., Sutcliffe, A.: Why people choose apps: an evaluation of the ecology and user experience of mobile applications. Int. J. Hum Comput Stud. **170**, 102965 (2023)
2. Assi, M., Hassan, S., Tian, Y., Zou, Y.: Featcompare: feature comparison for competing mobile apps leveraging user reviews. Empir. Softw. Eng. **26**(5) (2021)
3. Brown, T.B., et al.: Language Models are Few-Shot Learners (2020)
4. Chen, N., Lin, J., Hoi, S.C.H., Xiao, X., Zhang, B.: Ar-miner: mining informative reviews for developers from mobile app marketplace. In: ICSE 2014, pp. 767–778
5. Dabrowski, J., Letier, E., Perini, A., Susi, A.: Analysing app reviews for software engineering: a systematic literature review. Empir. Softw. Eng. **27**(2), 43 (2022)
6. Dalpiaz, F., Parente, M.: RE-SWOT: from user feedback to requirements via competitor analysis. In: REFSQ 2019, pp. 55–70
7. Deshpande, G., Rokne, J.: User feedback from tweets vs app store reviews: an exploratory study of frequency, timing and content. In: 2018 5th International Workshop on Artificial Intelligence for Requirements Engineering (AIRE), pp. 15–21 (2018). https://doi.org/10.1109/AIRE.2018.00008
8. Gao, C., Zeng, J., Lyu, M.R., King, I.: Online app review analysis for identifying emerging issues. In: ICSE 2018, pp. 48–58 (2018)
9. Genc-Nayebi, N., Abran, A.: A systematic literature review: opinion mining studies from mobile app store user reviews. J. Syst. Softw. **125**, 207–219 (2017)
10. Graham, R.: The great Twexit. Boston Globe (2022). https://www.bostonglobe.com/2022/11/08/opinion/great-twexit
11. Motger, Q., Franch, X., Marco, J.: Mobile feature-oriented knowledge base generation using knowledge graphs. In: ADBIS 2023, pp. 269–279
12. Salo, M., Makkonen, M.: Why do users switch mobile applications? trialing behavior as a predecessor of switching behavior. Commun. Assoc. Inf. Syst. **42**, 386–407 (2018)
13. Shah, F.A., Sirts, K., Pfahl, D.: Using app reviews for competitive analysis: tool support. In: WAMA 2019, pp. 40–46 (2019)
14. Stol, K.J., Fitzgerald, B.: The ABC of software engineering research. ACM Trans. Softw. Eng. Methodol. **27**(3) (2018)
15. Strønstad, G.H., Gerostathopoulos, I., Guzmán, E.: What's next in my backlog? time series analysis of user reviews. In: RE 2023 - Workshops, pp. 154–161 (2023)
16. Wohlin, C., Aurum, A.: Towards a decision-making structure for selecting a research design in empirical software engineering. Empirical Softw. Eng. **20**, May 2014

Exploring the Automatic Classification of Usage Information in Feedback

Michael Anders[✉][iD], Barbara Paech, and Lukas Bockstaller

Heidelberg University, Im Neuenheimer Feld 205, 69121 Heidelberg, Germany
{michael.anders,paech}@informatik.uni-heidelberg.de

Abstract. Context and motivation: User participation and involvement is important for system success. Communication between developers and users is an important part of participation. This communication influences user satisfaction and therefore system success. However, direct communication between users and developers is often not possible and thus developers need other information sources to understand how the users use the system and what improvements they want.

Question/problem: Feedback sources like app stores or online forums provide insights on the users' view. Due to its size this feedback has to be classified automatically. So far classification has focused on rough classification and the opinions of the users. Detailed usage information is more difficult to classify as the classes are more fine-grained.

Principal ideas/results: In this paper, we explore in how far it is possible to mine feedback for the usage information underlying the users opinions and wishes. We analyze multiple classification methods and investigate the transferability across different feedback sources. Additionally, we apply multi-stage classifications to improve classifier performance and we experiment with the granularity of the classes. Overall, BERT performs best in almost all experiments and multi-stage classification does not yield improvement. Improvements are possible with more coarse grained classes.

Contribution: To our knowledge, this paper is the first to explore the classification of usage information in explicit feedback. This is a first step towards bundling the usage information for individual functionalities.

1 Introduction

User participation and involvement is important for system success [2]. Communication between developers and users is an important part of participation and in particular influences user satisfaction [1]. This communication, of course, depends on the context. We have studied how to enhance user developer communication in large scale IT-projects [1,3]. Often, however, direct communication is not possible. This can lead to problems such as software behavior that might appear intuitive and self-explanatory for people experienced in software development, but is not as self-evident for the people that the software is actually developed for [24]. Without direct communication, developers need other information sources to understand the user view.

D. Mendez and A. Moreira (Eds.): REFSQ 2024, LNCS 14588, pp. 267–283, 2024.
https://doi.org/10.1007/978-3-031-57327-9_17

Users provide their view on an existing system and their needs for further development as feedback through channels such as app stores or online forums. There is a lot of research on mining such channels, but so far research has focused mainly on the opinions user have about software by classifying for example the features or bugs discussed therein and not the usage behavior underlying these opinions. An example taxonomy shows that product quality, user intention, user experience and sentiment are classified, but not concepts relating to usage information [28]. In a preliminary study we have shown that concepts relevant for usage information such as tasks or interactions or discussions of specific UI elements can be found in user feedback [6]. By providing this information about how users talk about the software and their use of it, developers could gain better understanding of the users. They may also be able to discover differences in how they envision a software and its functionalities being used as opposed how their actual user base does.

It is however, very time consuming to look for these concepts manually. In this paper we explore how well usage information can be automatically classified in user feedback. This extraction of information is difficult as the classes (e.g. aspects of the user interface (UI), actions performed or interaction data) are fine-grained. Also, it requires analysis on a token level rather than a sentence level, since a single sentence can contain different pieces of usage information.

We note that usage information in terms of interactions of the users with the software can also be gained from monitoring data as for example described in [19]. However, according to the ACM Code of Ethics [16]: "Only the minimum amount of personal information necessary should be collected in a system." Thus, due to data protection or ethical concerns it is not always viable to monitor users. Therefore, we rely on the feedback given explicitly.

In this paper, we experiment with multiple classification methods, multi-stage classification and different class granularities. Namely Stanford Named-Entity Recognition (SNER) [14], a Bi-LSTM based on [22] and a BERT [13] classifier. We also investigate the transferability across feedback sources.

The rest of the paper is structured as follows. Section 2 introduces the used classification framework and the classifiers with which the experiments are performed. Section 3 discusses related work. Section 4 details the design of our study, including the research questions, data collection and experiment configuration processes. The results of our experiments are presented in Sect. 5. Section 6 discusses our findings and ideas for future improvement. Lastly, in Sect. 7 we discuss threats to validity before concluding in Sect. 8.

2   Background

In this section we first introduce the framework with which we classify usage information and then the classifiers with which we perform our experiments.

2.1   TORE

Usage comprises the interactions with the software in terms of UI elements touched, data input and output and the functionalities used. These interactions are part of a usage context in terms of the user tasks and activities.

To classify the usage information, we are using the Task-oriented Requirements Engineering framework (TORE) which covers all these concepts [26]. TORE has been developed for requirements elicitation and specification. It has also been applied in different industrial development projects in the past [4] to guide requirements engineers in their communication and decisions while eliciting and specifying requirements.

TORE is comprised of two stages. The first stage consists of the TORE *levels*, e.g. *Interaction Level* and *System Level*. The levels are detailed by TORE *categories* which further specify the kind of usage information. The *Interaction Level* for example contains the *System functions* category. We also refer to these categories as the "second stage" in the TORE framework.

In one of our previous works we adapted the original TORE to analyze user feedback [6]. We simplified the model to include less categories, which did not fit the goal of analysing user feedback. The resulting levels and categories which are used in this paper are shown in Table 1. We combined the original first two levels (Goal & Task Level and Domain Level) into the *Domain level*. In addition, the distinction between As-is Activities, To-be Activities is not necessary for analyzing feedback. On the *Interaction Level* we replaced the original UI-structure category with a *Workspace* category. This captures all statements related to specific UI Elements which includes the GUI Elements (originally on the system level) as they refine workspaces. We performed further simplifications for the purpose of this paper. The *Goal* category was removed, because it is rarely ever found in feedback according to our previous study. Also, the *System Level* was unified into one category called *System* because users rarely have detailed knowledge of the inner workings of the software.

Table 1 provides a description of the resulting levels and categories and lists examples of usage information extracted from feedback (highlighted in bold). The feedback excerpts are largely extracted from our data sets introduced in Sect. 4.2 and talk about the popular hiking app Komoot. The *Stakeholder* example highlights two different roles which influence software. *Tasks* capture larger processes in the domain, which consist of multiple *Activities*. These *Activities* are individual steps in the domain. In the examples of Table 1 this is the Outdoor-Activity-Domain, for example hiking, biking. *Domain Data* captures all tokens that are relevant to the activities of the domain, in the example "camping grounds". *Interactions* describe actions performed by the user with the software. *System Functions* are functionalities that the software provides or according to the user should provide, as can be seen in the example. *Interaction data* comprise data provided and used in the interaction. *Workspace* highlights which specific view in the UI or which element in a view the user is discussing. Lastly, the *System* category captures mentions of the software as a whole (in the example "app") or users' discussions about the inner workings of the software.

Table 1. TORE Categories, their Definitions and Examples

Category	Definition	Example
Domain Level		
Stakeholders	Roles supported by or influencing the developed software (e.g. Users or Developers)	"I'm more of a **power user** while my friend is an **infrequent user**"
Tasks	Responsibilities of the Stakeholder as part of larger processes in the domain	"I frequently **plan outdoor routes** for my adventures"
Activities	Steps in the Stakeholders' Tasks	"This is the best app for **running, hiking** or **biking** out there"
Domain Data	Data relevant to an Activity	"The app keeps letting me down with bad information on **camping grounds** about 35% of the time"
Interaction Level		
Interaction	The interaction between a user and the software	"You can **track** your miles, **make** your own trails and even **get** directions"
System Functions	Functions executed by the software that consume, manipulate or produce data.	"It lacks **data sharing to Google Fit**"
Interaction Data	Data relevant for the System Functions or Interactions	"You deleted the **maps** I had, and now I can't even access **online maps**"
Workspace	Grouping of Interaction Data and System Functions which are relevant for one Task and specific UI elements	"The app shows completely different numbers than it records in the **Completed Rides Tab**"
System Level		
System	Components of the software as well as data and actions processed internally	"Komoot's **app** has become horribly slow and I'm not sure if that is because of **loading data** from some **server**"

TORE has similarities to other requirements models: Lauesen [21] introduces Goal, Domain, Product and Design Level. Goal and Domain level correspond to the original TORE levels. The Product level focuses on the functions on TOREs *Interaction Level*, and the Design level on the GUI details as in the original TORE *System Level*. Gorschek et al. [15] also introduce a requirements model with different abstraction levels. They use Product, Feature, Function and Component Level. The Product Level comprises *Goals* as in original TORE and the Features correspond to the *Activities*. The Function Level corresponds to our *Interaction Level* and the Component Level to our *System Level*. However, we use more categories to distinguish data and UI information in addition to

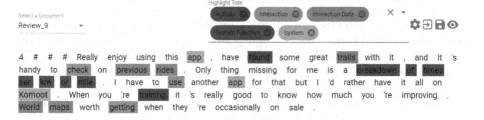

Fig. 1. TORE Annotation of App Review

functions on the *Interaction Level*. In our view the fine grained categories of TORE allow us to extract more detailed usage information. The setup of levels and categories also allows us to investigate the potential of automatic multi-stage classification.

When applying the adapted TORE framework to user feedback, individual tokens as opposed to complete sentences are coded. This very fine-grained approach is necessary as sentences can contain several different TORE categories. Figure 1 provides an example of the coding of an app review of Komoot. TORE categories are assigned to individual nouns and verbs to identify all mentioned usage information. Subsequent tokens that constitute a single term such as "World maps" as seen in Fig. 1 are assigned the same code.

2.2   Classification

We are performing our experiments in this paper with three separate classifiers. These are a BERT base classifier [13], the Stanford Named-Entity Recognition classifier (SNER) [14] and a Bi-LSTM based classifier [22]. They were selected because all three had been used in the past to perform token level classifications such as ours. They also cover three overlapping areas, namely pre-trained large language models (LLMs) with BERT, neural network deep learning (DL) with Bi-LSTM and more traditional machine learning (ML) with SNER. We decided to use BERT base as it is the most widely used in the related literature we found, namely three out of four approaches that use BERT employ BERT base.

We also combine these classifiers to perform a multi-stage classification. Here TORE levels (first stage) are classified by one of the three classifiers. The result of this is then used as a classification feature for the TORE category (second stage) classification by another classifier which takes the first stage into consideration when classifying the categories.

3   Related Work

We discuss related work regarding detailed user feedback analysis. This was identified by analyzing six existing literature reviews which deal with crowd sourced requirements and user feedback analysis. The literature reviews we looked at

[20, 23, 28–30], and [11] contained a total of 341 individual papers. We included only works that introduced an approach for the automatic classification of user feedback with pre-defined classes of *fine granularity*. That means classes referring to what users are specifically talking about. This goes beyond general requirements classes like "Feature" or "Bug" to include for example specific rationale users state in their feedback.

Based on 12 relevant papers we performed forward snowballing and found another 9 papers. The list of the papers along with the names of the classes they use as well as the used methods is provided on Zenodo[1]. These 21 papers can be sorted in three different groups: *Software Taxonomies* (5 papers) introduce classes that deal with specific aspects of a software such as *Android Version, UI, Battery and Price* e.g. [10]. *Rationale* approaches (4 papers) analyze the arguments and reasoning users give in their feedback and discussions. These papers include classes like *claim-supporting* and *claim-attacking*. The third and largest group *Requirements* (14 papers) is made up of papers looking at functional and non-functional requirements (NFRs). For example the work by Hadi et al. [17] introduces 16 classes among which are NFRs like *Usability* and *Reliability* as well as fine-grained functional requirements classes such as *Feature shortcoming* and *Aspect evaluation*. Some works include multiple data sets with different classification goals and as a result can be attributed to several groups, e.g, Khan et al. [5] focus on *Rationale* but also use a *Feature* class. We did not find any works that focus on usage information.

The number of classes classified by these papers ranges from 4 to 26. There is no apparent relation between the number of classes used and the methods employed for their classification. While for example Support Vector Machines are used by 4 out of 5 approaches with 5 classes, they are also used for approaches with for example 14 and 23 classes. In total, 21 different methods are used by the approaches, with 8 of those being used by more than one approach. We also looked at the evaluation results of the found approaches. It is difficult to name one specific value for each approach, given the difference in evaluation methodology. However, we can report the ranges in which the papers report their results. Papers within the *Software Taxonomy* group report highest F1 values between 0.64 and 0.95. Those within the *Rationale* group report their F1 values between 0.53 and 0.96. The range within the *Requirements* group is 0.64 to 0.98. When we only look at approaches that have a similar number of categories as us, i.e. those ranging from 7 to 11 classes, we find that all four fall into the *Requirements* group. These particular approaches span an F1 range between 0.64 and 0.93. This is very similar to the ranges of the other papers. We cannot directly compare our approach with theses results, since they use different data sets. However, in our view this gives indication for a range one would also expect for other, similar classification tasks.

---

[1] https://zenodo.org/records/10568458.

4   Study Design

In this section we first describe our research questions and the rationale behind them. Then we introduce our data sets and coding process and finally our experiment configuration.

4.1   Research Questions

For our analysis of the performance of the different classifiers, we focus on the following research questions:

RQ1:   *How well can automatic single-stage TORE classification be performed by SNER, Bi-LSTM and BERT and which classifier performs best?*

RQ2:   *Are classification results transferable across feedback sources?*

RQ3:   *Does multi-stage classification improve results compared to single-stage classification?*

RQ4:   *Does the granularity of the TORE categories influence classification results?*

RQ1 analyzes the performance based on a single-stage end-to-end classification where tokens in the feedback statements are directly assigned to TORE categories. As other studies have shown, classifier performance can be highly dependent on the feedback source [12,25]. With RQ2, we therefore analyze the transferability of our best performing classifier across feedback sources. Note that we do not investigate the generalizability across different software products, but how the platform through which the feedback is given affects classifier performance. Thus, we train on one or more of the sources and test on another source. As described in Sect. 2.2, TORE enables multi-stage classification. We investigate in RQ3 whether this improves the classification. In order to understand the limits of the granularity which our classifiers are able to tell apart we investigate in RQ4 the effects of combining classes which the classifier struggles to discern from one another.

We compare precision, recall and F1 measures of all classifiers.

4.2   Data Collection

We created three data sets from three different user feedback sources, an online questionnaire, app reviews and an online forum. All data sets used in this work can be accessed via the Zenodo Repository[2].

The data sets were coded using a custom coding tool designed specifically for TORE annotation called Feed. UVL[3] (shown in Fig. 1). The coding process was the same for all data sets. Multiple annotators coded a part of the data set with each part being coded by two annotators. Coding was done in multiple steps.

---

[2] https://zenodo.org/records/10568458.

[3] https://github.com/feeduvl.

Table 2. Data Set Sizes and Number of Words assigned to each Code

	Sentences	Words	Domain Data	Stake-holder	Activity	Task	Inter-action	Interaction Data	System Function	Work-space	System
Prolific	1146	26607	1097	385	350	249	1048	952	482	81	749
Forum	865	13775	280	33	85	1	572	443	5	345	678
App Review	901	14879	363	77	113	69	844	746	325	90	420

Each annotator coded a portion of the data set, and then the annotators met to resolve all disagreements between annotations. After all disagreements were resolved, coding rules were improved, after which another portion of the data set was coded. For each step, a Kappa value was calculated using Brennan & Perediger Kappa [8]. We call the resulting data sets the **Prolific**, **Forum** and **App Review** data sets. The only pre-processing performed on the data sets was the removal of web-links and emojis. An overview over the size of the final data sets can be seen in Table 2. The table also lists the number of words in each data set that are assigned each TORE code.

The Prolific data set was created through an online survey on the crowd platform Prolific[4]. A total of 100 participants took part in the survey. They were asked to answer four main questions about the Komoot[5] hiking app. Namely, "What is the Komoot App and what can the Komoot App do?", "What do you like about Komoot?", "What do you dislike about Komoot?", and "What could be improved about Komoot and why?". The survey was part of a prior research project. More information on the data set can be found in [7]. The participants' answers were then annotated by three PhD students (one being an author of this paper). The average Kappa value across the individual coding steps was 0.59.

The Forum data set consists of 98 threads including comments crawled from the online forum Reddit between May 2021 and May 2022. The threads were randomly sampled from the Google Chrome (40 threads), Komoot (20 threads) and VLC Video Player (38 threads) sub-forums on Reddit. Two master students, under the supervision of one of the paper's authors, annotated the dataset. The average Kappa value across the steps was 0.62.

The App Review data set was crawled from the Google Play Store. In total, 200 reviews submitted in July 2023 about the Komoot hiking app were gathered. Here, we excluded posts that were shorter than 100 characters since short posts are unlikely to contain relevant usage information. We also removed reviews that consisted purely of unrelated spam. This data set was annotated by a master student and a PhD student, both authors of this paper. The average Kappa value for this data set was 0.65.

The Kappa values for all three data sets range from 0.59 to 0.65. Annotators found that the uncertainty over which subsequent tokens to assign codes to and which not significantly increased the amount of disagreements, thus influencing the Kappa values.

---

[4] https://app.prolific.com/.
[5] https://play.google.com/store/apps/details?id=de.komoot.android.

Because we are dealing with three different feedback sources, we look at feedback in terms of documents within the data sets, i.e. one feedback equals one document. For the *Prolific* data set, we have 100 participants' answers to four questions, creating 400 individual documents. The *Forum* data set contains 98 threads, i.e. the original post by a user and all comments by other users there within. We treat these as 98 documents. The *App Review* data set contains 200 reviews, where each review is one document. Any document can contain an arbitrary number of sentences and each word in those sentences is one token. Because we purely code and classify tokens and never full documents, the different granularities between for example app reviews submitted by an individual user and the forum threads and comments submitted by multiple users do not affect our classification.

4.3   Experiment Configuration

We ran our experiments using the MLFlow machine learning platform[6]. The complete source code including the final hyperparameter setup and the code to reproduce the experiments is provided as part of an open-source project[7].

As introduced in Sect. 2 we performed our experiments on three separate classifiers. To optimize the parameters for each classifier we performed a grid search for both single and multi-stage classification. We ran experiments for each hyperparameter configuration with a data split of 80% training and 20% testing data across all three data sets. Experiments towards answering RQ1, RQ3, and RQ4 use 5-fold cross validation across all three data sets with the previously established best hyperparameter configuration. Experiments for RQ2 split training and test data sets among all combinations of feedback sources, training our best performing classifier on one feedback source (i.e. Prolific, App Review or Forum) and testing on another. Additionally, we perform leave-one-out-validation by training on two of our three data sets and testing on the third. For the multi-stage classification (RQ3) we experiment with all 3 classifiers as a first stage to classify TORE levels. These results are then fed as a classification feature towards a second-stage BERT classifier, which decides the specific TORE category. We additionally create Perfect-BERT which simulates a perfect first stage classifier, in order to see how dependent the second stage classifier is on the performance of the first stage. Lastly, for the TORE category granularity experiments (RQ4), we analyzed the individual class-performance of our best performing classifier to see which categories the classifier struggles to tell apart. We then combined the classes *Task & Activity* as well as *Interaction Data & Domain Data* and completely retrained the best single- and multi-stage classifier configurations with the combined classes using 5-fold cross validation.

The precision and recall values are calculated for each run on a per-class basis. Mean precision, recall, and F1 are averaged over all 5 folds on all data sets without any weights. Additionally the minimum and maximum precision, recall and F1 values across all folds are logged.

---

[6] https://mlflow.org/.
[7] https://github.com/feeduvl/uvl-tore-classifier-bert.

5   Results

In this section we provide the results for all research questions.

RQ1: *How well can automatic single-stage TORE classification be performed by SNER, Bi-LSTM and BERT and which classifier performs best?*
Table 3 shows the minimum, mean and maximum precision, recall and F1 values for all three classifiers in their single-stage configuration (The last row is addressed in RQ4). BERT performs best across all metrics achieving a mean F1 value of 0.65. SNER outperforms Bi-LSTM across all metrics.

Table 4 shows the individual class performance of the best performing BERT classifier, to analyze how well the classification of individual classes works. Because Max and Min values do not differ much, we only show the Mean values here. We can see that TORE categories have very different performance results. *Stakeholder* for example is the most accurately classified class with a mean value of 0.87 for all metrics. *Activity* has the lowest precision, recall and F1 with 0.42, 0.40 and 0.41 respectively.

Table 3. Single-Stage Classifier Results (RQ1)

	Min Prec	Mean Prec	Max Prec	Min Recall	Mean Recall	Max Recall	Min F1	Mean F1	Max F1
SNER	0.61	0.63	0.65	0.50	0.52	0.56	0.55	0.57	0.60
Bi-LSTM	0.55	0.58	0.61	0.48	0.51	0.53	0.53	0.54	0.56
BERT	0.65	0.66	0.67	0.61	0.64	0.66	0.63	0.65	0.67
BERT Comb. Class	0.70	0.73	0.76	0.68	0.70	0.73	0.70	0.72	0.74

Table 4. Individual Class Metrics for BERT Single-Stage (RQ1)

	Stakeholder	Task	Activity	Domain Data	Interaction	System Function	Interaction Data	Workspace	System
Mean P.	0,87	0,48	0,42	0,56	0,62	0,6	0,62	0,66	0,77
Mean R.	0,87	0,41	0,40	0,56	0,70	0,46	0,61	0,61	0,78
Mean F1	0,87	0,44	0,41	0,56	0,66	0,52	0,61	0,63	0,77

Answering RQ1: The TORE classification can be performed with a mean precision of 0.66 and mean recall of 0.64. This is achieved by BERT which overall shows the best performance. There is a large disparity between the performance of individual classes. *Stakeholder* performs best with a mean F1 value of 0.87 and *Activity* performs worst with a mean F1 value of 0.41

RQ2: *Are classification results transferable across feedback sources?*
Table 5 shows the precision, recall and F1 values for all transferability experiments of the best performing BERT single-stage classifier. Because training and test data are separate, there is no cross-validation for these and as such no Min or Max values to report. Results are consistently lower (F1 score 0.57) than those of the BERT classifier (F1 score 0.65) trained and evaluated on the combination of all data sets. The best performing combination of training and testing data by recall and F1 value are training on App Reviews and Forum and testing on Prolific data. There are only very small differences between the singular training

data experiments and the leave-one-out experiments with the latter achieving between 0 and 0.03 higher values compared to those without the second training data set. Consistently when only training and testing on one data set, all combinations including the Forum data set perform worse than their counterparts. When training on two data sets only testing on Forum shows worse performance. Training with forum and another data set results in increased performance compared to only training on the forum data set and in comparable performance to only training on App Review or Prolific and testing on the other.

Table 5. Single-Stage BERT Classifier Transferability Results (RQ2)

Training Data set	Test Data set	Precision	Recall	F1
Prolific	Forum	0.39	0.42	0.40
Prolific	App Review	**0.58**	0.42	0.49
App Review	Prolific	0.56	**0.57**	**0.56**
App Review	Forum	0.37	0.41	0.39
Forum	Prolific	0.46	0.43	0.44
Forum	App Review	0.46	0.37	0.41
Prolific & App Review	Forum	0.37	0.45	0.40
Prolific & Forum	App Review	**0.59**	0.45	0.51
App Review & Forum	Prolific	0.57	**0.57**	**0.57**

Answering RQ2: Classification does not transfer well across different feedback sources. The data sets transfer better, if Forum is not involved.

RQ3: *Does multi-stage classification improve results compared to single-stage classification?*

In Table 6 we report the results of the first-stage classifiers' TORE level classification in the first 3 rows. BERT 1st Stage outperforms the other first stage classifiers in all metrics and is only matched by SNER 1st Stage in minimum precision.

Table 6 also shows the results of the multi-stage classification (the last row is addressed in RQ4). Between the three classifier combinations SNER-BERT, Bi-LSTM-BERT and BERT-BERT, we see that the classifier with BERT as the first-stage outperforms the SNER and Bi-LSTM first-stages for all F1 and all but the minimum precision metrics. The highest recall is achieved by the Bi-LSTM classifier even though the Bi-LSTM 1st Stage does not outperform the other 1st Stage classifiers. SNER-BERT only shows the highest minimum precision, but is consistently lower than the other classifiers. As with first stage classifiers, results are again relatively close between the three. While the BERT first-stage classifier out-performs the other first-stage classifiers in all metrics, BERT-BERT does not do that for all multi-stage classifiers.

We do not see any remarkable differences between the multi- and single-stage performance of the classifiers. All metrics between the best performing

Table 6. First-Stage and Multi-Stage Classifier Results (RQ3)

	Min Prec	Mean Prec	Max Prec	Min Recall	Mean Recall	Max Recall	Min F1	Mean F1	Max F1
SNER 1st Stage	**0.75**	0.76	0.78	0.67	0.70	0.71	0.71	0.73	0.74
Bi-LSTM 1st Stage	0.68	0.72	0.77	0.62	0.65	0.67	0.65	0.68	0.77
BERT 1st Stage	**0.75**	**0.78**	**0.80**	**0.76**	**0.78**	**0.80**	**0.75**	**0.78**	**0.80**
SNER - BERT	**0.61**	0.63	0.64	0.54	0.56	0.58	0.57	0.60	0.61
Bi-LSTM - BERT	0.53	0.55	0.56	**0.65**	**0.67**	**0.70**	0.59	0.60	0.62
BERT - BERT	0.60	**0.64**	**0.65**	0.62	0.65	0.67	**0.61**	**0.64**	**0.66**
Perfect - BERT	0.83	0.85	0.86	0.83	0.84	0.86	0.83	0.84	0.86
BERT - BERT Comb. Class	0.70	0.71	0.73	0.73	0.73	0.73	0.72	0.72	0.73

multi-stage BERT-BERT and best-performing single-stage BERT only have an average difference of 0.02. The Table also reports the metrics for the Perfect-BERT classifier, which simulates a perfect first stage classifier as input for the second stage BERT classifier. We see remarkably higher values when compared to the other multi-stage as well as the single-stage classifiers, with a mean F1 value of 0.84.

Answering RQ3: Multi-stage classification with the achieved performance of the first-stage classifiers does not result in improvement over single-stage classification. Perfect-BERT, however, shows the potential of multi-stage given further improvement in first-stage classifiers.

RQ4: *Does the granularity of the TORE categories influence classification results?*

Figure 2 shows the normalized confusion matrix for the best performing BERT single-stage classifier on the left. Each row represents the percentages of predicted labels for every true label. *Activity* for example was correctly labeled 40% of the time and mislabeled as *Task* 17% of the time. The brighter the color in the off-diagonal elements the higher the confusion.

Ignoring the default label 0, where no class is assigned to a token, the *Stakeholder* is the least confused class in the data sets. We see higher confusion between the *Activity, Interaction* and *Task* classes as well as the *Interaction Data* and *Domain Data* classes. The matrix indicates that the classifier is not able to consistently tell these classes apart. Therefore, we combined the *Activity* and *Task* categories as well as the *Interaction Data* and *Domain Data* categories and re-run the 5-fold cross validation experiments for BERT single-stage and BERT-BERT multi-stage. We did not combine the *Interaction* category with *Activity* despite the higher confusion because we see this distinction as essential when extracting usage information.

The right side of Fig. 2 shows the resulting confusion matrix of BERT single-stage, where the classes have been combined. The "Data" category represents the combined *Interaction Data & Domain Data* categories. The combined categories show an increase in the number of correctly assigned labels. The *Data* class sees no confusion higher than 0.04 with any of the other classes. The *Task & Activity* class sees a 14% improvement compared to *Activity* and *Task* as separate classes.

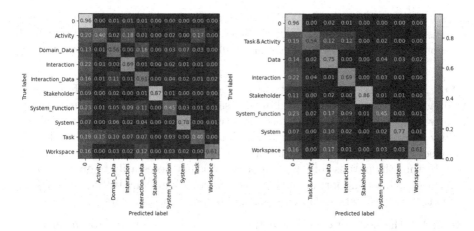

Fig. 2. Normalized Confusion Matrix BERT Single-Stage original (left) and combined classes (right)

The metrics achieved by the BERT single-stage classifier with the combined classes can be seen in the last row of Table 3 and those of the BERT-BERT multi-stage in the last row of Table 6. The combined class classifiers outperform the others (with the exception of Perfect-BERT) in all metrics. When comparing combined class BERT single-stage in Table 3 with BERT-BERT multi-stage in Table 6 we see almost identical precision and F1 and slightly higher recall on the multi-stage.

Answering RQ4: The results are affected by the granularity of the classes. Less fine-grained data and domain level categories improve classification results overall and reduce confusion between classes.

6   Discussion and Future Work

In this section we discuss the results and corresponding future work.

RQ1 & RQ4: Our highest mean F1 value achieved is 0.71. Comparing this to the F1 ranges of 0.64 and 0.93 discussed in Sect. 3 for the works with a similar amount of classes, we see a performance slightly below the average of 0.76. Note, however, that we compare our mean F1 value to their highest F1 scores because mean F1 is not presented by every paper. Our highest F1, achieved by the BERT single-stage with combined classes on the *Stakeholder* category, is 0.87 which comes much closer to the high values in the established ranges. This comparison gives us a first indication whether the performance of our classifiers is on par with those of other existing fine-grained automatic classifiers.

Interestingly, both our highest and all the lowest performing classes are on the *Domain Level*, while the other two levels are much closer to the mean F1 values. The *Domain Level* appears to be the hardest level to accurately classify. In the future, we want to investigate the reasons for this in more detail.

As discussed in Sect. 2.1 TORE shares similarities with other requirements models. These models use only levels. Since RQ4 indicates that less fine grained categories have better performance, one could explore our classifiers for their use cases.

RQ2: As expected, transferability across feedback sources is low. Differences in feedback sources are also reflected in the usage information. The Forum data set is very different from the others as evidenced by both Table 2 and the transferability analysis using only one training data set. These differences have less of an effect when training with Forum and another data set. Part of the difference may be that the purpose of a post on a forum is not necessarily to give feedback but rather ask for support or to have a discussion. Additionally, the Forum data set is the only one which contains posts about multiple software products.

RQ3: For the multi-stage classification, we do not see any improvement compared to single-stage. Nonetheless, Perfect-BERT demonstrates the potential for improvement compared to single-stage classification. It was reported by the manual annotators and can be seen in the results of the first-stage classifiers in Table 6 that the first-stage classification is much easier than the second-stage because of the reduced number of classes (3 compared to 9). This could offer an avenue for future improvement, by using a manually annotated first-stage classification to then automatically extract detailed usage information in the second stage.

Overall we conclude that our results are in the range of similar fine-grained classification approaches, but our classifiers need improvement to be useful. We aim to conduct more experiments with our data by applying even newer classifiers such as the ChatGPT LLM which seems promising because of its ability to understand contextual information within sentences [31]. Different preprocessing methods, such as spell checkers or resolving abbreviations might also help in overcoming some of the shortcomings of online crowd-based feedback [27]. Data balancing methods such as SMOTE [9] might counteract data imbalance in our data sets. Additionally, using inside-outside-beginning (IOB) tagging could help improve results by allowing a classifier to detect commonly terms consisting of multiple words (e.g. "data sharing" in Table 1). Currently, each token is treated individually and assigned its own code, meaning the classifier might only correctly label parts of a term resulting in a decrease in performance.

Another area for future improvement is the chosen BERT model. We chose BERT base because it was the most widely used model in the related literature. In the future we want to investigate versions of BERT that are specifically trained for named entity recognition, such as the bert-base-NER model provided by the hugging face library[8].

With improved classification, our future work is going to focus on connecting feedback and usage information contained therein with the existing requirements of a software. The goal is to bundle usage information for every requirement to support a focused access to the usage information for the developers. Then

---

[8] https://huggingface.co/dslim/bert-base-NER.

developers can match their expectations on the system usage with actual usage information. In our opinion, a promising approach for connection is to adapt approaches which connect feedback with bug reports (e.g. [18]).

7   Threats to Validity

Construct Validity: The main threat to construct validity concerns the Prolific data set and is one inherent to online survey platforms. Users are anonymous participants who are paid for their participation. This could result in them feeling the need to give feedback that is not representative to other forms of online feedback. To alleviate this to some degree, answers were checked for originality. Answers found to be copied from online sources (even those lightly changed from these sources) were excluded from the data set.

Internal Validity: One threat to the internal validity is the implementation of our classifiers. To alleviate this, we have used widely used libraries such as Huggingface and performed our testing with an established machine learning experiment management platform in the form of MLFlow. We also made our code publicly available so results can be replicated. We alleviated the internal threat due to the manual coding by employing multiple coders for every data set and ensuring that every document was independently coded by two people after which inter-rater agreement was established.

External Validity: The main threat to the external validity concerns the contents of our data sets. Almost all data therein is concerning one software product, Komoot. For unrelated reasons Komoot was chosen during our initial design of the questionnaire for the Prolific data set [7]. While this helped us to better analyze the transferability across feedback sources in RQ2, it poses a threat towards the generalizabilty of our classifiers on unseen data of different software products. Given the domain dependence of the classification especially on the Domain level, we don't expect similar performance of our classifier in different domains. This would require retraining on data from unseen domains. We are currently working on another data set and want to provide an analysis of the generalizabilty in the future. The Forum data set partially contains feedback from other software products. This however, introduces another threat especially for the transferability investigation in RQ2, as it is the only one that is not solely about Komoot. We made our data sets publicly available to allow independent investigation and replication.

8   Conclusion

In this work, we conducted experiments in order to extract usage information from user feedback gathered from three different sources. We investigated how well usage information can be extracted automatically using different classifiers, stages and granularities. Our results indicate that extraction is possible, but further improvements are needed. We already performed experiments with multi-stage classification and class granularity. While more coarse-grained classes yielded better results, multi-stage classification with the current performance of the first-stage classifier did not. In the future we aim to further improve these results. We also want to improve the usefulness of the usage information in the feedback for developers. For this, we aim to connect existing requirements for a software with related feedback and thus the usage information contained therein.

References

1. Abelein, U., Paech, B.: State of practice of user-developer communication in large-scale IT projects. In: Salinesi, C., van de Weerd, I. (eds.) REFSQ 2014. LNCS, vol. 8396, pp. 95–111. Springer, Cham (2014). https://doi.org/10.1007/978-3-319-05843-6_8
2. Abelein, U., Paech, B.: Understanding the influence of user participation and involvement on system success-a systematic mapping study. Empir. Softw. Eng. **20**, 28–81 (2015)
3. Abelein, U., Paech, B., et al.: Evaluation of the simulated application of the udc-lsi method: the ipeople case study. In: IEEE/ACM Workshop on Cooperative and Human Aspects of SE, pp. 22–28. IEEE (2015)
4. Adam, S., Doerr, J., Eisenbarth, M., Gross, A.: Using task-oriented requirements engineering in different domains - experiences with application in research and industry. In: IEEE Requirements Engineering Conference, pp. 267–272 (2009)
5. Ali Khan, J., Liu, L., Wen, L.: Requirements knowledge acquisition from online user forums. IET Softw. **14**(3), 242–253 (2020)
6. Anders, M., Obaidi, M., Paech, B., Schneider, K.: A study on the mental models of users concerning existing software. In: Gervasi, V., Vogelsang, A. (eds.) REFSQ 2022. LNCS, vol. 13216, pp. 235–250. Springer, Cham (2022). https://doi.org/10.1007/978-3-030-98464-9_18
7. Anders, M., Obaidi, M., Specht, A., Paech, B.: What can be concluded from user feedback?-an empirical study. In: IEEE Requirements Engineering Workshops CrowdRE, pp. 122–128. IEEE (2023)
8. Brennan, R.L., Prediger, D.J.: Coefficient kappa: some uses, misuses, and alternatives. Educ. Psychol. Measur. **41**(3), 687–699 (1981)
9. Chawla, N.V., Bowyer, K.W., Hall, L.O., Kegelmeyer, W.P.: Smote: synthetic minority over-sampling technique. J. AI Res. **16**, 321–357 (2002)
10. Ciurumelea, A., Schaufelbühl, A., et al.: Analyzing reviews and code of mobile apps for better release planning. In: SANER, pp. 91–102. IEEE (2017)
11. Dabrowski, J., Letier, E., et al.: Analysing app reviews for software engineering: a systematic literature review. Empirical SE **27**(2), 1–63 (2022)
12. Devine, P., et al.: Evaluating software user feedback classifier performance on unseen apps, datasets, and metadata. Empirical SE **28**(2), 26 (2023)
13. Devlin, J., Chang, M.W., Lee, K., Toutanova, K.: Bert: Pre-training of deep bidirectional transformers for language understanding. arXiv preprint arXiv:1810.04805 (2018)
14. Finkel, J.R., Grenager, T., Manning, C.D.: Incorporating non-local information into information extraction systems by gibbs sampling. In: Annual Meeting of the Association for Computational Linguistics (ACL), pp. 363–370 (2005)
15. Gorschek, T., Wohlin, C.: Requirements abstraction model. Requirements Eng. **11**, 79–101 (2006)

16. Gotterbarn, D., et al.: Acm code of ethics and professional conduct (2018)
17. Hadi, M.A., Fard, F.H.: Evaluating pre-trained models for user feedback analysis in software engineering: a study on classification of app-reviews. Empirical SE **28**(4), 88 (2023)
18. Haering, M., Stanik, C., Maalej, W.: Automatically matching bug reports with related app reviews. In: IEEE/ACM International Conference on Software Engineering, pp. 970–981 (2021)
19. Johanssen, J.O., Kleebaum, A., Bruegge, B., Paech, B.: Feature crumbs: adapting usage monitoring to continuous software engineering. In: Product-Focused Software Process Improvement, pp. 263–271. Springer (2018)
20. Khan, J.A., Liu, L., Wen, L., Ali, R.: Crowd intelligence in requirements engineering: current status and future directions. In: Knauss, E., Goedicke, M. (eds.) REFSQ 2019. LNCS, vol. 11412, pp. 245–261. Springer, Cham (2019). https://doi.org/10.1007/978-3-030-15538-4_18
21. Lauesen, S.: Software requirements: styles and techniques. Pearson Education (2002)
22. Li, N., Zheng, L., Wang, Y., Wang, B.: Feature-specific named entity recognition in software development social content. In: IEEE Conference on Smart Internet of Things, pp. 175–182 (2019)
23. Lim, S., Henriksson, A., Zdravkovic, J.: Data-driven requirements elicitation: a systematic literature review. SN Comput. Sci. **2**, 1–35 (2021)
24. Mann, J.E.C.: It education's failure to deliver successful information systems: Now is the time to address the it-user gap. J. Inf. Technol. Educ. Res. **1**, 253 (2002)
25. Novielli, N., Calefato, F., et al.: Can we use se-specific sentiment analysis tools in a cross-platform setting? In: Mining Software Repositories, pp. 158–168. ACM (2020)
26. Paech, B., Kohler, K.: Task-Driven Requirements in Object-Oriented Development, pp. 45–67. Springer (2004)
27. Pagano, D., Maalej, W.: User feedback in the appstore: an empirical study. In: IEEE Requirements Engineering Conference, pp. 125–134. IEEE (2013)
28. Santos, R., Groen, E.C., Villela, K.: A taxonomy for user feedback classifications. In: REFSQ, vol. 2376, CEUR-WS (2019)
29. Wang, C., Daneva, M., et al.: A systematic mapping study on crowdsourced requirements engineering using user feedback. J. Softw. Evol. Process **31**(10), e2199 (2019)
30. Zhang, T., Ruan, L.: The challenge of data-driven requirements elicitation techniques. master thesis, Blekinge Institute of Technology (2020)
31. Zhou, J., Ke, P., Qiu, X., Huang, M., Zhang, J.: Chatgpt: Potential, prospects, and limitations. Frontiers of IT & Electronic Engineering, pp. 1–6 (2023)

Channeling the Voice of the Crowd: Applying Structured Queries in User Feedback Collection

Leon Radeck[(⊠)] and Barbara Paech

Institute for Computer Science, Heidelberg University, 69120 Heidelberg, Germany
{radeck,paech}@informatik.uni-heidelberg.de

Abstract. **[Context/Motivation]** Crowd-based Requirements Engineering (CrowdRE) promises to overcome limitations of traditional requirements engineering by actively involving a larger number of users. A common theme in CrowdRE research is the development and deployment of online platforms where users can autonomously provide their feedback about software. **[Problems]** Current platforms only provide minimal guidance regarding the topics mentioned in the feedback. Therefore, it is difficult to map the feedback to existing requirements and identify gaps. **[Principal ideas]** To address the problems, we present a systematic process for collecting feedback through a variety of structured questions and an online platform which supports the process. We evaluate the process and the platform in the large-scale interdisciplinary research project SMART-AGE. **[Contribution]** To our best knowledge, this is the first platform that gathers feedback from a large number of users simultaneously through structured queries. Our contributions include a tailored list of questions for feedback collection, an analysis of how the question characteristics influence the answers and an evaluation of the user acceptance. Our findings indicate that both the quality and quantity of feedback are high, with variations based on the characteristics of the questions. Additionally, our platform is well accepted by its users.

Keywords: requirements engineering · crowd · user feedback collection

1 Introduction

User feedback is essential for the continuous development of software, because it contributes substantially to the elicitation of requirements. Traditional methods of collecting user feedback, such as interviews or workshops, are only feasible with a limited number of users as they are very time-consuming. Crowd-based requirements engineering (CrowdRE) is an umbrella term for automated or semi-automated approaches to collect and analyze feedback from a large number of users, also called "crowd", to derive validated user requirements [5]. In recent years two main approaches for collecting text-based feedback have been studied: on the one hand all users can provide feedback through an open channel like the AppStore [8]. On the other hand, users give feedback on a dedicated platform offered for a specific product. In this paper we focus on the latter. Wouters et al. [26] have presented a platform and the method CREUS which focuses on the generation of ideas. Based on three case studies they show that CREUS can elicit new ideas. They also conclude that steering the crowd is very important to get feedback. In their approach however, users are only given little guidance to which part of the product they should generate ideas. This means that the requirements engineers (ReEng) have to check in detail whether and which existing requirements are associated with the feedback. However, there is a semantic gap between how users express themselves and how requirements are formulated [6]. Therefore, we want to explore whether it is possible to collect feedback through guided questions about specific system functions and non-functional requirements which can provide the context for the mapping to requirements.

In this paper we present in detail the platform smartFEEDBACK (SF) and the process CREII (Crowd-based Elicitation with Integrating Implicit Feedback) which applies structured queries to collect feedback in terms of opinions, problems and improvement ideas in general and on specific product features. The paper builds on earlier publications [16, 17]. The platform and process are developed in the project SMART-AGE [17]. The SMART-AGE project is a comprehensive 5-year intervention trial conducted in Heidelberg and Mannheim, Southwestern Germany, aimed at enhancing the quality of life of older adults (OA) through app technology. It involves approximately 500 participants who use three specific tablet-based apps. The first app, smartVERNETZT (SV), is designed to encourage social networking and participation. The second, smartIMPULS (SI), offers health advice targeted at the key health and functional areas of OA. The aforementioned SF collects user feedback based on structured queries. Due to project constraints in SMART-AGE direct interaction between users is not allowed. Thus, in contrast to CREUS and other approaches, only the ReEng can read the feedback and react accordingly in the communication with the users. Typically, less communication is detrimental during requirements engineering (RE) and the motivation of the users to participate [14]. In our process this is counteracted by two means: In the first phase the ReEng asks structured questions to stimulate the feedback. In the second phase the ReEng derives requirements from the feedback and elicits feedback about these requirements. In this paper we present details about the first phase. We contribute a customized set of questions for gathering feedback, an assessment of the level of user engagement on our platform, an analysis of the impact of question characteristics on answers, a comparison of the quality of feedback with other platforms, and an evaluation of user acceptance of our platform.

The paper is structured as follows: Sect. 2 presents related work and Sect. 3 presents our research questions. Section 4 describes our approach, Sect. 5 the design of our evaluation study and Sect. 6 the results. In Sect. 7 we discuss the results and we conclude in Sect. 8.

2 Related Work

We have identified several platforms which collect user feedback. All have different features. We compare SF in the following regarding the most frequent features. Like the platforms CrowdCore [24], CRUISE [22], REfine [23], iThink [4], SKLSEForum [10], WikiWinWin [27] and WPFSRE [12] we collect feedback in the form of freetext. All other platforms (Tournify [13], GARUSO [7], KMar-Crowd [25], Requirements Bazaar [18], Athena [11], WikiReq [1] and Winbook [9]) collect feedback by using structured templates like user stories or use cases. The reason is that we want to minimize the barriers for OA in providing feedback. Since our users are OA, we also do not integrate gamification features like in [4, 7, 23]. Further design considerations relating to OA are discussed in [17]. We also do not use voting or other rating or prioritization mechanisms integrated in all platforms except [11]. The reason is that direct interaction between users is not allowed in SMART-AGE. All platforms give the users the possibility to comment on existing feedback from other users, but we allow comments only for one's own feedback. Since we want to ask questions at specific times, we could not base our app on forums [10] or wikis [1, 12, 27]. We did not find any platform which asks structured questions. Additionally, in terms of user engagement, none of the platforms measure the time users spend on them, and only a few assess how frequently the platform is used. Table 2 of [26] gives an overview of 5 evaluations studies and their results. We extend this in our evaluation (see Table 9).

3 Research Questions

We want to understand whether our approach is effective and whether it is accepted by the users. We therefore define the research questions (RQ) as described in Table 1.

Table 1. Research questions

RQ1 Is it feasible to collect feedback with our approach?
RQ1.1 How much feedback is given?
RQ1.2 How much time do users spend on our platform and how often do they use it?
RQ1.3 How do the answer options of a question influence its skip rate and the average text length included in its answers?
RQ1.4 How do the characteristics of a question influence its skip rate?
RQ2 Can we collect high quality feedback with our approach?
RQ2.1 Are the answers comprehensible?
RQ2.2 Does the feedback contain ideas?
RQ3 Do users accept our platform?
RQ3.1 How satisfied are the users with the platform overall?
RQ3.2 How satisfied are users with the system functions of the platform?

[RQ1] We study feasibility of feedback collection in terms of the quantity of the resulting feedback (RQ1.1) and the time and frequency of the platform usage (RQ1.2). Further indicators for the user engagement are the skip rate of questions and the length of the feedback provided. A low skip rate and a high average length of freetext motivation of the users. This is important, because freetext is essential for requirements derivation. In RQ

1.3. we analyse the effects of the answer options of a question on these indicators. With RQ1.4 we analyse the effects of a question's characteristics on its skip rate. For example, questions that are more abstract, such as those regarding non-functional requirements, may be seen as too complicated to answer, leading to a higher skip rate. Also, adaptive questions that probe into reasons for inactivity could be perceived as confrontational, resulting in avoidance and an increasing skip rate. We analyse the average text length contained in answers solely in relation to the answer options of a question, and not in relation to the question characteristics, as certain characteristics mix different answer options with and without free text.

[**RQ2**] For quality we look at comprehensibility of the feedback (RQ2.1) as well as the ideas contained in the feedback (RQ2.2). We use the number of ideas as a quality measure, because ideas are key for creating new requirements. This is also used by other approaches as discussed in related work. [**RQ3**] For user acceptance we measure how the users are satisfied with the platform overall (RQ3.1) and we analyze how users are satisfied with the system functions of the platform (RQ3.2).

4 Approach

In this section we describe our approach to collect feedback which includes the process and the platform that supports the process.

4.1 Process

Figure 1 gives an overview of the process CREII. While we focused in [16] on the usage data, we focus in this paper on the answers of the users to our questions. The subsequent process of feedback collection about requirements and requirements derivation and refinement will be presented and evaluated in future research work.

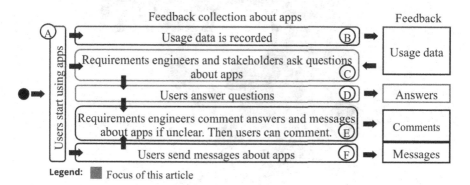

Fig. 1. Process CREII to collect feedback about apps and requirements and derive requirements

After the users begin to use the apps (A), they can provide feedback about the apps by sending messages to ReEng (F) or they can answer questions about the apps (D), that the ReEng and stakeholders (responsible persons for the apps SV and SI) have asked (C). Some of the questions are asked based on the usage behaviour of the users. This is why the usage data is recorded (B). The ReEng can also comment the answers to questions and messages about apps to ask for clarification (E). The users can comment back on messages that they sent (E). In general, users only see their own feedback and not that of other users. In the following we describe the questions that are used to ask for feedback and their rationale.

Question Characteristics. Figure 2 visualizes the different question characteristics and how many questions are asked per characteristic. We have 202 questions altogether. The figure shows that the approach is mainly based on scheduled questions, asked by the ReEng to derive requirements. The questions are grouped according to the following characteristics: *owner, purpose, type, aspect, category* and *app*. Each characteristic is mapped to a ring in Fig. 2 starting with *owner* as the innermost ring and *app* with the outermost ring. One can see that all questions are asked with respect to the three different apps (SV, SI, SF), but with different amounts. The *owner* of a question represents the person that is responsible for that question. Most of the questions have been defined by us as ReEng, but some questions also were defined by the stakeholders.

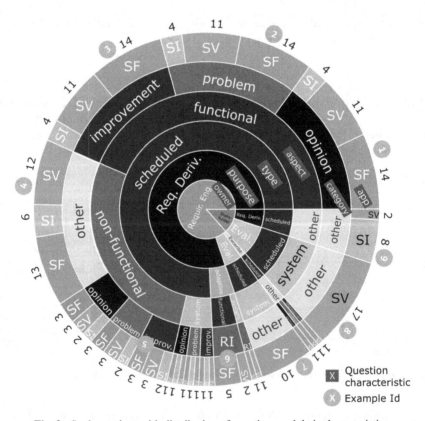

Fig. 2. Sunburst chart with distribution of questions and their characteristics

The *purpose* of a question can either be to derive requirements, to evaluate the app or it could be an exercise question (used during app introduction). For the Evaluation questions we use the System Usability Scale (SUS) questionnaire [2]. The *type* of a question can either be scheduled or adaptive. Scheduled means that the question is asked after a fixed number of days relative to the start of SF. Adaptive means that the question is asked depending on the usage behavior of the user. We detail in [16] the reasons for adaptive questions and how they are derived. The *category* represents whether the question asks about opinions, problems or improvements (OPI) or about reasons and improvements (RI). We ask for OPI in isolation to obtain more specific answers. The category RI is used for adaptive questions. There we ask why the users didn't use the app or a specific functionality of it for some time and we ask how the functionality can be improved so it is used more often. Questions of category *other* a) are asked by the stakeholders (as they are not interested in a combination of OPI), b) are asked for the purpose of evaluation taken from the questionnaire or c) address non-functional aspects often need to be formulated in a specific way. The characteristic *aspect* distinguishes whether the question refers to the system (that means one of the apps) as a whole or to functional aspects or non-functional aspects. Questions with a functional aspect refer to system functions and questions with a non-functional aspect refer to quality in use and product quality. Questions with aspect *other* cannot be uniquely mapped to either the system as whole or a specific functional or non-functional requirement.

Table 2. Example questions and answer options for example id from Fig. 2

Id	Example Questions	Answer options
1	How do you like the history function in smartFEEDBACK? Why?	Likert scale selection, Freetext
2	Are there any problems with displaying the history in smartFEEDBACK ? If yes, which ones?	Yes/No selection Freetext
3	Can the display of the history in smartFEEDBACK be improved? If yes, how?	Yes/No selection Freetext
4	Are you concerned about the security of your data in smartVERNETZT? Why?	Yes/No selection Freetext
5	Can smartVERNETZT be improved to make it particularly good for users over 67? If yes, how?	Yes/No selection Freetext
6	What is the reason that you have not looked at a question in SF in the last week? How could the app be improved so that you use it more often?	Freetext Freetext
7	I can well imagine using smartFEEDBACK regularly. Why is that?	Likert scale selection, Freetext
8	Using smartVERNETZT helps me to cope with everyday life.	Likert scale selection Freetext
9	Do you find the questions in the smartIMPULS app easy to understand and clear? What contributes to this?	Yes/No selection Freetext

Example Questions and Answer Options

Figure 2 contains green circles with numbers for specific combinations of question characteristics. For example, the green circle with the number 8 refers to example questions with owner "Stakeholder", purpose "Evaluation", type "scheduled", aspect "system", category "other" and app "SV". These are used in Table 2 where example questions and their answer options are listed. Not all questions feature a question mark. For instance, some questions taken from questionnaires are asked in a neutral manner, as illustrated by question 8. Often, we combine two sub-questions in one question, such as "How do you like the history function in SF? Why?". Here the first part of the question can be answered by selecting a value from a likert scale and the second part can be answered by freetext. For questions with category OPI we use a combination of selection and freetext. In category other, we mix questions with the answer option combinations: freetext – selection, selection – only, freetext – freetext and freetext only.

Asking Questions. When the users start with an app, we deliberately postpone posing questions to them. A grace period of seven days is given to become acquainted with the apps. This decision is guided by literature [3] which indicates that posing questions too early can be disruptive to users. Additionally, we limit the number of questions to a maximum of five per day. This is supported by the same literature [3] which suggests that an excessive number of questions can be overwhelming. Out of these five questions, three are specifically designed to gather feedback on OPI, while the other two questions are selected randomly to provide a variation. When we ask for OPI, we always ask for the opinion first, then for problems and then improvements. We strategically schedule the questions regarding functionality and user experience. We ask questions about less prominent system functions later in the process, allowing users ample time to explore and familiarize themselves with all aspects of the app. After three months of app usage we ask questions regarding the overall evaluation. Questions that are asked to the users do not expire. If the users don't answer the questions they receive on one day, they can answer these questions on the next day together with the new questions of that day. We provide a repository which includes the detailed plan when which questions are asked [15].

4.2 Platform

SF is a web application that is developed for the use by OA on a tablet. We describe in [17] how we tailor SF to the needs of OA. For example, we describe how we adjust the user interface (UI) and we describe our decision against using gamification. Screenshots of SF and its UI are shown in Fig. 3 and Fig. 4. SF allows the users to answer or skip questions (Fig. 3). The users are notified about new questions in form of red circles with white numbers in the navigation sidebar in the tab "Questions". The white numbers represent the number of new questions. Furthermore, we remind every user who has more than five unanswered questions once per week, that there are new questions. The users can see a history of all answers given (Fig. 4).

The SF platform is developed using Vue.js for the front end and Java's Spring Boot for the backend. A PostgreSQL database is used for data persistence. The backend employs a GraphQL API for communication, integrating WebSocket technology to enable real-time updates like displaying new questions to users without the need to reload the webpage. The deployment of SF is streamlined through containerization, with both the frontend

Fig. 3. "Question" page of SF   **Fig. 4.** "History" page of SF

and backend packaged as docker containers. These containers are then deployed within a multi-node Docker Swarm cluster to ensure high availability. For network management, the reverse proxy Traefik is utilized to manage HTTPS encryption.

5 Evaluation Study

In the following we describe the design and execution of our first evaluation study and the corresponding threats.

5.1 Data Collection and Analysis

We collect the data within the SMART-AGE study.

Recruiting. We recruit the OA in collaboration with local municipalities. Each week we send an invitation letter to a fixed number of OA, and upon successful answer, we conduct a telephone screening to check for their exclusion criteria. The major reason for this screening is that we exclude candidates who cannot benefit from our app, such as severely diseased candidates, individuals with probable cognitive impairment, candidates who are already physically very active or particularly socially involved, candidates with a very low technical affinity, candidates who do not have an internet connection at home, and candidates who live in a nursing home. After acceptance, we schedule a first appointment for a home visit.

Introduction of the Technology. At the first home visit, we equip the OA with a tablet containing the three apps. These apps are initially not accessible. The project process consists of several home visits where the different apps are introduced step by step to the OA.

Datasets. Table 3 gives details on our data sets. The coding dataset is a subset of the complete dataset and contains coded answers. For both datasets we only report about the OA which used the apps on their own after the first week, since during the first week the OA are supervised during their app usage. Due to the gradual recruiting the number of OA increase with time, leading to a higher number of OA in their early user weeks.

Table 3. Datasets

Measurement	Complete dataset	Coding dataset
Time range	24.05.23 – 24.10.23	24.05.23-10.09.2023
Number of questions asked	12,717	5,394
Number of answers	11,358	4,969
Number of answers including freetext	3,878	1,821
Number of OA after week 1	111	72

Coding. In the coding dataset we code the answers of the OA by classes. We base part of our classes on [20]. These classes include *Information Giving, Feature Shortcoming, Feature Strength, New Feature*, and *Problem*. Additionally, we include classes like *Cannot Answer Question* for situations where OA either misunderstand the question or cannot answer it because of a lack of the necessary experience, as well as *Reference to Other Answer* for instances where an OA references another answer. Examples for each class can be found in Table 4. If an answer contains multiple classes, we split the answer into statements, so that each statement corresponds to one class. In preparation for the systematic requirements derivation and refinement process which is not covered in this paper, the answers are also coded for *semantic comprehensibility*. If the coder cannot understand an answer, the answer is marked as incomprehensible. An example for an incomprehensible answer to the question "Are there any problems with adding private links in SV? If yes, which ones?" is "like the pocket". The coding was done by the first author of the paper and a master student who communicated on a regular basis.

Table 4. Examples for answers that correspond to the classes

Class	Example 1	Example 2
Information Giving,	I like it as simple as possible	I am still learning
Feature Shortcoming	It would be very good if the question were written in a larger font.	I don't find the black and white symbols very appealing.
Feature Strength	Private links enable to avoid chaos, this is good.	The time overview is practical!
New Feature	I would prefer the option of placing frequently used apps (icons) on the home screen to save clicks	A dictation function for text input is better because it is clear and correctable
Problem	The link is not up to date	My camera isn't working
Cannot answer question	Everything can always be improved but wouldn't know in which direction	I don't understand the question.
Reference to other question	See previous answer	I have already described this before

5.2 Threats to Validity

We identify potential threats across various dimension according to [3].

Construct Validity. The question order, ambiguous wording or leading questions are a threat to construct validity. The sequence in which questions are presented can impact how the OA interpret and respond to subsequent ones, possibly leading to biased feedback. Additionally, ambiguous wording in questions might result in varied interpretations, which could mean the answers don't accurately capture the user perspective. Moreover, the use of leading questions can bias answers towards a particular viewpoint. To alleviate this, we sequence questions about the system or about the same functional or non-functional aspect. We also use consistent wording throughout these questions and try to ask questions that are not leading.

Internal Validity. The positive relationship between project personnel and the OA, especially during home visits, might influence the answers, leading to more favourable feedback. Also, providing free tablets as incentives could influence motivation and feedback. To mitigate potential biases in the answers, we ensure, especially during home visits, conversations remain focused on study-related topics and consciously avoid delving into personal matters.

External Validity. The exclusion of OA who meet specific criteria may limit the generalizability of the findings. This selection bias restricts our ability to apply conclusions to a broader population. Furthermore, the lack of visibility of others' feedback and the absence of crowd interactions can limit external validity. The feedback in this isolated environment might not accurately reflect the feedback that users would give in a more interactive setting. Furthermore, we coded only part of the feedback. To mitigate concerns regarding external validity, we draw OA from two highly diverse cities (Heidelberg and Mannheim) and plan to share individual feedback among all OA during the requirements derivation process.

Reliability. There is also a reliability threat in the coding as the main inventor of the process and developer of the platform was one of the coders. Thus, the coding might be biased by a positive view on the feedback. Furthermore, apart from the regular communication between the coders, we did not conduct an additional interrater agreement.

6 Results

In this section, we present the results of our study.

RQ1: Is it feasible to collect feedback with our approach?
RQ1.1. How much feedback is given?
12,717 questions were asked. 1,359 questions were skipped and 11,358 answers were given. 3,878 answers included freetext (see Table 3).

RQ1.2. How much time do users spend on our platform and how often do they use it?
We distinguish three kinds of SF users: *Inactive* users who did not start the app. *Accessing* users who started the app, but did not give feedback. *Contributing* users who gave an answer. *Active* users are the sum of accessing and contributing users. We include accessing users in the active users, because these users might also be interacting in the

app apart from answering questions or sending messages, e.g. when there are no open questions for them to answer.

Considering the 111 users over the whole time period, there were 6 users which never were active and 68 users which were not active in at least one week. 43 users were accessing at least once, but none of these was only accessing the whole time. 97 users were contributing at least once, with 35 contributing every week.

Figure 5 presents a detailed stacked bar graph depicting usage of SF in *user weeks* 2 to 12. User week n is week n after SF was introduced to the user. The number is decreasing, because each week some new users are added. So there are much more users in their week 2 than in week 12. Figure 6 normalizes the distribution according to user numbers. This shows that roughly 30% of the users in a user week are inactive. They do not answer the questions in this week, and therefore the questions are postponed to the following week.

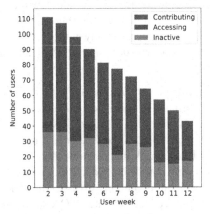

Fig. 5. Absolute Distribution of users (n = 111)

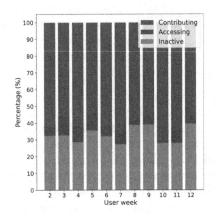

Fig. 6. Relative distribution of users (n = 111)

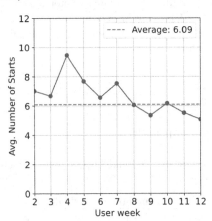

Fig. 7. Average number of starts for active users with minimum 12 week usage (n = 41)

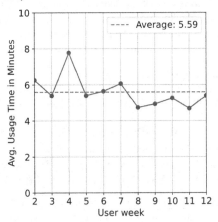

Fig. 8. Average usage time for active users with minimum 12 week usage (n = 41)

To understand the usage patterns over the user weeks in more detail we looked at the 41 users who were active for 12 weeks. Figure 7 displays the average number of starts of the users and Fig. 8 the average usage time in minutes. The red dashed lines represent the overall average for each metric. The number of starts is declining during the weeks from roughly 7 to 5. There is a little less decrease in usage time. The users take around 6 min to give feedback in each week. Both graphs exhibit a peak in week 4. We also analyzed for this user group whether the usage of SF depends on the usage of the apps. So we related the starts of SV with the answers in SF regarding SV. The calculated pearson correlation coefficient of approximately 0.385 shows that there is tendency that increased starts of SV leads to more answers. This is statistically significant, as the p-value 0.012 is less than the common alpha level of 0.05. We also analyzed the relation to user characteristics age, gender and mobile device proficiency (measured with the MDPQ-Score [19]), but could not find any statistically significant relationships.

RQ1.3. How do the answer options of a question influence its skip rate and the average text length included in its answers?

The skip rate for a question is calculated by dividing the number of times the question is skipped by the total number of times the question is asked. The average length of freetext contained in an answer is calculated by dividing the sum of characters of all answers for that question by the total number of times the question is asked. Table 5 presents the relation of the answer options with the skip rate and the average length of freetext contained in the answers. Questions which require only a selection are skipped much less (4.6%) than questions asking for freetext (11.4%). On average an answer to a "freetext only" question is 25 characters. This length is doubled, if a reason for a selection is asked (selection freetext). Questions with two freetext answers result in 5 times longer answers.

Table 5. Differences between answer options regarding number and length of answers and their skip rate

Answer option 1	Answer option 2	#Questions asked	#Skips	Skip rate	Avg. length of freetext
Selection	Freetext	10,888	1,246	11.4%	57.9
Selection	-	1,318	61	4.6%	-
Freetext	Freetext	398	42	10.6%	138.2
Freetext	-	113	10	8.8%	25.3
	Total:	12,717	1,359	10.7%	73.8

RQ1.4. How do the characteristics of a question influence its skip rate?

Table 6 shows the skip rate and the number of questions asked per question characteristic.

Table 6. Skip rate per question characteristic

owner		purpose		type		subject				category				
Requirement engineer	Stakeholder	RD	Evaluation	Scheduled	Adaptive	System	Functional	Non-Functional	Other	Opinion	Problem	Improvement	RI	Other
Skip rate: 10.2	8.4	10.3	7.4	10.1	9.4	7.8	12.7	9.1	11.4	9.0	10.0	14.2	9.1	8.6
#Questions 11,222	1,495	11,135	1,582	12,426	291	2,260	7,369	2,911	177	3,102	2,921	2,919	291	3,484

The volume of questions exhibits a large variation from over twelve thousand ("Scheduled") to fewer than two hundred ("Other"). This must be kept in mind when interpreting the skip rates, because the skip rate is less meaningful when the number of questions asked is low. The biggest difference in skip rate overall is between questions with purpose "Evaluation" (7.4%) and questions with category improvement (14.2%).

To examine how various combinations of question characteristics affect the skip rate, Table 7 presents the skip rates associated with each specific combination of these characteristics. The biggest overall difference in skip rate is between evaluation questions from the stakeholders (4.6%) and improvements questions for specific functions (17,0%).

Table 7. Skip rate for each question characteristic combination

owner	purpose	type	subject	category	Skip rate	#Questions asked
Requirements engineer	Requirements derivation (RD)	Scheduled	System	opinion	7.2	209
				problem	8.7	207
				improv.	14.0	207
			Functional	opinion	10.7	2,463
				problem	10.7	2,336
				improv.	17.0	2,334
			Non-functional	opinion	9.0	430
				problem	9.0	378
				improv.	10.6	378
				other	8.7	1,725
		Adaptive	System	RI	7.0	55
			Functional	RI	10.7	236
	Eval.	scheduled	System	other	8.4	264
Stakeholder	RD	scheduled	Other	Other	10.6	177
	Eval.	scheduled	System	Other	4.6	1,318

Total: 12,717

Answering RQ1: We conclude for RQ1 that it is feasible to collect feedback with structured questions, and that we see differences with respect to question characteristics and answer options.

RQ2: Can we collect high quality feedback with our approach?
RQ2.1. Are the answers comprehensible?
The feedback is very comprehensible, since out of the 1,821 coded freetext answers, 1,801 are comprehensible (98.9%).

RQ2.2 Does the feedback contain ideas?
Table 8 shows the classes of the 1,884 statements that could be extracted out of the 1,821 answers of the coded feedback subset.

Table 8. Classes of statements of coded answers with freetext

Class	#Statements
Information giving	700 (37.2%)
Feature strength	433 (23.0%)
Cannot answer question	305 (16.2%)
Feature shortcoming	254 (13.5%)
Reference to other answer	95 (5.0%)
New Feature	57 (3.0%)
Problem	40 (2.1%)
	Total: 1,884

The largest category is *Information Giving* (37.2%). *Feature Strengths* are provided almost double (23%) than *Feature Shortcomings* (13.5%). *New Features* are contained in 3% of the answers. To understand whether this amount of new features is high, we compare our platform in Table 9 with other platforms.

Table 9. Comparison of SF to other platforms

Platform	SF	Tournify	Kmar-Crowd	REfine	GARUSO
Product	SMART-AGE	Tournify	S-Sys	V-Sys	Qubus 7	Smart living
Active users	62	157	135	385	19	726
Contributing users	55	39	60	130	19	32
Contributing users/total users	**76%**	25%	44%	34%	100%	4%
Ideas	57	57	32	78	21	56
Ideas/active user	**0.92**	0.36	0.24	0.20	1.11	0.08
Ideas/contributing user	1.03	1.46	0.53	0.60	1.11	1.75

Table 9 is an extended version from [26] which compares Tournify [13], Kmar-Crowd [25], REfine [23] and GARUSO [7]. This table shows for SF our coded sample of 62 active users and 55 contributing users (see Table 3). We count new features as ideas. Values where SF outperforms the other platforms are bold. We list REfine, but do not compare our numbers with it as this was a very small study where all users had to contribute. Compared to the other platforms we have much more contributing users

(71% vs 44%) and therefore also a higher idea rate per active users (0.92 vs. 0.36). Tournify (1.46) and GARUSO (1.75) have a higher idea rate per contributing user than SF (1,03).

Answering RQ2: We conclude for RQ2 that our feedback has quite high quality.

RQ3: Do users accept our platform?
We have two sources for user satisfaction with our platform.

RQ3.1: How satisfied are the users with the platform overall?
We applied the *System Usability Scale (SUS) Score* [2][1] in that we asked the corresponding questions through SF. There are 22 users who answered all the SUS questions. Based on the SUS questions we calculated a system usability score as described in literature. Our score of 76.81 is well above the usual threshold of 68 [21] indicating above-average-usability.

RQ3.2: How satisfied are users with the system functions of the platform?
We asked the users directly about the different system functions. Figure 9 shows the results. The likert-scale answers are visually represented using a color-coded system. For opinions, dark green is mapped to the highest level of satisfaction, followed by green, gray, orange, and red for the lowest level. Black means the question was skipped. For problems and improvements, "Yes" is mapped to red and "No" is mapped to dark green. The number of answers differ per question, because we ask some questions in early user weeks and some in late user weeks. Combined with the fact that we have more users in the first user weeks, there are more answers for questions that are asked early. Overall the opinions are positive to neutral and the users see few problems and some improvements.

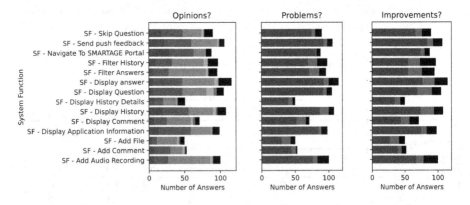

Fig. 9. Answers to opinion, problem and improvement questions per system function

Answering RQ3: We conclude that the users accept the platform.

---

[1] The questions of the SUS questionnaire (*purpose* = Evaluation, *app* = smartFEEDBACK) can can be found in the repository.

7 Discussion

In the following we discuss some of the results in detail.

[RQ1.1] We see that less than 10% of our questions get skipped, this is a sign that the users really care about answering the questions.

[RQ1.2] We can observe a peak in number of starts and usage time in usage week 4. This could be due to a third home visit around 4 weeks after the introduction of SF so that the users might start using the apps again for preparation. After the peak we see a slight but consistent decline in number of starts and usage time. This could be due the lack of interaction in SF. We also observe that users with more starts in SV tend to answer more questions in SF related to it. However, as our current dataset mainly consists of newer users with fewer user weeks, this pattern could change as we include more users with a longer history of using the SV and SF.

[RQ1.3] Questions which only ask for a selection are skipped much less than questions including freetext. This is understandable as this is less effort for the users. While selection-only questions are well-suited for quantitatively gathering user opinions, they do not provide enough detailed information to use them as a basis for creating or modifying requirements.

[RQ1.4] It is interesting that improvement questions are skipped most often. This is understandable as it is more difficult to come up with new ideas than to report problems. It might also be due to the fact we observe in coding that improvement ideas are also reported in the answers to opinions and problems questions which are asked before the improvement questions. The skip rates of scheduled and adaptive questions are very similar indicating that the questions based on monitored user behaviour are not more effective in triggering feedback. However, the volume is very different.

[RQ2.1] The comprehensibility of our feedback is very high, this reenforces the point that the users put a lot of effort in their feedback.

[RQ2.2] We see a higher percentage of contributing users and therefore a higher idea rate per active user compared to other platforms. This is very likely due to our study setting where users agree to participate. Platforms like Tournify and GARUSO have higher idea rates per contributing user. However, we only categorized feedback as an idea when it suggests new features ("New feature"). Feedback that we code with "Feature shortcoming" often contains not only a critique to a feature, but also a proposal for modification. We will explore in the future the value of this feedback. Also, the processes of the other platforms are focused on generating ideas only. We think that collecting details regarding OPI can inspire the ReEng for new ideas. We assume that we are able to collect even more user ideas in the second phase of the process CREII, when we ask questions regarding existing feedback and requirements.

[RQ3.1] Our SUS score is well above the usual threshold, potentially owing to SF's intuitive layout and straightforward navigation, simplifying user interactions.

[RQ3.2] The users have mainly positive and neutral opinions about the system functions of SF. The high number of neutral opinions could be explained by the fact that providing feedback on SF offers little benefit to the users themselves.

Overall, our results show that useful feedback can be gathered by structured questions. We cannot make any judgement so far whether our approach outperforms approaches with less structured questions.

8 Conclusion and Future Work

We presented our process CREII and the platform SF. The results show that our platform is accepted and it is possible to collect high quality feedback with structured queries. Our next step is to derive requirements. We will do this based on a larger coded dataset. Reflecting the discussion on RQ2.2, we will separate "Feature shortcoming" into "Problems" and "ChangeRequests" and include the "New Features" in the latter. If necessary, we will the ask the user base to clarify the "ChangeRequests". Based on the "ChangeRequests", we will derive requirements and ask the user base for feedback on the requirements.

References

1. Abeti, L., Ciancarini, P., Moretti, R.: Wiki-based requirements management for business process reengineering. ICSE, 14–24 (2009)
2. Brooke, J.: SUS: a "quick and dirty" usability scale. usability evaluation in industry, 207–212 (2020)
3. Easterbrook, S., Singer, J., Storey, M.-A., Damian, D.: Selecting empirical methods for software engineering research. In: Shull, F., Singer, J., Sjøberg, D.I.K. (eds.) Guide to Advanced Empirical Software Engineering. pp. 285–311. Springer, London (2008). https://doi.org/10.1007/978-1-84800-044-5_11
4. Fernandes, J., Duarte, D., Ribeiro, C., Farinha, C., Pereira, J.M., Da Silva, M.M.: IThink: a game-based approach towards improving collaboration and participation in requirement elicitation. Procedia Comput. Sci. 15, 66–77 (2012)
5. Groen, E.C., et al.: The Crowd In Requirements Engineering: The Landscape And Challenges. IEEE Softw. 34, 44–52 (2017). https://doi.org/10.1109/MS.2017.33
6. Kifetew, F.M., Perini, A., Susi, A., Siena, A., Muñante, D., Morales-Ramirez, I.: Automating user-feedback driven requirements prioritization. Inf. Softw. Technol. 138, 106635 (2021). https://doi.org/10.1016/j.infsof.2021.106635
7. Kolpondinos, M.Z., Glinz, M.: GARUSO: a gamification approach for involving stakeholders outside organizational reach in requirements engineering. Requirements Eng. 25, 185–212 (2020)
8. Kujala, S.: Effective user involvement in product development by improving the analysis of user needs. Behav. Inf. Technol. 27, 457–473 (2008). https://doi.org/10.1080/01449290601111051
9. Kukreja, N., Boehm, B.: Integrating collaborative requirements negotiation and prioritization processes: a match made in heaven. In: ACM International Conference Proceeding Series, pp. 141–145 (2013)
10. Lai, H., Peng, R., Sun, D., Liu, J.: A lightweight forum-based distributed requirement elicitation process for open source community. Int. J. Adv. Comput. Technol. 4, 138–145 (2012)
11. Laporti, V., Borges, M.R.S., Braganholo, V.: Athena: a collaborative approach to requirements elicitation. Comput. Ind. 60, 367–380 (2009)

12. Lohmann, S., Dietzold, S., Heim, P., Heino, N.: A web platform for social requirements engineering. Lect. Notes Inform. (LNI), **P-150**, 309–315 (2009)
13. Menkveld, A., Brinkkemper, S., Dalpiaz, F.: User story writing in crowd requirements engineering: the case of a web application for sports tournament planning, RE **19**, 174–179 (2019)
14. Nuseibeh, B., Easterbrook, S.: Requirements engineering: a roadmap. ICSE **2000**(1), 35–46 (2000)
15. Radeck, L.: Appendix (Repository). https://github.com/lradeck/appendix
16. Radeck, L., Paech, B.: Integrating implicit feedback into crowd requirements engineering – a research preview. In: REFSQ. ACM (2023)
17. Radeck, L., et al.: Understanding IT-related well-being, aging and health needs of older adults with crowd-requirements engineering. In: REWBAH, pp. 57–64. IEEE (2022)
18. Renzel, D., Behrendt, M., Klamma, R., Jarke, M.: Requirements bazaar: social requirements engineering for community-driven innovation. RE **5**, 326–327 (2013)
19. Roque, N.A., Boot, W.R.: A new tool for assessing mobile device proficiency in older adults: the mobile device proficiency questionnaire. J. Appl. Gerontol. **37**, 131–156 (2018). https://doi.org/10.1177/0733464816642582
20. Santos, R., Groen, E.C., Villela, K.: A taxonomy for user feedback classifications. In: CEUR Workshop Proceedings, vol. 2376 (2019)
21. Sauro, J., Lewis, J.r.: quantifying the user experience (2012)
22. Sharma, R., Sureka, A.: CRUISE: a platform for crowdsourcing requirements elicitation and evolution. In: ICCC, pp. 1–7 (2018)
23. Snijders, R., Dalpiaz, F., Brinkkemper, S., Hosseini, M., Ali, R., Özüm, A.: REfine: a gamified platform for participatory requirements engineering. CrowdRE, 1–6 (2015)
24. Vogel, P., Grotherr, C., Semmann, M.: Leveraging the internal crowd for continuous requirements engineering - lessons learned from a design science research project. In: ECIS 2019 (2020)
25. Wouters, J., Janssen, R., Van Hulst, B., Van Veenhuizen, J.: CrowdRE in a governmental setting: lessons from two case studies. RE (2021)
26. Wouters, J., Menkveld, A., Brinkkemper, S., Dalpiaz, F.: Crowd-based requirements elicitation via pull feedback: method and case studies. RE (2022)
27. Yang, D., Wu, D., Koolmanojwong, S., Brown, A.W., Boehm, B.W.: WikiWinWin: a Wiki based system for collaborative requirements negotiation. In: ICSS, pp. 1–10 (2008)

Emerging Topics and Challenges
in Requirements Engineering

Requirements Information in Backlog Items: Content Analysis

Ashley T. van Can[(✉)][iD] and Fabiano Dalpiaz[iD]

Utrecht University, Heidelberglaan 8, 3584 Utrecht, CS, The Netherlands
{a.t.vancan,f.dalpiaz}@uu.nl

Abstract. [**Context and motivation**] With the advent of agile development, requirements are increasingly stored and managed within issue tracking systems (ITSs). These systems provide a single point of access to the product and sprint backlogs, bugs, ideas, and also tasks for the development team to complete. [**Question/problem**] ITSs combine two perspectives: representing requirements knowledge and allocating work items to team members. We tackle a *knowledge problem*, addressing questions such as: How are requirements formulated in ITSs? Which types of requirements are represented? At which granularity level? We also explore whether a distinction exists between open source projects and proprietary ones. [**Principal ideas/results**] Through quantitative content analysis, we analyze 1,636 product backlog items sampled from fourteen projects. Among the main findings, we learned that the labeling of backlog items is largely inconsistent, and that user-oriented functional requirements are the prevalent category. We also find that a single backlog item can contain multiple requirements with different levels of granularity. [**Contribution**] We reveal knowledge and patterns about requirements documentation in ITSs. These outcomes can be used to gain a better empirical understanding of Agile RE, and as a basis for the development of automated tools that identify and analyze requirements in product and sprint backlogs.

Keywords: Agile Requirements Engineering · User Stories · Backlog Items · Issue Tracking Systems · Content Analysis

1 Introduction

Agile software development emerged as a means for many software companies to stay competitive by improving market responsiveness, gaining the ability to continuously and quickly define and re-prioritize software requirements based on the ever-changing stakeholder needs [11].

In agile software development, requirements are defined in an iterative and incremental fashion. Documenting and discussing requirements in agile practices is commonly carried out by writing and discussing user stories [4], which are key elements of the product and sprint backlogs [9].

While researchers have explored and reviewed the benefits and challenges of agile RE [11] and the nature of the product backlog [23], there is limited research

© The Author(s), under exclusive license to Springer Nature Switzerland AG 2024
D. Mendez and A. Moreira (Eds.): REFSQ 2024, LNCS 14588, pp. 305–321, 2024.
https://doi.org/10.1007/978-3-031-57327-9_19

that analyzes how requirements are documented in real-world product and sprint backlogs, which are often stored in an issue tracking system like JIRA.

In this paper, we conduct a preliminary analysis of 1,636 issues from fourteen JIRA repositories (seven open source, seven proprietary), aiming to address our main research question **MRQ**: *How are requirements documented within backlog items?*. Since we focus on requirements that are stored in product and sprint backlogs, we examine only those JIRA repositories that we know are used or that are likely to serve as a representation of these backlogs[1]. Methodologically, we employ content analysis [14, 27] as the lens through which we analyze a sample of the issues in these repositories, leading to a bottom-up understanding that is rooted in empirical data.

Studies that adopt a similar research method to analyze data from product or sprint backlogs include the investigation of architectural knowledge in JIRA issues [25], the analysis of emotions in backlog items [20], and linking JIRA issues to software life-cycle activities [18].

Given the informal nature of issue tracking systems as documentation tools, in this paper we use the term *requirement* to denote a variety of textual fragments, without limiting ourselves to the use of specific templates such as the 'shall' format, user stories, or the like. This is in line with the CPRE Glossary by IREB [7], which states that a requirement is "A documented representation of a need, capability or property".

This paper makes two contributions to the state-of-the-art:

- Through content analysis of a large sample of backlog items from fourteen projects, we present first insights on how requirements are documented in agile RE via issue tracking systems;
- As a byproduct of the analysis, we share a coding scheme that can be used for the analysis of additional datasets.

The rest of the paper is structured as follows. Section 2 refines the MRQ into five research questions. Section 3 presents our research method, including the sampled dataset and the coding scheme. Section 4 shows our results. Section 5 contrasts our work with related work. Section 6 discusses the results in terms of the research questions. Finally, Sect. 7 draws conclusions and sketches future directions.

2   Research Questions

In order to address the main research question stated in the introduction, we put forward five more specific research questions: RQ1 through RQ5.

RQ1. To what extent do the backlog item labels chosen by practitioners reflect the requirements expressed in the items?

---

[1] We do not make a distinction as to whether the items belong to the product backlog or to the sprint backlog; in the remainder of this paper, we therefore use the term 'backlog items' to refer to the components of either backlog.

RQ1 focuses on whether the practitioners use suitable issue labels (e.g., 'Story' or 'Epic' in JIRA) to distinguish requirements from issues that represent other aspects, such as bugs or tasks. In particular, we define one hypothesis that emerges when separating issues with labels about requirements from issues with labels that concern the execution of tasks:

H1. Backlog items with requirement-related labels contain requirements more often than backlog items with task-related labels.

In addition to analyzing backlog items based on the appropriateness of the labels, we dig deeper into the understanding of the requirements categories that are represented. A full explanation of the categories we employ is presented later in the paper (see Table 1); for now, an example is the classic distinction between functional and non-functional requirements [3]. This leads to RQ2:

RQ2. What categories of requirements information are more commonly used?

One of the key properties of requirements, when stored in a requirements management system, is for them to be uniquely identifiable [10]. We want to study whether this is the case also with issue tracking systems, or if multiple requirements co-occur in the same backlog item (in issue tracking systems, backlog items are the smallest identifiable piece of information). This entails RQ3:

RQ3. How often does a single backlog item include multiple requirements?

We can further refine RQ3 based on the question regarding categories (RQ2), leading to studying whether – when multiple requirements appear in the same backlog item – certain combinations of different categories are prevalent:

RQ3.1. What different requirements categories do co-occur more often in a backlog item?

Given the importance of justifying the *why* in requirements engineering [28], we put forward RQ4 that explores whether the requirements in backlog items also include information concerning why they are needed.

RQ4. To what extent are requirements complemented by a motivation for their existence?

Finally, we aim to conduct a preliminary analysis on whether open source software (OSS) projects, which are easier to retrieve and access, are representative of how requirements are represented in proprietary projects, which are subject to confidentiality rules. To this extent, we introduce RQ5:

RQ5. Can we identify differences w.r.t. RQ1–RQ4 when comparing proprietary projects and open-source projects?

3   Research Method

We apply content analysis to gain insight into the requirements information present in issue tracking systems. Content analysis is a "research technique for making replicable and valid inferences from texts (or other meaningful matter) to the contexts of their use" [14]. In our case, in line with the distinction by White and Marsh [27], we make use of quantitative content analysis: after defining our research questions and hypotheses when suitable, we collect data, determine the collection unit (issues organized into projects), define the coding scheme, tag the data, analyze, and write up the results.

One important deviation from the classic approach is that, while we started from literature knowledge to define basic codes (we performed *inductive coding*), we have used a subset of the data to refine the coding scheme by introducing specific sub-categories (*deductive coding*). Our process is visualized in Fig. 1.

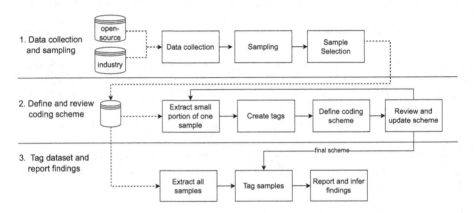

Fig. 1. The research method describing how we 1) collected and sampled the data, 2) defined and reviewed the coding scheme and 3) tagged the data.

3.1   Data Collection and Sampling

The topmost row of Fig. 1 illustrates how we collected data and we sampled the backlog items to analyze.

Data Collection. We analyze data extracted from issue tracking systems of several projects, aiming to identify backlog items. We also collected seven internal development projects from a collaborating company: the low-code development platform provider Mendix. The other projects were sampled from two large public datasets of issue tracking systems for OSS projects [18, 26].

From a quick scan, we could determine that many of the issue tracking systems for the open-source projects are employed mainly for reporting bugs rather than for documenting requirements. We automatically excluded projects with ≥80% issues labeled as 'bug' and <10% labeled as 'story'. This screening allowed us to filter out the projects whose dataset is likely not to serve as sprint or product backlog, retaining 88 projects.

In addition, since user stories (arguably the main requirements artifact in Agile RE [4,13,16]) are user-oriented requirements, we decided to focus on projects with a clear user interaction component, eliminating projects without a clear user interface. The first author assessed the 88 retained projects based on their UI component and excluded irrelevant projects.

This led to 16 candidate open-source projects. Since we knew that the JIRA issues from all seven industry projects were used to represent the sprint backlog, we did not exclude any industry projects.

Sampling. Since the projects varied in size (from 57 to 5,750 issues), we selected a sample from each project. We defined an initial sample size of 100 items, as we found that a sample selection based on a time period would result in a large variation in sample sizes (as each project varies greatly in how often items are uploaded). We aimed to select a sample that represents well the backlog items that the development teams worked upon in a given time period. Therefore, we randomly selected one issue and included the 99 subsequent issues (not counting bugs) in order of creation time, excluding the last 100 issues from the random selection phase. In other words, considering the dataset as a set of sliding windows of size 100 (and sliding interval of 1), we randomly selected one sliding window per project.

Links between issues provide additional context regarding how various backlog items relate and depend on one another [17]. Thus, we decided to include this information when extracting samples of the projects. We considered each issue as part of a *cluster* of linked issues. After selecting a sample S of 100 issues, we added all issues that were directly and transitively linked to a particular issue in S, excluding those created later than the last issue in S.

Sample Selection. Before the different OSS projects[2] could be utilized further, the two authors of this paper independently evaluated the first 12 issues of each sample in order to verify if the initial selection contained a sufficiently high volume of relevant requirements information and were not merely exploited for bug reports and task lists. Each of the 12 issues per project was classified as to whether it contained information related to requirements. After an independent tagging, the authors reviewed the classification together and subsequently discussed any differences. Finally, the projects with over 50% (of the 12 issues) marked as requirements relevant were shortlisted (details in our online appendix[3]). This resulted in 7 open-source projects and 7 private projects.

---

[2] Thanks to our collaboration with Mendix, we knew those projects were using the issue tracking system to represent the backlog items the teams would implement in the various sprints.

[3] Online appendix: https://doi.org/10.5281/zenodo.10643450.

3.2   Coding Scheme Construction

The coding scheme was constructed through the analysis of two of the 14 project samples, one OSS (QT Design Studio) and one proprietary (Portfolio). We defined the coding scheme iteratively and performed the tagging using the software Nvivo. In each iteration, one tagger examined a small additional section of data to identify a variety of information in the backlog items, focusing specifically on content and writing patterns. After creating the codes, the first tagger grouped related codes or adjusted codes to construct a coding scheme. When the first tagger was unsettled about certain scenarios, the first and second taggers discussed the situation, adjusting the scheme accordingly. The rest of the process was repeated each time adding new data to adjust the scheme until the scheme no longer required adjustment.

After the two projects were all tagged based on the scheme, the second tagger checked the scheme and the tagged dataset, resulting in some minor final adjustments, leading to the coding scheme that is available in our online appendix and that is summarized in Sect. 3.3.

The first tagger applied the scheme to 50% of the remaining projects. The second tagger independently tagged a random 20% of each project. We subsequently compared the tagged items and discussed any conflicts. We consider tagging difference as a conflict when a specific text is tagged with a different granularity level or type. This resulted in a percent agreement rate of 65%, after which we made a few minor adjustments to 6 projects.

Next, the first tagger completed the remainder of the projects, of which the second tagger independently tagged 20%. After comparing the tags, we agreed on 71% of the items, which is an improvement over the first 6 projects. Based on these minor mismatches, we adjusted the dataset.

3.3   Coding Scheme

The coding scheme (Table 1) distinguishes two characteristics on which a requirement can be classified: a) the requirement *type* and b) the *granularity* level.

The requirement type indicates whether it is a functional or non-functional requirement. For functional requirements, we have defined two possible subcategories: a) user-oriented (indicating that the user directly experiences the added functionality, and b) system-oriented (the added functionality is not directly experienced by the user but is necessary for the system to function as desired). The granularity level denotes the level of refinement of the requirement, where we distinguish between low-level (e.g., acceptance criteria), medium-level (e.g., user stories), and high-level requirements (e.g., epics). Recognizing the importance of the reason for the requirement, we also tag whether there exists a *motivation* for the requirements in that backlog item. The complete tagging guidelines are available in our online appendix.

Table 1. Overview of our coding scheme.

Characteristic	Category	Description
Requirement type	User-oriented functional	Functionality directly experienced by the user
	System-oriented functional	Functionality that the system will implement but that is not directly experienced by the user
	Non-functional	Requirement that constrains or sets some quality attributes upon functional requirements [5]
Granularity level	Low	Requirement that is directly verifiable (e.g., acceptance criterion)
	Medium	Requirement that refers to one specific functional or non-functional aspect of the system (e.g., user story)
	High	Requirement that encompasses multiple aspects or functionalities of the systems (e.g., epic or theme)

Table 2. The projects selected in this study, showing the total size, sample size, number of items with requirements labels (*E*: Epic, *F*: Feature, *US*: User Story, *SU*: Suggestion), number of task-labelled issues (*T*: Task, *ST*: Sub-task, *TT*: Technical task, *ST*: Support ticket), and other issues.

Project	Size		Req-labeled				Task-labeled				Other
	Total	Sample	E	F	US	SU	T	ST	TT	ST	
Control	738	120	15	0	90	0	0	14	0	1	0
Service	173	100	6	0	57	0	37	0	0	0	0
Store	634	109	12	0	69	0	0	0	0	21	7
Company	29	29	0	0	29	0	0	0	0	0	0
Portfolio	97	97	4	0	84	0	0	9	0	0	0
Data	57	57	8	0	27	0	20	2	0	0	0
Learn	994	143	15	0	116	0	0	5	0	7	0
Cost Management	2,038	179	15	8	99	0	28	29	0	0	0
Jira Performance Testing Tools	777	105	2	0	26	57	12	8	0	0	0
Lyrasis Dura Cloud	1,125	113	0	0	105	0	7	0	0	0	1
Network Observability	137	102	2	0	99	0	1	0	0	0	0
OpenShift UX Product Design	369	130	3	0	113	0	3	11	0	0	0
Qt Design Studio	4,983	180	5	0	51	6	11	21	86	0	0
Red Hat Developer Website	5,750	172	21	0	151	0	0	0	0	0	0
Total	17,901	1,636	108	8	1,116	63	119	99	86	29	8

3.4 Selected Projects

Table 2 shows the different projects included in this study, indicating the number of items in the original dataset, the sample size and the labels used. Each backlog item consists of a label, which is specified by one of the team members and should reflect the content of the item. In addition to the label, we examine the description, summary (i.e., title) and cluster to which each item belongs.

4 Results

We present the findings of our content analysis for RQ1–RQ5. We address RQ1–RQ4 in four sections, each of which ends with a reflection on RQ5: we split the projects population between proprietary and OSS projects in order to assess if differences exist. Given the small sample size for RQ5 (7 projects per group), we do not run statistical significance tests as their reliability would be low. Due to space limitations, the raw data are available in our online appendix.

4.1 Do Practitioners Choose Accurate Backlog Item Types (RQ1)?

RQ1 examines whether practitioners' labeling accurately reflects the content of backlog items and to identify any inconsistencies. We expect backlog items of types 'Epic', 'Feature', 'Story' and 'Suggestion' to contain requirements information, while items with types 'Task', 'Technical task', 'Sub-task' and 'Support-tickets' not to. We exclude 8 issues with a rarely occurring type that are hard to relate to requirements or tasks. Our hypothesis H1 is that the the first kind of backlog items will more frequently contain requirements-related tags.

Table 3 shows, for each project, the ratio of items that (i) are labeled as tasks and include at least one requirement, and that (ii) are labeled as requirements and include at least one requirement. The columns on the left focus on proprietary projects, the ones on the right on OSS projects.

Table 3. Ratio of items task-labeled and requirements-labeled items that include at least one requirement according to our tagging.

Project (proprietary)	Task	Req	Project (OSS)	Task	Req
Control	0.2	0.62	Cost Management	0.28	0.58
Service	0.14	0.92	Jira Performance Testing Tools	0.15	0.64
Store	0	0.49	Lyrasis Dura Cloud	0.29	0.75
Company	0	0.66	Network Observability	0	0.42
Portfolio	0.22	0.51	OpenShift UX Product Design	0.79	0.58
Data	0.73	1	Qt Design Studio	0.30	0.52
Learn	0.25	0.45	Red Hat Developer Website	0	0.44
Macro-average	0.22	0.66		0.26	0.56
Std- dev	0.25	0.22		0.27	0.11
Macro-average (all)	0.24	0.61			
Std-dev (all)	0.25	0.17			

To confirm H1 statistically, given the limited sample size of n=14, we choose a robust non-parametric test: Mann-Whitney's U, verifying whether the ratio of requirements in the requirements-labeled items is greater than that in the task-labeled items (H1), or alternatively if they can be considered equal (H0). The Mann-Whitney U test results in a test statistic of 173 with a p-value of 0.0001.

At a significance level of α >0.05, we can reject the claim H0 that the two rates are equal. The effect size is *large*, as $d_{cohen} = 1.716$.

When comparing proprietary and OSS projects, task-labeled items exhibit similar results: for proprietary projects we obtain an average $\overline{x} = 0.22$ ($\sigma = 0.25$), and for OSS projects an average $\overline{x} = 0.26$ ($\sigma = 0.27$). Similarly, when comparing the items labeled as requirements, the proprietary projects yield an average $\overline{x} = 0.66$ ($\sigma = 0.22$) while the OSS projects result in $\overline{x} = 0.56$ ($\sigma = 0.11$). The results indicate a slight difference, with the proprietary projects having on average more items correctly labeled as requirements in comparison to OSS projects, although more investigations are necessary to draw robust conclusions.

4.2 What Are the Most Commonly Used Categories (RQ2)?

RQ2 aims to reveal what type of requirements are frequently present in backlog items and with what degree of granularity they are formulated. Figure 2 shows, for each combination of type and granularity (see Table 1), the occurrence across the 14 projects, distinguishing between proprietary from OSS projects.

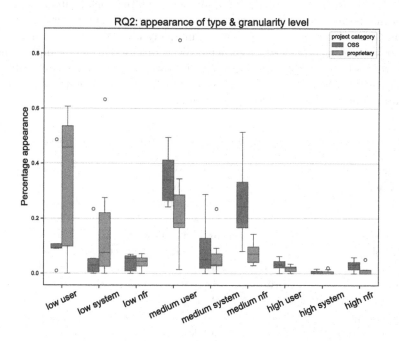

Fig. 2. Percentage of occurrence of the type-granularity combinations.

When examining the medians per combination, medium-level user-oriented requirements appear to be most prevalent in all projects with an overall median of 0.266 ($\overline{x} = 0.313$, $\sigma = 0.199$). The second more frequent ones are low-level user-oriented requirements with a median of 0.106 ($\overline{x} = 0.237$, $\sigma = 0.228$). As can be seen by the relatively high standard deviation (compared to the $\overline{x}$) and the gap between the median and $\bar{x}$, low-level user-oriented requirements also exhibit the largest fluctuations across projects.

Figure 2 shows that certain combinations exhibit a considerable difference between the OSS and proprietary projects. In particular, for low-level user-oriented requirements, the large variation for proprietary projects contrasts with a nearly nonexistent one for OSS projects. The results also show the higher percentage of medium-level requirements for OSS projects compared to proprietary projects. For high-level requirements and low-level non-functional requirements, only a slight variation exists between project types.

4.3   Do Backlog Items Include Multiple Requirements (RQ3)?

Since requirements are expected to be uniquely identifiable [10], RQ3 examines whether backlog items comply with this property. We explore how many of the backlog items with requirement-related information have more than one requirement (RQ3). Then, we examine which combinations of tags (type and granularity level) are most prevalent when 2+ requirements per item are identified (RQ3.1).

Table 4 shows the percentage of requirements-related backlog items consisting of multiple requirements. The results show that the projects in our sample does not only comprise requirements that are uniquely identifiable. Nonetheless, the standard deviation indicates large per-project variations. For example, among the items containing requirements in Jira Performance Testing Tools, only 14% contain items with multiple requirements, while the project Company has almost 95% items containing multiple requirements.

Table 4. Presence of multiple requirements in a single issue.

Project (proprietary)	Multiple (%)	Project (OSS)	Multiple (%)
Control	0.632	Cost Management	0.540
Service	0.048	Jira Performance Testing Tools	0.140
Store	0.550	Lyrasis Dura Cloud	0.235
Company	0.947	Network Observability	0.286
Portfolio	0.745	OpenShift UX Product Design	0.538
Data	0.549	Qt Design Studio	0.164
Learn	0.419	Red Hat Developer Website	0.434
Macro-average (propr)	0.556	Macro-average (OSS)	0.334
Std-dev (propr)	0.280	Std-dev (OSS)	0.170
Macro-average (all)	0.445	Std-dev (all)	0.251

Table 5 shows the most common combinations of different tags that co-occur in an issue (RQ3.1) having at least 10 total occurrences. The most frequent combination is having a medium requirement to be refined into one or more low-level requirements of the same type. In addition to its frequency, this combination appears in 12 of our 14 projects. The second most common combination is two medium-level requirements: one non-functional and one functional user-oriented: this occurs in 13 projects. The third row is complementary to the first one and it shows that in several projects, medium-level system-oriented functional requirements are refined into low-level requirements of the same type.

Table 5. Most frequent combinations of different tags in the same issue, showing both the total and the per-project counts.

Combinations	Total	Control	Service	Store	Company	Portfolio	Data	Learn	Cost mgmt	JIRA Perf	Lyrasis	Network Obs	OpenShift	QT Design	RH Developer
low user, medium user	96	16	0	1	1	17	17	17	2	1	0	6	1	2	15
medium nfr, medium user	61	2	3	3	0	0	2	0	6	2	15	1	23	2	2
low system, medium system	34	3	0	4	12	6	0	0	8	0	0	0	0	0	1
low user, medium user, low system	16	7	0	2	0	3	0	3	0	0	0	0	0	0	1
low user, low nfr, medium user	13	4	0	1	0	1	2	1	1	0	0	0	1	0	2
low user, medium nfr	13	2	0	2	0	3	3	0	0	0	0	0	1	1	1
low nfr, medium nfr	11	1	0	2	0	2	0	0	0	0	0	0	4	0	2
low user, medium nfr, medium user	10	0	0	0	0	0	4	0	2	0	0	0	1	1	2

Table 4 also compares the two types of projects (RQ5) in terms of the presence of multiple requirements. On average, the proprietary projects hold more items with multiple requirements ($\bar{x} = 0.556$, $\sigma = 0.280$) than the OSS projects ($\bar{x} = 0.334$, $\sigma = 0.170$). The standard deviations likewise show larger fluctuations in these percentages among proprietary projects compared to OSS projects.

4.4   Are Requirements Complemented by a Motivation (RQ4)?

RQ4 examines whether a backlog item containing requirements includes an associated motivation. Table 6 shows the percentage of backlog items where we identified at least one requirement that contain at least one justification. The overall macro-average shows that the motivation behind many requirements is not present in the backlog. The standard deviation of 0.169 also indicates that the percentages fluctuate only slightly across the projects. In ten of the fourteen projects, less than 50% of the requirements-containing items have an associated justification. The other four projects have less than 60% of their backlog items with motivations.

Table 6 compares the type of projects (RQ5) in terms of the presence of motivation. The macro average reveals only a small difference, with OSS projects having slightly more frequent motivations for their backlog items.

Table 6. Backlog items with at least one requirement that also have a motivation.

Project (proprietary)	Yes (%)	Project (OSS)	Yes (%)
Control	0.088	Cost Management	0.540
Service	0.524	Jira Performance Testing Tools	0.649
Store	0.525	Lyrasis Dura Cloud	0.383
Company	0.368	Network Observability	0.238
Data	0.529	OpenShift UX Product Design	0.487
Portfolio	0.319	Qt Design Studio	0.269
Learn	0.194	Red Hat Developer Website	0.171
Macro-average (propr)	0.364	Macro-average (OSS)	0.391
Std-dev (propr)	0.176	Std-dev (OSS)	0.175
Macro-average (all)	0.377	Std-dev (all)	0.169

4.5   Threats to Validity

We discuss threats to validity, ranging from internal to external factors, and discuss how we mitigated these to preserve the credibility of the study.

Regarding project selection, only one tagger eliminated irrelevant projects. Since this evaluation was conducted by a single tagger, it is possible that some interesting projects in the OSS datasets were excluded. Moreover, we did not tag all issues in the projects, but only a subset. We reduce this vulnerability by randomly selecting a subset of a representative size.

In addition, a single tagger created the initial coding scheme. To eliminate bias, the second tagger reviewed all the data on which the initial coding scheme was built. The feedback from the second tagger was used to adjust the coding scheme. Additionally, only one tagger coded the full sets of remaining projects for the final tagging of the projects. We mitigated these biases by including a second tagger who randomly tagged 20% of the issues in the sample.

For some open source projects, we could not ascertain whether the analyzed issues are part of a sprint or product backlog, due to the absence of such details in the datasets. We employed filtering mechanisms to only retain projects whose issue tracking systems are likely to be used to support sprint backlogs or to serve as a product backlog; however, we cannot be certain.

Furthermore, we have excluded issues labeled as "bug". Although it is possible that bug issues exist that contain requirements information, based on an initial exploration phase, we have determined that this situation is improbable.

For the proprietary projects, we examine projects from one company. Selecting only projects from one specific source could lead to a number of threats. We reduced this vulnerability by using projects from different teams and including a wide variety of OSS projects from different companies, but we reckon that future work needs to use data from multiple companies.

5   Related Work

Lüders and colleagues conducted research on the visualization and automated categorization [17] of links between issues. Our approaches are complementary. While they are concerned with the relationships (including dependencies and hierarchies) between requirements and other issue types, we offer an in-depth analysis of the *contents* of the issues.

Rath *et al.* [22] explored the effectiveness of automated traceability by assessing the ability of machine learning in recovering trace links between code commits and JIRA issues. Similar studies have been conducted, in the context of model-driven engineering, by van Oosten *et al.* [19]. Although these studies also analyzed issues in JIRA repositories, their focus is on repairing trace links, while we examine the requirements information in the issues.

Interview studies have been often employed to learn about the practices of documenting requirements in agile development. For example, Behutiye *et al.* [2] conducted fifteen interviews with practitioners from four companies using agile software development and they studied how quality requirements are documented. Their findings showed that in certain cases issues and epics are used to represent quality requirements, but also that prototypes and face-to-face communication are very important. A similar analysis was conducted by Alsaqaf *et al.* [1] in the context of large-scale, distributed settings. Their exploratory study reveals fifteen challenges, several of which are related to the minimal documentation principle in agile development. The study by Franch and colleagues [6] on requirements specification shows that, in agile contexts, project management tools are commonly used to document requirements. These studies are based on interviewing practitioners, while we focus on obtaining insights through the content analysis of backlog items.

Some research groups have collected and shared collections of issues extracted from publicly available repositories. The two largest and renowned datasets are the TAWOS dataset by Tawosi *et al.* [26] and the 'alternative' one by Montgomery and colleagues [18]. We make use of six projects from Montgomery's dataset and one from TAWOS, and we contrast these with seven proprietary projects.

Some studies applied content analysis to backlog items. For instance, Soliman *et al.* [25] investigated where architectural knowledge is located in JIRA issues, Ortu and colleagues [20] have studied the emotions that are included in the issues, and Montgomery *et al.* [18] linked issues to software life-cycle activities. We conduct a more in-depth analysis of requirements within backlog items.

Content analysis has been used extensively in software engineering for the analysis of communication within instant messaging systems as well as chat rooms. For example, Parra *et al.* [21] compare the contents present in Slack and Gitter in terms of Bin's categorization [15]: do the messages fulfill a developer's personal needs, team-wide purposes, or community support? Silva *et al.* [24] conducted thematic on a large number of Slack and Gitter chatrooms to identify what developers talk about. In this paper, we also apply content analysis but we focus on backlog items rather than messaging systems.

6   Discussion

We address each research question on the basis of the findings reported in Sect. 4. While doing so, we highlight remarkable results and provide additional likely explanations for certain phenomena.

RQ1: To what Extent do the Backlog Item Labels Chosen by Practitioners Reflect the Requirements Expressed in the Items? The hypothesis H1 formulated for this research question tested whether the percentage of requirements in requirements-labeled items is higher than in task-related items. The test results reported in Sect. 4.1 confirm the hypothesis, indicating it is more likely to find requirements in items labeled as such than in items labeled as tasks. This result provides empirical evidence in support of a straightforward conjecture. The average percentages shown in Table 3 are, instead, more surprising. On average, over 20% of the items labeled as tasks do in fact contain requirements. In addition, on average, more than 30% of the items labeled as requirements contain no requirements at all. These results show that practitioners do inconsistently label the items; therefore, in order to locate requirements within backlog items, it is not sufficient to simply display the issues that are labeled as requirement (here: epics, features, user stories, and suggestions).

RQ2: What Categories of Requirements Information are More Commonly Used? Fig. 2 visualizes the occurrence of different types of requirements with specific granularity levels in backlog items. The figure shows a high fluctuation between certain categories, especially for low and medium-level user-oriented functional requirements and medium-level non-functional requirements. This reflects the different usage patterns of teams managing their product or sprint backlogs. For example, the combination low-level functional requirements ('low user' in Fig. 2) for proprietary projects shows high variability; this happens because some teams include acceptance criteria in the same issue where a user story is written, while others do not specify them, or store them in a different environment.

RQ3: How Often does a Single Backlog Item Include Multiple Requirements? For each project, Table 4 shows how many of the items containing at least one requirement also contain multiple requirements. In general, the results show that many of these backlog items contain multiple requirements, making them no longer uniquely identifiable. In addition, Table 5 indicates which different categories of requirements often occur together, revealing that low-level requirements most often occur in combination with medium-level requirements. This is a generalization of the refinement pattern where a user story is refined into acceptance criteria. In addition, the results show that non-functional requirements often co-occur with functional requirements. This could be interpreted in terms of the refinement of non-functional requirements into functional requirements that are closer to system design [8]. More frequently and surprisingly, however, we found functional requirements that also specified a non-functional aspect of the system. An example from the Cost Management project is 'As a user, I want to quickly filter my tags based on tag key or value, [...]', which points to the functionality of a filtering option and also to the quality of performing it quickly.

RQ4: To what Extent are Requirements Complemented by a Motivation for Their Existence? We found that more than half of the backlog items that contain at least one requirement did not include any justifications for those requirements. These results show the lack of recognition of the importance [28] and recommendations [10] for expressing the 'why' behind requirements. This may be due to agile software development practices, where requirements are formulated in a concise manner, as they are intended to support and foster the conversation within the team [12], rather than acting as a precise and complete specification. In addition, this table considers the presence of justifications for all categories of requirements, while justifications are not equally essential for all categories, particularly when we look at low-level requirements. Nevertheless, the project Service, for example, contains no low-level requirements and still only includes justifications for roughly 40% of the required items.

RQ5: Can we Identify Differences w.r.t. RQ1-RQ4 when Comparing Proprietary Projects and Open-Source Projects? For each of the research questions above, we distinguish the open source projects from the proprietary projects. One of the most notable discrepancies between the open source projects and the proprietary dataset is the difference in labeling the backlog items, with the open source projects showing more inconsistencies (see Table 3). This lack of consistency in open source projects may be due to the low level of oversight or the varying experience level of contributors. In contrast, industrial projects in our sample show large variations between teams in their structuring of the project backlog (e.g., what type of requirements they include), and more often document multiple requirements within a single backlog item (Table 4).

7   Conclusion and Future Work

We performed content analysis on collections of backlog items to better understand the occurrence of requirements-related information in backlogs. For this purpose, we collected, tagged, and analyzed fourteen samples of open-source and proprietary projects, summing up to a total of 1,636 items.

Our results show that backlog item labeling is applied inconsistently and in a misleading manner by practitioners. In addition, teams may use one backlog item to document multiple requirements, making requirements within backlogs not uniquely identifiable. Both aspects pose challenges for those who need to retrieve requirements information. Furthermore, the most common are medium- and low-level user-oriented requirements, which also occur together in one item and mirror the refinement pattern of user stories in acceptance criteria.

In general, we find that item labels chosen by practitioners do not fully represent the content of requirements, especially when a backlog item contains multiple requirements, possibly of different types. Our most immediate future work aims to build a prototype tool to help practitioners automatically extract and classify requirements from collections of backlogs items.

Although backlog items contain a significant amount of requirements, they may not represent all the requirements. Especially in agile development, scenarios may arise where developers discover the need for a new feature during system development without specifying the implementation in the backlogs. Therefore, future work could focus on examining the completeness of sprint and product backlogs as a requirements specification artifact, or whether other documents (e.g., user journeys and vision documents, as indicated by our industrial partner Mendix) should be considered to obtain a fuller picture.

Acknowledgements. This research is partially funded by the Dutch Research Council (NWO) through the Open Technology Programme 2021-II TTW, project AUTOLINK (19521). We would like to thank Mendix, and especially Toine Hurkmans, for providing us with the proprietary datasets used in this study.

References

1. Alsaqaf, W., Daneva, M., Wieringa, R.: Quality requirements challenges in the context of large-scale distributed agile: an empirical study. Inf. Softw. Technol. **110**, 39–55 (2019)
2. Behutiye, W., Seppänen, P., Rodríguez, P., Oivo, M.: Documentation of quality requirements in agile software development. In: Proceedings of the 24th International Conference on Evaluation and Assessment in Software Engineering, pp. 250–259 (2020)
3. Cleland-Huang, J., Settimi, R., Zou, X., Solc, P.: Automated classification of non-functional requirements. Requirements Eng. **12**, 103–120 (2007)
4. Cohn, M.: User Stories Applied: For Agile Software Development. Addison-Wesley Professional, Boston (2004)
5. Cysneiros, L.M., do Prado Leite, J.C.S., de Melo Sabat Neto, J.: A framework for integrating non-functional requirements into conceptual models. Requirements Eng. **6**, 97–115 (2001)
6. Franch, X., Palomares, C., Quer, C., Chatzipetrou, P., Gorschek, T.: The state-of-practice in requirements specification: an extended interview study at 12 companies. Requirements Eng., 1–33 (2023)
7. Glinz, M.: A glossary of requirements engineering terminology. Standard Glossary of the Certified Professional for Requirements Engineering (CPRE) Studies and Exam, Version 2.0.1 (2022)
8. Gross, D., Yu, E.: From non-functional requirements to design through patterns. Requirements Eng. **6**, 18–36 (2001)
9. Hess, A., Diebold, P., Seyff, N.: Understanding information needs of agile teams to improve requirements communication. J. Ind. Inf. Integr. **14**, 3–15 (2019)
10. IEEE: Systems and software engineering - life cycle processes -requirements engineering. ISO/IEC/IEEE 29148:2018(E) (2018)
11. Inayat, I., Salim, S.S., Marczak, S., Daneva, M., Shamshirband, S.: A systematic literature review on agile requirements engineering practices and challenges. Comput. Hum. Behav. **51**, 915–929 (2015)
12. Jeffries, R.E., Anderson, A., Hendrickson, C.: Extreme Programming Installed. Addison-Wesley Longman Publishing Co., Inc., Boston, MA, USA (2000)

13. Kassab, M.: The changing landscape of requirements engineering practices over the past decade. In: Proceedings of the EmpiRE, pp. 1–8. IEEE (2015)

14. Krippendorff, K.: Content Analysis: An Introduction to Its Methodology. Sage publications, California (2018)

15. Lin, B., Zagalsky, A., Storey, M.A., Serebrenik, A.: Why developers are slacking off: Understanding how software teams use slack. In: Proceedings of the CSCW Companion, pp. 333–336 (2016)

16. Lucassen, G., Dalpiaz, F., Werf, J.M.E.M., Brinkkemper, S.: The use and effectiveness of user stories in practice. In: Daneva, M., Pastor, O. (eds.) REFSQ 2016. LNCS, vol. 9619, pp. 205–222. Springer, Cham (2016). https://doi.org/10.1007/978-3-319-30282-9_14

17. Lüders, C.M., Pietz, T., Maalej, W.: On understanding and predicting issue links. Requirements Eng., 1–25 (2023)

18. Montgomery, L., Lüders, C., Maalej, W.: An alternative issue tracking dataset of public Jira repositories. In: Proceedings of the MSR, pp. 73–77 (2022)

19. van Oosten, W., Rasiman, R., Dalpiaz, F., Hurkmans, T.: On the effectiveness of automated tracing from model changes to project issues. Inf. Softw. Technol. 160, 107226 (2023)

20. Ortu, M., et al.: The emotional side of software developers in JIRA. In: Proceedings of the MSR, pp. 480–483 (2016)

21. Parra, E., Alahmadi, M., Ellis, A., Haiduc, S.: A comparative study and analysis of developer communications on Slack and Gitter. Empir. Softw. Eng. 27(2), 1–33 (2022)

22. Rath, M., Rendall, J., Guo, J.L., Cleland-Huang, J., Mäder, P.: Traceability in the wild: automatically augmenting incomplete trace links. In: Proceedings of the ICSE, pp. 834–845 (2018)

23. Sedano, T., Ralph, P., Péraire, C.: The product backlog. In: Proceedings of the ICSE, pp. 200–211. IEEE (2019)

24. Silva, C.C., Galster, M., Gilson, F.: A qualitative analysis of themes in instant messaging communication of software developers. J. Syst. Softw. 192, 1–15 (2022)

25. Soliman, M., Galster, M., Avgeriou, P.: An exploratory study on architectural knowledge in issue tracking systems. In: Biffl, S., Navarro, E., Löwe, W., Sirjani, M., Mirandola, R., Weyns, D. (eds.) ECSA 2021. LNCS, vol. 12857, pp. 117–133. Springer, Cham (2021). https://doi.org/10.1007/978-3-030-86044-8_8

26. Tawosi, V., Al-Subaihin, A., Moussa, R., Sarro, F.: A versatile dataset of agile open source software projects. In: Proceedings of the MSR, pp. 707–711 (2022)

27. White, M.D., Marsh, E.E.: Content analysis: a flexible methodology. Libr. Trends 55(1), 22–45 (2006)

28. Yu, E.S., Mylopoulos, J.: Understanding "why" in software process modelling, analysis, and design. In: Proceedings of the ICSE, pp. 159–168. IEEE (1994)

Requirements Engineering for No-Code Development (RE4NCD)

A Case Study of Rapid Application Development During War

Meira Levy[1,2](✉) ⓘ and Irit Hadar[2] ⓘ

[1] Shenkar, College of Engineering, Design and Art, 5252626 Ramat Gan, Israel
lmeira@shenkar.ac.il
[2] University of Haifa, City Campus, 3303220 Haifa, Israel

Abstract. Context and motivation: In recent years, a new development approach has emerged, for rapid application development (RAD) supported by platforms that enable low or no-code development (NCD). This approach is designed for developers with limited or no coding expertise and for achieving a very short time-to-deployment. Question/problem: This research explores the process of RAD as performed with the Monday NCD platform. It focuses on the phases of requirement engineering (RE) and design, which are typically omitted during RAD, posing challenges in ensuring a rigorous, sustainable, and flexible application. **Principal ideas/results:** Addressing this gap, the paper introduces a proposed RE for NCD (RE4NCD) method, based on a case study in which a civilian management system was rapidly developed during a time of war, and proposes a research preview for further exploration and development of this research direction. **Contribution:** The paper highlights the theoretical and practical implications of RE4NCD, underscoring the potential transformative impact of NCD on the software development industry. It further proposes future research aimed at refining and validating the RE4NCD method, tracking the adoption and evolution of applications in diverse organizations, and applying the method to additional case studies for comprehensive evaluation and validation.

Keywords: Requirements Engineering · No-Code Development · Rapid Development · War · Crisis

1 Introduction

Research of requirements engineering (RE) in times of crisis, such as fire, flood, tsunami, or terrorist attack, is an emerging field that copes with the unique challenges of developing software systems that can deliver information for managing such events [13]. Rapid application development (RAD) and no-code development (NCD) platforms aim to provide drag-and-drop tools to enable users in a constrained domain to develop applications by assembling pre-coded components [5]. While this prototyping approach is very appealing for rapid development in a bottom-up manner by both expert and non-expert developers, it requires applying the necessary rigorous techniques and tools for ensuring the requirements completeness and adaptation [5, 6].

© The Author(s), under exclusive license to Springer Nature Switzerland AG 2024
D. Mendez and A. Moreira (Eds.): REFSQ 2024, LNCS 14588, pp. 322–329, 2024.
https://doi.org/10.1007/978-3-031-57327-9_20

This paper proposes a first edition of requirements engineering for no-code development (RE4NCD) method, which can be used for any NCD platform for developing software systems and templates that can be further used in similar challenges. The RE4NCD method was developed and practiced in a case study where a system for managing civilian efforts was rapidly developed, during the Israel-Hamas war following the Oct. 7th terror attack on Israel. The first author was involved in this case study as an RE expert, enabling participatory action research (PAR) [7, 10]. Requirements were elicited and a system was developed and deployed over the course of two weeks. This paper describes this experience and the resulting developed method and proposes a research preview for further developing this research direction.

The rest of the paper is organized as follows. Section II presents the theoretical background of rapid application development (RAD) focusing on NCD. Section III describes the case study settings, where the RE4NCD method was developed and practiced, and Section IV follows with the description of the RE4NCD principles and how they were applied to the case study. Finally, Section V presents conclusions and future research directions that target to refine and enhance the proposed RE4NCD method.

2 Rapid Application Development (RAD via No-Code Development (NCD)

A recent approach of low/no-code software development has received much attention in the context of RAD [3]. Low and no-code software development has changed the dynamics of software engineering by introducing the capability of nonprofessional developers to rapidly build tools without having a programming background [9]. The reduction in code development provides us with several benefits, including reduction of software development cost, enhanced business agility [1], rapid iterations [11], and improved manufacturing capability [9]. While demonstrating these advantages, such development techniques suffer from several shortcomings. For example, the developed systems may be difficult to update when new technical requirements are needed [4], and quality concerns, such as security, are yet to be fully addressed [1].

The growth and increasing demand of digitization of businesses, many of which not having sufficient software engineering capacity to meet the demand, has led to an increasing adoption of the solutions of NCD platforms, with an estimation to be used in 65% of application development work by 2024 [11].

While NCD tools contribute to the rapid development of software [2], formalization and representation of the requirement specifications of the software still remain a significant challenge [3]. Several former studies have addressed the need for conceptual modeling during NCD [1, 3]. However, a literature review we performed in the field of RE did not yield research regarding the RE process of NCD. Our study aims at bridging this gap with the developed RE4NCD method, for facilitating rapid requirements process during RAD, in times of crisis, when emergency occurs and there is a demand for an urgent deployment of a software system that can facilitate information sharing and processing for crisis management.

3 The Case Study

3.1 Settings

The case study took place in a small countryside town in the north of Israel. During the COVID'19 pandemic, a social initiative entitled "Kulanu Yachad" (meaning, all of us together) has emerged for supporting isolated people with community services such as transportation to hospitals, food delivery, home repairs, and more, all based on volunteers. Following the terror attack on Israel, on October 7, 2023, and during the Israel-Hamas war it initiated, which necessitates civilians living in bombarded areas to stay at home close to their shelters, these social initiatives have resumed.

The community services were tracked and managed via Excel. A call to develop a software application that will handle the volunteering activities was distributed via the WhatsApp group which initiated the intensive work of a dedicated team of volunteers. This team included one of the managers of the volunteering organization ("Kulanu Yachad"), a student who works in the organization as an operator, who receives the requests via phone calls and matches them with volunteers, and the first author, who joined the team as an RE expert. As part of this team, the first author acted as a researcher following the principles of the participatory action research approach [7, 10].

In addition, the team had at its disposal a limited capacity of several hours of a volunteer for developing the application. The team chose monday[1] as the NCD platform. Monday is a cloud-based system that enables teams and organizations to efficiently manage work, tasks, and projects. It provides a centralized platform for collaboration, task tracking, and project management.

3.2 Research Method

Participatory action research (PAR) is a research method in which the researcher takes an active part as a community or organization member in a certain studied case. The researcher collects and analyses data for the purpose of influencing the situation while generating knowledge [7]. PAR aims to solve a real-world problem by advanced research understanding, while ownership and control of the research process are transferred from an outsider's view to a collaboration with the research participants [10].

In the reported case study, the first author joined the team as a volunteer, and gathered data during the researched process. The data underwent an iterative thematic analysis [14] for identifying emergent themes which were then compared with existing literature. This resulted in identifying the gap and in the proposed RE4NCD method addressing it.

3.3 Development Procedure and Resulting Application

The first meeting of the team started by introducing the main concepts of the volunteering organization, which includes the initiatives from the COVID pandemic, such as transportation to medical care, as well as many additional ones, addressing new challenges introduced by the war, such as finding accommodations for civilians evacuated

---

[1] Monday.com.

from their homes. As a first step, the team had to come to an agreement about a core terminology. For example, currently there are formal organizations and social initiatives. Social initiatives can be initiated by formal organization, e.g., a school sending pupils to agricultural missions, or by groups of people who do not belong to an organization e.g., a group of people forming a joint cooking effort for evacuees. Thus, it is possible, but not necessary, to connect a *social initiative* to an organization, while having volunteers involved in different roles with various initiatives. Another important concept is *request*, which represents needs that can be linked to a certain social initiative. A request can be split into several *activities*, each of which can be performed by different volunteers. For example, if a request deals with a house that needs to be renovated, there could be several activities related to this request, such as painting, floor tiling, and more.

After the team agreed on the main concepts and respective terms, it was agreed that the system should be developed as a prototype for further usage by other volunteering initiatives in the same town, as well as in other volunteering organizations in Israel and abroad. Therefore, wherever flexibility is required there should be data items that can be modified and extended after the development and deployment. For example, roles in a social initiative should be a separate table, enabling roles' modifications and additions.

After a trial to start building the system in the drag and drop manner facilitated by the NCD platform, the team understood that this process: (a) is highly time consuming, and (b) will not lead to a robust system. Having the monday developer available for a very limited time and the need to develop a robust and flexible application required a more formal RE and design approach prior to the drag and drop development. Thus, the team started over, first developing an ERD and a Use Case diagram, describing the data and processes of the system.

Based on the diagrams, the monday developer created a test environment with all the tables and processes as designed. The developed system was tested by the operator in the team, who ran it with usage scenarios of incoming requests. Following the testing results, several improvements were made, and the test environment was delivered and deployed within the volunteering organization and became the official system with which the volunteering operation is managed.

The operator took over as the technical professional of the application and learned how to configure it and how to use the dashboarding options as well as the manipulation of the presentation and search options. Monday has a special language that enables presenting data in groups. For example, the presentation screen can be divided into activities sent via WhatsApp for seeking volunteers, activities that are already allocated to volunteers, and activities that are done. The dashboard is very useful and provides statistics according to different queries, the volunteering activities statuses, and more. The manipulation of the dashboard is very easy to learn and apply. The system also provides tools to create forms for data entering, either by the operator or through a link to the form, which makes it very useful and convenient in case we want people to fill in the form from a distance.

The entire development process took two weeks, and the system has been successfully used since its deployment. New users learn how to work with the system within one hour, and we received good feedbacks about its usability and usefulness. Shortly after deployment, a new social initiative was added to the application, which required additional data items. The implementation of this addition took about three hours, demonstrating the flexibility and robustness of the system.

Several volunteering organizations, from Israel and abroad, were exposed to the application and expressed their wish to adopt it to their own needs. These organizations include, for example, a larger organization that handles volunteering activities all over Israel, and a North American organization that supports Israeli social initiatives. It should be noted that enhancements to the application can be made by each organization separately, storing the enhanced application and their data in their own data space.

4   The Proposed RE4NCD Method

Based on the experience and insights we gained during the case study, Table 1 presents the steps and activities of RE4NCD, for achieving improved performance and resulting artifact of rapid NCD.

Table 1. RE4NCD: Main Steps and Activities as Emerged from the Case Study

No.	Step	Activities	Case study data
1	*Create a diverse team with shared goal and terminology*	Identify the most capable contributors through an open call via any communication platform (e.g., WhatsApp)	The call for volunteers to join an effort of creating the application was sent through WhatsApp to the group of active volunteers
		Establish a team, including people who: relate to the challenge, practice the current non-digital process to be replaced with the developed application, are familiar with the NCD platform, are familiar with RE and design practices, can form collaboration with other communities that face similar challenges	The team included 3 people: a manager from the organization, an operator, and an RE expert. An additional volunteer joined for a very short time for building the application on the NCD platform
		Develop a common terminology among the team members, so that each member understands the process, users, data items, and jargon	Achieving shared understanding regarding terms, distinguishing for example between organization and social initiative, and request and activity
2	*Rapid RE & Design*	Identify the processes and data that should be digitized and prioritize them according to urgent needs	Handling the requests and matching them to volunteers was identified as the most important process to be supported, while supporting operator shifts handling, for example was backlogged

(continued)

Table 1. (*continued*)

No.	Step	Activities	Case study data
		Create an Entity Relational Diagram (ERD) of the data entities to ensure logical correctness	ERD was created by the RE expert in collaboration with the other team members
		Identify necessary tables that will enable extension and flexibility after application deployment and for various communities	The tables were planned and created beyond addressing the immediate requirements, for facilitating flexibility (i.e., roles, status, etc.)
		Create Use Case diagrams for representing the usage scenarios that should be supported by the application. Make sure to identify several users with different permissions to keep data protection, privacy and security	Use case diagram that describes various roles and their activities was developed
3	*Development*	Select a NCD platform according to participants' capabilities, and develop the application based on the design	The application was developed using monday
		Create a test environment and test the application according to the usage scenarios	Several scenarios, described by the operator, were used to the test system
		Migrate the existing data to the newly developed application	The data from the old Excel file was automatically inserted to the application. Whenever missing data was identified, it was inserted manually
		Create users and authorizations Nominate a team member that best knows the NCD platform for further enhancements over the application usage lifecycle	Users were defined for the operators of the organization. The operator who learned the monday platform, took responsibility for future enhancements
4	*Production*	Start working with the application and improve when needed	After deployment all the operators worked with the system. Several improvements have already been made, mainly adding sub-groups of activities
		Establish collaboration with other communities that face similar challenges. Identify additional requirements that can serve you as well and improve your application accordingly	The organization had already several meetings with other social organizations in Israel and abroad. Several of them are interested to reuse our design and the monday platform with a few adjustments

This method is the result of the case study described in this paper, and we believe it shows much promise, based on the success of its outcome. It is, however, important to acknowledge that the method construction is based on a single, limited case study. We plan to apply this method to additional varied domains and settings, in order to further develop and refine it toward its generalization as a new RE method that will support the emerging approach of NCD.

5 Conclusion and Future Research Directions

No-Code development (NCD) is a software development method that provides non-professional users with a platform for visually creating applications with little or no coding. NCD gained rapid growth of its adoption in industry and is expected cause a turnaround in the software development industry and foster digital transformation [12].

Our research was initiated following a realization that there are both theoretical and a practical gaps in the rapid application development (RAD) method via NCD. The case study presented in this paper serves as the first step in this research, in which the NCD practice of drag and drop, and connect components for building mobile or web applications [8] was applied. The time and resource availability constraints of this case, set during a crisis, created an opportunity for us to rethink the development process, which eventually resulted in the proposed RE4NCD method. This method was successfully practiced in the case study, as described above.

While our study examined one particular platform in a unique case study, the RE4NCD method does not refer to any particular characteristics of the platform, thus we believe it is applicable to any other NCD platform.

Future research is needed for the refinement, validation, and generalization of the proposed method. An immediate direction we plan to pursue is following up on future adoption of the application developed in this case, and the ways in which it is adapted and enhanced throughout its lifecycles in different organizations, and specifically in organizations that develop software systems based on the Monday platform. In parallel, this method can be further refined and validated by applying it to other platforms and case studies where RAD is required and NCD platforms are used. While the RE4NCD was developed during a participatory action research, the following case studies can be based on non-participatory observations on other development teams, starting with qualitative analyses and following with the determination of quantitative KPIs toward further evaluation and validation of the method.

References

1. Al Alamin, M.A. et al.: An empirical study of developer discussions on low-code software development challenges. In: Proceedings - 2021 IEEE/ACM 18th International Conference on Mining Software Repositories, MSR 2021, pp. 46–57. Institute of Electrical and Electronics Engineers Inc. (2021). https://doi.org/10.1109/MSR52588.2021.00018
2. Caldeira, J., et al.: Unveiling process insights from refactoring practices. Comput. Stand. Interfaces **81** (2022). https://doi.org/10.1016/j.csi.2021.103587

3. Hossain, B.A., et al.: Natural language-based conceptual modelling frameworks: state of the art and future opportunities. ACM Comput. Surv. **56**(1), 1–26 (2024). https://doi.org/10.1145/3596597
4. Khorram, F., et al.: Challenges & opportunities in low-code testing. In: Proceedings - 23rd ACM/IEEE International Conference on Model Driven Engineering Languages and Systems, MODELS-C 2020 - Companion Proceedings, pp. 490–499. Association for Computing Machinery, Inc (2020). https://doi.org/10.1145/3417990.3420204
5. Kramer, J.: RE runtime: the challenge of change RE'20 Conference Keynote. In: Proceedings of the IEEE International Conference on Requirements Engineering, pp. 4–6. IEEE Computer Society (2020). https://doi.org/10.1109/RE48521.2020.00012
6. Li, J., Tei, K.: Done is better than perfect: iterative adaptation via multi-grained requirement relaxation. In: Proceedings of the IEEE International Conference on Requirements Engineering, pp. 288–294. IEEE Computer Society (2022). https://doi.org/10.1109/RE54965.2022.00043
7. Macdonald, C.: Understanding participatory action research: a qualitative research methodology option. Can. J. Act. Res. **13**(2), 34–50 (2012)
8. Mclean, A.: Software development trends 2021. Can. J. Nurs. Inform. **16**, 1 (2021)
9. Waszkowski, R.: Low-code platform for automating business processes in manufacturing. In: IFAC-PapersOnLine, pp. 376–381. Elsevier B.V. (2019). https://doi.org/10.1016/j.ifacol.2019.10.060
10. Widjaja, A., Matitaputty, S.: Empowerment of small medium enterprises through student participatory action research in implementation of accounting information system. SHS Web Conf. **59**, 01002 (2018). https://doi.org/10.1051/shsconf/20185901002
11. Woo, M.: The rise of no/low code software development—no experience needed? Engineering **6**(9), 960–961 (2020). https://doi.org/10.1016/j.eng.2020.07.007
12. Yan, Z.: The impacts of low/no-code development on digital transformation and software development (2021). https://arxiv.org/abs/2112.14073v1
13. Yang, L., et al.: GDIA: eliciting information requirements in emergency first response. Requir. Eng. **20**(4), 345–362 (2015). https://doi.org/10.1007/s00766-014-0202-2
14. Yin, R.K.: Case Study Research: Design and Methods. Sage Publications, Thousand Oaks (1984)

Behavior-Driven Specification in Practice: An Experience Report

Joel D. Allred⬤, Simon Fraser⬤, and Alessandro Pezzoni$^{(\boxtimes)}$⬤

Anaplan Limited, York, UK

joel@allred.ch, {simon.fraser,alessandro.pezzoni}@anaplan.com

http://www.anaplan.com

Abstract. Agile methods are now widely used in software engineering organizations, whereas most formal methods are limited to niches and are perceived as inadequate in the context of agile development. This paper presents a case study of the innovative practices used at Anaplan, a financial planning and analysis software provider, to integrate formal specification within an agile process. The results show how Behavior-Driven Specification (BDS), by documenting behavior using executable acceptance criteria (EAC), is used to validate the design and implementation of calculation functions in Anaplan's sparse calculation engine, while keeping all stakeholders aligned on the requirements. We also show that the interaction between the specifiers and the developers allows catching implementation issues at early stages of the development, while allowing the specification to remain amenable to emerging implementation constraints. The validated requirements have enabled the development of a framework to automatically generate extensive test coverage that is used to verify the implementation. As a result, over 200 bugs were caught in the production code before release, not counting the hundreds of issues that BDS allowed developers to detect earlier in the process. We show that BDS leads to high levels of confidence in the behavioral correctness of software while being fully aligned with agile practices, and proves to be a significant evolution in the field of software development.

Keywords: Behavior-driven specification · Agile · Software development · Requirements engineering

1 Introduction

Agile methods [3] are becoming ubiquitous in modern software engineering companies. Meanwhile, formal methods are seen as having limited use, with the perception being that they can only provide value in niche industries, such as safety-critical system or chip development. In our experience, many software developers maintain a view that formal methods are incomprehensible, are tied to linear development practices, and are incompatible with agile methods. Changing requirements is often inevitable in the world of software, and recent efforts – based on incrementality and an iterative design loop – have been made to marry agile processes with model-checking [9].

D. Mendez and A. Moreira (Eds.): REFSQ 2024, LNCS 14588, pp. 330–343, 2024.
https://doi.org/10.1007/978-3-031-57327-9_21

Anaplan is a Software as a Service company which produces an enterprise planning platform using an agile development process. The complexity of planning models has led Anaplan to develop an alternative calculation engine suited to notionally large models that are sparsely populated with data. Development had to satisfy a consistency requirement, where the sparse engine had to be functionally equivalent to the existing one, except for intentional differences that had to be documented. When adding new functionality to the sparse engine, flawlessness also had to be ensured, because modifying the behavior of calculation functionality that has already been shipped to customers would cause significant reputational damage, as organizations rely on these numbers to make critical strategic decisions.

To satisfy these goals, we devised Behavior-Driven Specification (BDS). This approach and its technical process were introduced by Fraser and Pezzoni [8]. We hereby elaborate on how the approach was implemented from a process and team point of view for the delivery of the sparse calculation engine, from its inception to General Availability (GA), and conduct a thorough evaluation of the effectiveness of BDS at Anaplan based on quantitative quality metrics and stakeholder feedback.

2   Background

Anaplan is renowned for its enterprise planning management platform. It has been widely adopted for business planning purposes, including budgeting, forecasting, financial planning, and analysis, as well as for operations and supply chain management. Anaplan provides an environment that customers use to model the various aspects of their business [15]. At the center of the Anaplan platform is a robust calculation engine. This engine oversees the management of OLAP (Online Analytical Processing) cubes [5], which are multidimensional arrays of data representing various business measures. Anaplan hosts a rich modeling language, offering extensive slicing, dicing, and calculation capabilities. Crucially, Anaplan's aggregation capabilities allows the construction of models that take trillions of raw data points and produce simple dashboards that provide health indicators and insights that are critical to business planning. To ensure efficient and consistent model operations, all cube data is kept in memory for rapid retrieval and calculation. When any cell's value changes within a single model, Anaplan instantly recalculates all dependent figures, thereby offering a dynamic, real-time view of business operations.

2.1   Diversifying the Engine Portfolio

The Anaplan calculation engine was optimized for the dense population of a model, making use of large chunks of contiguous memory for cube data, leading to efficient calculation. However, when running in the public cloud, this approach is physically limited by the size of available hardware. Hence, a calculation engine optimized for the sparse population model was developed and offered alongside the existing engine.

This new engine was developed in parallel to the dense engine and continuously deployed to production environments while ensuring functional consistency with the dense engine, with necessary differences documented. Due to the rapid and organic growth of the software in earlier years, the dense engine had been developed without a precise specification of the requirements. The dense engine now had to act as the 'source of truth', and its behavior needed to be captured and used as a specification for the sparse engine. This had to be done in an agile way where the sparse development could start without having a full understanding of the dense behavior, and be revisited as the specification is created. In this asynchronous way of working, implementation could even start before specification, in which case feature flags would be used to hold off the release of a feature until behavior is validated and verified.

2.2 Validating New Functionality

In addition to the formalization of the existing behavior, the introduction of new features such as calculation functions necessitates meticulous design and testing. Users depend on the consistency of numeric calculations embedded in their models, and thus, rolling out a new version of the Anaplan software that alters these computations, even for rectifying a prior bug, would prove significantly disruptive, principally because small changes at raw data level can lead to substantial changes in aggregated figures.

Therefore, before any new functionality can be released to users, Anaplan must:

1. validate the requirements — ensure the feature aligns with the users' needs
2. validate the implementation — confirm the feature satisfies the requirements
3. verify the implementation — ensure the implementation does not contain any fault.

2.3 Timeline

The engineering process described in this article was instrumental in the successful delivery of the sparse calculation engine, which was an ambitious undertaking that mobilized several engineering teams[1] for over three years.

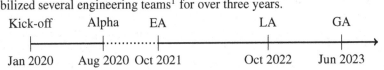

The development work for the sparse engine was started in January 2020. In March 2020, a steel thread [1] was merged into the main codebase and deployed to production for selected internal users to evaluate. An alpha version with limited functionality was made available in August 2020 to selected customers and for internal use. In October 2021, a feature-incomplete version of Anaplan with the sparse engine was released in Early Access (EA) to selected customers for experimentation. A feature-complete Limited Availability (LA) version was released in October 2022 for use on customer production models. Finally, GA was announced in June 2023 and the engine is now being widely distributed to customers who have a need for high dimensionality in their models.

---

[1] Approximately 50 engineers were involved over the lifetime of the project.

3 Leveraging Behavior-Driven Specification

Behavioral correctness of the Anaplan calculation engines being essential, the engineering team has embraced formal methods while also adopting agile development practices. The task of ensuring correctness can be divided into *verification* — 'doing things right' — and *validation* — 'doing the right thing'. Initially, the task was to ensure consistency between the two engines. Subsequently, there emerged a requirement to validate and verify new functionality introduced in the sparse engine. During the lifetime of the project, this latter aspect grew in importance relative to replicating the behavior of the dense engine. To realize these objectives, a formal specification was instituted to encapsulate the logic inherent in the modeling language. The formal specification is written in VDM-SL [14], a specification language that has tooling with animation capabilities.

Precisely documenting the behavior is of little use if the product requirements are unknown or misunderstood. Although a formal specification is useful for recording expected behavior in an implementation-agnostic way, it is not an adequate tool to be discussing requirements. Evans [6] describes how design and development can be problematic when language is fractured.[2] Despite the existence of a large amount of research discussing the agile engineering of requirements, the software engineering community still lacks appropriate processes to work with iteration-based specification [11], and missing or incomplete requirements is a cause of project delays [16].

In an effort to integrate the specification work with well-proven agile processes, Anaplan implemented BDS [8], a formal specification approach aligned with agile principles. The BDS process resembles Behavior-Driven Development (BDD) [17], an agile technique that stimulates collaboration, communication, and understanding between the project stakeholders, and like the Cucumber framework that supports BDD, requires a domain-specific language (DSL) for writing scenarios. However, unlike most BDD use cases, BDS clearly separates verification from validation.

The descriptive specification language introduced by Fraser and Pezzoni [8] and materialized in the form of EACs, was brought forth to facilitate discussions and validate requirements coherently among all stakeholders. This language was formulated to ensure the convergence of understanding, enabling precise, unambiguous communication and agreement on the requirements and functionalities discussed. Each EAC is structured as a Hoare Triple [10] of the form:

- *given* an initial context (precondition)
- *when* the stakeholder performs an action (command)
- *then* the outcome is as expected (postcondition)

These scenarios are written using a DSL, the execution of which is enabled by a set of implementation and specification adapters. Examples of some simple EACs are given in listings 1.1 and 1.2.[3,4]

---

[2] Language fracture exists when stakeholders use different formalisms.

[3] Note that the keyword whenever is used in the scenario as when is a Kotlin keyword.

[4] We are unable to provide more accurate examples due to the proprietary nature of the work, but the simplicity of these scenarios is comparable to that of many of our actual EACs.

```
@Eac("A created list is empty")
fun create() {
  whenever {
    createAList("list")
  }
  then {
    listContains("list")
  }
}
```

Listing 1.1. EAC for creating a list

```
@Eac("When an entity is added,
      it is contained by the list")
fun addToAList() {
  given {
    thereIsAList("list", entities = "a")
  }
  whenever {
    addEntityToList("list", "b")
  }
  then {
    listContains("list", entities = "a", "b")
  }
}
```

Listing 1.2. EAC for adding to a list

EACs are abstract and can be validated against any system which provides an adapter for the abstract modeling language used. Adapters were developed for both the dense and sparse engines, as well as the formal specification. Importantly, this means that the EAC is — by default — validated against all three systems every time it is run.

Since requirements engineering is inherently a human-centric process [2], maintaining the requirements specification of the software in a readable form allows stakeholders to continuously refer to an up-to-date expected behavior. This living documentation acts as a unique source of truth and provides a starting point to discuss behavior while avoiding redundant exchanges that would arise from incomplete or outdated information, thus improving the shared understanding of requirements [4, 12]. The various artifacts of BDS, as well as the framework [7] that allows their integration in the development process are all maintained by a dedicated team of engineers called *System Specification*.

3.1   The Agile Process for New Functionality

On the sparse engine project, the traditional requirements engineering process of elicitation, analysis, documentation, and validation [13] of a newly introduced feature is adapted to follow an agile workflow. In the typical initiation phase of an agile project, user stories are created and prioritized, forming an initial backlog under the guidance of a Product Owner (PO). Once the user story is prioritized, the development process starts with conception and concludes with the feature's release to the user, as depicted in Fig. 1.

The *conception* phase involves interactions between the PO, user group representatives, and the System Specification team to outline the product requirements for the feature. This stage also addresses how the new feature aligns with the existing constructs from a modeling standpoint, and whether it is more appropriate to introduce a new concept or expand existing functionality. The product requirements at this stage are written in natural language.

Once the high-level concept is approved by all stakeholders, the feature advances to the *refinement* stage, during which development teams investigate the feasibility of the feature and estimate the implementation effort. This constitutes a decision point to reevaluate the technical approach or potentially the development priority of the feature. Concurrently, the specifiers formalize the product requirements. For each behavior, the initial step involves writing a natural language description of the acceptance criteria

(AC). For each AC, a scenario that exemplifies it is then crafted, potentially extending the DSL if new language expressions are needed. Each scenario then serves as an EAC used to validate the criteria against both the formal specification and the target implementation. The Azuki framework [7, 8] allows scenarios to execute automatically against any given implementation or specification for which a DSL adapter is available. It is not necessary for every adapter to implement every available DSL feature, as the framework is designed to automatically skip unsupported scenarios, but at this stage we expect that at least the adapter for the formal specification will implement all the DSL features required to animate the new scenario. The specifier therefore adjusts the formal specification, and potentially its adapter, as needed to satisfy the EAC, taking care not to invalidate other EACs. The specifier can provide a reference implementation as a separate library for any functionality that would be prohibitive or impossible to specify in pure VDM, which is then automatically used by the VDM interpreter when animating unspecified functions while still checking any specified conditions and invariants [8] — this property of VDM proved extremely useful, for example, when we needed to introduce a Rust-based library to overcome Java's limitations with Unicode characters outside the Basic Multilingual Plane. This process for writing and specifying EACs is then repeated until all requirements have been captured.

EACs keep being updated as they are handed to developers for behavior implementation. Iterative validation and refinement are necessary to keep stakeholders aligned on changing requirements [13]. During the *implementation* phase, as developers add new functionality the implementation adapter is extended as needed to allow running the new EACs, and the development teams run the EACs and ensure that they pass before considering the implementation complete. Note that, while in our instance the System Specification team currently owns and maintains the DSL and all of its adapters, it would be equally feasible for a developer team to own the adapter for their implementation.

The agility of the process makes

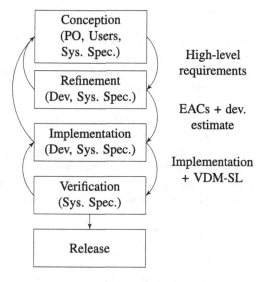

Fig. 1. Development process for new functionality. The labels on the right describe the inputs into the next phase.

it possible for the implementation to start before the specification is fully agreed, or even started. Importantly, the System Specification team remains separate from the development teams and has no involvement in the implementation.

A powerful complement to BDS is the *verification* framework, which is another feature of Azuki [7]. This is used to generate regression tests that combine many features.[5] Many levers are offered that enable targeted testing, but the basic process involves:

- generation of a scenario using the DSL of the EACs
- validation of the scenario using the specification; i.e., is the scenario meaningful?
- querying of the implementation for scenario results
- verification of said results by ensuring they satisfy the specification.

Once verified the scenario is easily transformed into a test case that can be added to a suite of regression tests. The whole chain, from generating a scenario down to adding the resulting test to the regression suite, is entirely automated and is driven by manually written scripts.[6] To date, these scripts have been mainly provided by the System Specification team, but recently an outsourced team was quicly able to start contributing to the verification process with little direct oversight. This process has now generated hundreds of thousands of tests that provide significant coverage, but which take significantly less time to execute than mechanisms used with the dense engine.

4 Evaluation

Since its introduction in early 2020, the engineering process described in this article has allowed the product and engineering teams to find an efficient and safe path to releasing the Anaplan platform whose functional correctness is critical. Leveraging BDS offers many advantages: it sharpens the clarity of the product requirements, proactively identifies deviations from these requirements, effectively catches bugs pre-production, and reduces dependence on traditional verification methods like manual quality assurance (QA) and exhaustive user model testing.

In this section, we recall the known benefits of BDS [8] and elaborate using recent experience, examples, and metrics from our issue-tracking system. We also give the results of a survey carried out among the different stakeholders involved in the BDS process and gather insights that pinpoint the various aspects that differentiate BDS from other development methods.

4.1 Requirements Clarification

The fact that functional requirements are embodied in a unique source of truth – the EAC – is a fundamental shift in how features are designed and engineered. The precise documentation of the behavior makes discussing the requirements more straightforward than in areas where this process has not been applied yet.

Previously, when the PO wished to add a feature to the product, the development teams would have had to rely on the natural language requirements given by the PO, which were usually not precise enough to fully drive the implementation, leaving room

---

[5] The optimization of said combination being a common source of bugs.

[6] The inner workings of the verification framework are complex and will be the subject of an upcoming paper.

for interpretation. Resolving these imprecisions would lead to long discussions because of the lack of a precise specification language. Furthermore, the resulting decision may have not always been clear to all stakeholders, introducing a risk of decisions that later need to be reverted.

Using BDS, the fact that requirements are immediately translated to acceptance tests means that our developers have been able to remove any ambiguity that would have existed. It also means that limitations coming from the implementation — for instance, choices that were made for calculation performance reasons — were surfaced and discussed appropriately and promptly and the requirements were adapted accordingly.

4.2   Early Detection of Requirements Deviation

Having a formal specification with EACs means that the implementation can be accurately validated against the requirements before a feature is released, even if the release cycle is short. The sparse calculation engine is released every two weeks, and ensuring that each version of the code satisfies the high-level requirements of the product would be very hard to achieve without this automatic validation step.

Crucially, BDS ensures that implementation and specification stay consistent with each other. As a point of comparison, during the analysis of the behavior of the dense engine, 218 difference (Diff) tickets were raised that identified a defect in the dense engine. Each defect was then corrected in the sparse engine and the Diff tickets kept as documentation of differences between the two engines. Diff tickets also document the justification for the differences, as those rationales are useful for later reference when discussing adjacent topics, or when explaining behavior differences to customers.

It is worth emphasizing that we use *defect* in the broad sense. Some defects are simply errors in the code, stemming from misunderstandings of the requirements, but bugs can exist in the specification itself. Analyzing the behavior of the dense engine using BDS led us to uncover many issues in the requirements that were caused by not having an independent source of truth.

4.3   Reduced Reliance on Manual Testing

Currently, the formal specification spans 303 specification modules and 149 test modules, comprising about 36,000 and 47,000 lines of VDM-SL respectively, plus 8 backing Java libraries and 1 Rust library. There are about 3,500 EACs, while the verification framework has allowed the generation of hundreds of thousands of test cases.

Extending this suite to test specific aspects of the functionality requires very little work. This innovative tool ensures a regression coverage that would be unthinkable to achieve with manually-written tests. During the development of the project, the regression suite has caught dozens of bugs that would have otherwise been deployed to production.[7] Also, having an extensive regression suite allows the development teams to keep developing at speed to implement improvements in the engine while keeping the

---

[7] Each development team has its own unit and functional test suites, so the regression suite catches issues that the existing testing mechanisms have failed to detect.

risk of introducing regressions low. Compute performance is a critical aspect of the Anaplan platform and the engineering roadmap for the sparse engine contains much performance enhancement work. This means that the code will incur many changes in the future and the regression suite will ensure that functionality remains unaffected.

Parts of the functionality that have been fully specified can be released without any manual checks. The existence of the specification and the fact that the EACs and the generated regression suite pass remove the need for manual testing at release time. QA resources can thus be redirected to areas that are not yet covered by specification.

4.4   Catching Code Defects: A Quantitative Evaluation

During development of the sparse engine, a number of defects were discovered in the implementation. We call *defect* an observable deviation from the intended behavior. In total, 307 calculation logic defects have been identified in the implementation over the course of the development.[8] These defects are partitioned into three categories, depending on which agent identified them:

1. System Specification process and tooling
2. Internal stakeholder (e.g. Customer Support)
3. Customer

In addition, we distinguish whether the defect was found in a released or an unreleased version of the software. Table 1 shows how many issues were detected in each category between September 2020 and August 2023.

Table 1. Calculation issues detected between September 2020 and August 2023

	Unreleased	Released
System Specification Process and Tooling	296	11
Internal Stakeholder (e.g. Customer Support)	0	2
Customer	–	1

It is worth mentioning that the above only counts bugs that have been caught after the code was pushed to the production branch of the repository, either because the specification was written after the implementation was completed, or because the issue was caught by the regression suite. Many more bugs, perhaps the majority, are caught by EACs before the implementation reaches production because EACs run on the continuous integration system and can immediately be fixed by the developers. In sum, in the vast majority of cases, existing defects were caught before users had a chance of detecting them.

For reasons of commercial sensitivity, we cannot disclose specifics around customer adoption nor corporate strategy, but there were many model builders extremely keen to

---

[8] To underline the significance of this number, fewer than 500 calculation bugs have been raised against the dense engine over the past 5 years.

evaluate and adopt the new engine, specifically looking to tackle the most complex use cases. As soon as it was possible to do so, at the start of LA in October 2022, customers adopted the new calculation engine, and had already begun to deploy applications based upon it to their own end-users before the end of the LA phase. Given, the behavioural similarity to the existing engine, customers expected and treated the new system as a fully mature product. With Anaplan being a cloud based system, we were able to observe similar resource usage between sparse and dense engine instances.

4.5 Qualitative Results

A central aspect of BDS is the cross-team interactions and the various stakeholder's perception of the specification process. To understand people's experience with the process, a questionnaire was sent to representative stakeholders to understand how people interact with the specification and what kind of value they see in the process. This was not a widespread survey, but rather a collection of insights from a selection of people in each functional area of the project. We interrogated the PO, the Engineering Lead in charge of the project, two Engineering Managers, two Technical Leads, and four Software Engineers, of which two are in the System Specification team. This non-anonymous survey consisted of a mix of graded and open questions. Responses were analyzed and collated to give a synthesized account of everyone's experience with the process.

Awareness and Understanding. To understand the extent to which people are exposed to and understand BDS, we asked stakeholders to self-report whether they were aware of the process and whether they understood it. One respondent was unaware that BDS was being used. The other stakeholders understood the process with various degrees of familiarity. Engineers outside of the System Specification team reported that they had a practical understanding – allowing them to contribute to the decision-making and the implementation – but not a formal one, which demonstrates that all feel the benefits despite different levels of understanding.

Communication and Collaboration. All respondents agreed that BDS improves communication between teams. It is considered that the document provided by BDS is far superior to ad-hoc requirements documents as it removes much of the guess-work around areas that have not been defined with sufficient clarity. EACs are considered a useful tool as they are much more readable for engineers than the formal specification and require no knowledge of any particular code. Even junior engineers can translate EACs into appropriate unit tests at the appropriate level of abstraction. Having a common language and process has helped understand and document the differences between the sparse and dense engines. Bug reports are also enhanced as they are supported by clear and unambiguous test cases that also use the EAC formalism. The clarity allowed many issues to be agreed as 'working as intended' very early in the bug triage process, concluding that there were missing or incorrect requirements.

Crucially, the PO noted that BDS provides much greater definition of detailed requirements, as subject matter experts can own the precise requirements specification.

In addition, there were many instances where the PO had to think through the implications of their initial proposals and revise them as part of the process. This forced the PO to make explicit decisions about the product where, without BDS, the developer would have had to hazard a best guess.

Exposure to EACs. The members of the System Specification team are the only stakeholders engaged in writing EACs. The technical leads and one engineer are able to understand the meaning of most EACs. One manager has not had any exposure to EACs. All other managers, engineers, and PO have had some exposure to EACs but were less familiar with the syntax. Nonetheless, the PO was able to grasp the syntax as necessary and make informed decisions based on the EACs.

Engineers working on the kernel have little familiarity with EACs because they are written using the Anaplan concepts whereas the kernel is a much more generic OLAP engine. Also, stakeholders mainly involved in non-calculation aspects of Anaplan are unfamiliar with EACs which, for now, only cover calculation functionality. However, engineers working on the translation layer above the kernel regularly look at failing EACs to debug the implementation.

Quality and Outcomes. All respondents considered that BDS significantly increased the quality of the software produced. Since the sparse engine was made available, a single bug was identified by a customer, while hundreds of issues — and possibly thousands if we count the occurrences of developers running EACs before pushing their code to production — were caught internally by the process and the tooling. Another outcome is that the process led to the identification and documentation of over 300 behavioral differences between the dense and the sparse engine. This documentation enables the creation of precise user-facing documentation that explains the intended differences to a user that encounters a behavior in the sparse engine that is unexpectedly different from the dense behavior they are familiar with. This improves the feeling that a given feature is working as intended, rather than being mistaken for a bug. Finally, the extensive regression suite that is generated from the specification gave the team the confidence to transition safely to fortnightly releases.

Development Speed. When asked what effect BDS had on the overall development speed, 80% of respondents judged that BDS made development faster, whereas 20% considered it made it slower. Most people consider that BDS increases the wall time to release features because the specification work, although executed by a team separate from the development team, can become a bottleneck due to resourcing. Also, when writing EACs, there is occasionally the need to introduce some DSL to support the constructs used in the scenario.

However, respondents acknowledge that the reduced back-and-forth achieved by getting the implementation *right the first time* contributes to lower overall effort because fewer bugs are shipped. In addition, the cost of bug investigation, model breaks, management of dissatisfied customers and Anaplan stakeholders, fixes, and release is reduced, freeing resources that enable the development of more features.

Challenges. Practicing BDS effectively requires effort and it takes time to get a feel for what constitutes a good EAC. There is also a potential interlocking between developer repositories and specification test repositories, since if either the implementation or the EACs are not ready, the test suite will not pass, which can block development.[9] This was resolved by introducing a form of deferred compliance where EACs are temporarily skipped until compatibility is achieved.

Engineers also found that the lack of a naming convention between EACs and implementation could make EACs difficult to read for engineers using different names internally. For various reasons, Anaplan constructs can have different names at different levels of abstraction. One engineer found the EAC syntax confusing because the DSL varies from the testing constructs they use in their team.

The issue of workload synchronization was also raised. When planning to implement a feature, managers need to ensure that enough workforce capacity is available in the System Specification team as well as the development teams, which can complicate planning.

General Feedback. All stakeholders believe that BDS should be extended beyond calculation to other areas of the platform. Among the most cited suggestions for expansion are exports, imports, filters, and generally all the grid functionality that the user can interact with. Overall, all respondents are very satisfied with the process and consider that the BDS process is essential to the quality of the Anaplan offering and is a cost-effective way of catching issues early in the process. It has been instrumental in the ability to release the calculation engine at a fast cadence with a high confidence in its correctness, which is essential to the mission-critical operations of Anaplan's customers.

4.6 Key Takeaways

The following prerequisites are crucial for implementing BDS successfully:

– The subject matter must be of a nature that can be formalized. For instance, an OLAP system has well-defined states and transitions, making it a good candidate.
– BDS is a cross-functional operation. Buy-in from all teams involved is necessary.

When the above conditions are satisfied, and the correctness of the software is critical, a BDS approach is clearly valuable:

– Features are validated early in the process, preventing the user from experiencing unnecessary changes.
– The overall development effort is reduced because the released features are (logically) flawless and the cost of defect management is removed.
– Compared to specification-first approaches, the lead time to releasing features is much reduced due to the agility of the process.

---

[9] This only applies to changes to existing behavior such as bug fixes. New behavior can be implemented and specified in parallel.

Since the inception of BDS, some lessons were learned:

– The choice of a Kotlin DSL for the EACs, rather than a natural language like Gherkin, was driven by the need to keep the verification scenario generator implementation-agnostic. This was achieved by making extensive use of the DSL in the generator code, which would not have been feasible with a distinct natural language. While the Kotlin DSL could be perceived to raise the barrier of entry to reading and understanding EACs, we found the effect to be minor and strongly counterbalanced by the greater ease of reasoning about more complex scenarios afforded by a more terse language. However, it should be pointed out that OLAP is in itself a technical topic, so the balance could be different for another application.
– Although it is tempting to specify the low-level implementation requirements, the specification should formalize the behavior of the system as a whole, to be closely matched to the product requirements.
– When analyzing a feature, it is easy to fall into the trap of leaking unnecessary implementation details into an EAC. Keeping the EACs abstract and user-centered allows them to be a form of implementation-agnostic documentation that is accessible to a broader audience.
– Because of the agility of the interaction, an effective setup requires a high level of coordination between the product team, the technical leads, and the specifiers. Weekly meetings with all stakeholders have been found to be highly beneficial, especially in the earlier stages of the project.

5   Conclusion

As a methodology to efficiently design, implement, and release software whose correctness is critical, BDS has proved to be remarkably effective in the development of Anaplan's sparse calculation engine. Our analysis shows that it is not only possible to integrate formal methods in an agile context, but implementing BDS brought levels of confidence in the correctness of the software that would not have been achievable by other means. From a team perspective, BDS has been successfully integrated without alienating staunch agile advocates, some of whom are now promoters of the process.

Through the establishment of EACs, the functionality validation process ensures that implementation remains constantly in line with the expected behavior as understood by all stakeholders. More importantly, the requirements specification is agile and can never go stale because it evolves with the code. Coordination is achieved by discussing requirements during their inception, as well as during subsequent modifications. Adherence to the requirements specification is enforced by the EACs acting as acceptance tests. The VDM-SL formal specification allows the automatic generation of an extensive regression suite that detects faults that are not captured by the acceptance criteria, for instance when features are combined.

As our evaluation shows, BDS can be considered to have decreased overall feature development effort by catching nearly 300 issues before they reached production, thus avoiding expensive rework and impact on customer relations. To date, only a single calculation bug was ever detected by an external user. The process was instrumental in bringing an entirely new engine to general availability through frequent and drama-free releases.

Disclosure of Interests. The authors have no competing interests to declare that are relevant to the content of this article.

References

1. Alkobaisi, S., Bae, W.D., Narayanappa, S., Debnath, N.: Steel threads: software engineering constructs for defining, designing and developing software system architecture. J. Comput. Methods Sci. Eng. **12**(s1), S63–S77 (2012). https://doi.org/10.3233/JCM-2012-0437
2. Alwidian, S., Jaskolka, J.: Understanding the role of human-related factors in security requirements elicitation. In: International Working Conference on Requirements Engineering: Foundation for Software Quality, pp. 65–74. Springer, Cham (2023). https://doi.org/10.1007/978-3-031-29786-1_5
3. Beck, K., et al.: Manifesto for agile software development (2001). http://www.agilemanifesto.org/
4. Buchan, J.: An empirical cognitive model of the development of shared understanding of requirements. In: Zowghi, D., Jin, Z. (eds.) Requirements Engineering. CCIS, vol. 432, pp. 165–179. Springer, Heidelberg (2014). https://doi.org/10.1007/978-3-662-43610-3_13
5. Codd, E.F., Codd, S.B., Salley, C.T.: Providing OLAP (on-line analytical processing) to user-analysts. An IT Mandate. White Paper. Arbor Software Corporation 4 (1993)
6. Evans, E.: Domain-Driven Design: Tackling Complexity in the Heart of Software. Addison-Wesley Professional (2004)
7. Fraser, S., Pezzoni, A.: Azuki framework to assist with behavior-driven specification. https://github.com/anaplan-engineering/azuki. Accessed 01 Oct 2023
8. Fraser, S., Pezzoni, A.: Behaviour driven specification. In: Proceedings of the 19th International Overture Workshop, pp. 5–20 (2021). https://doi.org/10.48550/arXiv.2110.09371
9. Ghezzi, C.: Formal methods and agile development: towards a happy marriage. In: The Essence of Software Engineering, pp. 25–36. Springer, Cham (2018). https://doi.org/10.1007/978-3-319-73897-0_2
10. Hoare, C.A.R.: An axiomatic basis for computer programming. Commun. ACM **12**(10), 576–580 (1969). https://doi.org/10.1145/363235.363259
11. Inayat, I., Salim, S.S., Marczak, S., Daneva, M., Shamshirband, S.: A systematic literature review on agile requirements engineering practices and challenges. Comput. Hum. Behav. **51**, 915–929 (2015). https://doi.org/10.1016/j.chb.2014.10.046
12. Jebreen, I., Awad, M., Al-Qerem, A.: A propose model for shared understanding of software requirements (SUSRs). In: Information Science and Applications (ICISA) 2016, pp. 1045–1056. Springer, Singapore (2016). https://doi.org/10.1007/978-981-10-0557-2_100
13. Kotonya, G., Sommerville, I.: Requirements Engineering: Processes and Techniques. Wiley (1998)
14. Larsen, P.G., Battle, N., Ferreira, M., Fitzgerald, J., Lausdahl, K., Verhoef, M.: The overture initiative integrating tools for VDM. ACM SIGSOFT Softw. Eng. Notes **35**(1), 1–6 (2010). https://doi.org/10.1145/1668862.1668864
15. Sidorova, M.I., et al.: Strategic planning software as a tool for improvement of business information space. In: European Proceedings of Social and Behavioural Sciences (2022)
16. Wang, X., Zhao, L., Wang, Y., Sun, J.: The role of requirements engineering practices in agile development: an empirical study. In: Zowghi, D., Jin, Z. (eds.) Requirements Engineering. CCIS, vol. 432, pp. 195–209. Springer, Heidelberg (2014). https://doi.org/10.1007/978-3-662-43610-3_15
17. Wynne, M., Hellesoy, A., Tooke, S.: The Cucumber Book: Behaviour-Driven Development for Testers and Developers. Pragmatic Bookshelf (2017)

The Return of Formal Requirements Engineering in the Era of Large Language Models

Paola Spoletini[1]([✉])[ID] and Alessio Ferrari[2][ID]

[1] Kennesaw State University, Kennesaw, GA, USA
pspoleti@kennesaw.edu
[2] ISTI-CNR, Pisa, Italy
alessio.ferrari@isti.cnr.it

Abstract. [*Context and Motivation*] Large Language Models (LLMs) have made remarkable advancements in emulating human linguistic capabilities, showing potential in executing various traditional software engineering tasks, including code generation. [*Question/Problem*] Despite their generally good performance, utilizing LLM-generated code raises legitimate concerns regarding its correctness and the assurances it can provide. [*Principal Idea/Results*] To address these concerns, we propose turning to formal requirements engineering—a practice currently predominantly used in developing complex systems where adherence to standards and accountability are required. [*Contribution*] In this vision paper, we discuss the integration of automatic formal requirements engineering techniques as a complement to LLM code generation. Additionally, we explore how LLMs can facilitate the broader acceptance of formal requirements, thus making the vision proposed in this paper realizable.

Keywords: Formal Requirements · Formal Methods · Model Checking · Large Language Models

1 Introduction

Large Language Models (LLMs) [30,31] have demonstrated exceptional proficiency in replicating human linguistic capabilities. These models can comprehend and generate human-like text, finding significant applications within many fields, including software engineering (SE). In SE, LLMs can contribute to different activities, such as code synthesis, documentation generation, and ideation, making some processes more efficient, especially those that traditionally require intensive human labor. Comprehensive overviews of the work currently done to explore the use of LLMs in SE have already been published, despite the recent development of the field [14,17,26].

Among the different uses of LLMs in SE, code generation is one of the most studied areas and has yielded promising results (e.g., [4]). The idea is to generate code through a prompt representing the user's needs, often expressed in Natural Language (NL). This represents a paradigm shift in how code is conceptualized

© The Author(s), under exclusive license to Springer Nature Switzerland AG 2024
D. Mendez and A. Moreira (Eds.): REFSQ 2024, LNCS 14588, pp. 344–353, 2024.
https://doi.org/10.1007/978-3-031-57327-9_22

and developed, thus creating new challenges for the SE community, and in particular for the requirements engineering (RE) community. Indeed, given that the specification is expressed through an NL prompt and the code is generated by LLMs, questions arise on the reliability, correctness, and interpretability of the code generated by these models. To address these concerns, we propose using formal requirements and formal verification techniques to guarantee that the generated code satisfies the user's needs.

Formal requirements represent a structured and unambiguous specification of the expected behavior and constraints of a system [16]. They precisely describe system functionality, performance, and design criteria. Unlike informal requirements, which may be prone to ambiguity, formal requirements employ mathematical or logical language, minimizing the risk of misinterpretation. Formal requirements allow the adoption of formal methods, a set of mathematical techniques for verifying systems. These methods, including model checking [9], theorem proving [1], and SMT [11], aim to enhance the rigor and correctness of the software development process. Because of the expertise needed to apply formal requirements and associated verification techniques, their use is currently mainly limited to the development of critical and complex systems (e.g., [12]). Given the proposed central role of formal RE in software development, our vision also includes steps to make formal RE more accessible. To this end, we propose to use LLMs to support the formalization of requirements, explain formal requirements, and help interpret the results of formal tools.

Section 2 analyzes the challenges of using LLMs in software development to generate code. Section 3 presents our vision of how formal RE can be used in support of LLM-generated code and how LLMs can be used to make formal requirements more accessible; it also discusses how formal techniques can be used to guarantee the reliability and consistency of LLMs output for these activities. Section 4 provides some final remarks.

2 Open Challenges on the Use of LLMs and Formal Requirements in Software Development

The use of LLMs to support code generation has the potential to simplify and improve software development but also introduces many challenges to the RE community because it changes the software development paradigm. In particular, in this setting, the requirements specification becomes the NL prompt, and the implementation happens as a black-box operation.

A primary challenge is

C1.a: the interpretability of LLM-generated code.

While these models showcase impressive language comprehension, the intricacies of code semantics and the rationale behind certain decisions remain obscured. Understanding the black-box nature of and the rationale behind LLM-generated code becomes a crucial hurdle and an imperative for robust software engineering practices. This also makes it necessary

C1.b: ensuring the reliability and correctness of code

through suitable techniques that, starting from the user's prompt, can guarantee that the generated code is semantically correct. In addition, once LLMs are used in software development, given LLMs probabilistic behavior, guaranteeing

C2: the reliability and consistency of the output,

also assuring that **hallucinations** [29] **are avoided** becomes fundamental.

Another major challenge is

C3: the introduction of biases [2],

a critical concern when deploying LLMs in software development. Biases embedded in the training data can manifest in the generated code, potentially leading to unintended consequences. For example, when making decisions on how to distribute aid using income and gender as parameters, the generated code can create inequality by assigning aids only to a specific gender [18]. The RE community faces the challenge of identifying and mitigating biases in LLM-generated code to avoid perpetuating discriminatory and unethical patterns.

In addressing the challenges posed by LLMs in software development, our vision involves the strategic incorporation of formal requirements and methods. This vision aims to exploit the strengths of formalism to enhance the reliability, interpretability, and ethical soundness of LLM-generated code. The extensive proposed adoption of formal RE, though, presents many challenges. The major challenge connected to Formal RE is

C4: the accessibility of formal languages and techniques.

In particular, it has been shown that mastering formal specification languages, such as temporal logic [25], requires training and expertise [3] and, the lack of adequate knowledge often results in incorrect specifications [10]. Finally, other challenges are connected to the verification of these specifications. In particular, it is important to

C5: correctly interpreting the results and easing the interaction with tools,

especially those requiring manual interactions, e.g., interactive theorem provers.

3   Formal RE 2.0: Vision and Roadmap

Figure 1 represents some of the main contributions of our vision. Box 1 describes how LLMs can be used to generate code (yellow box—yellow is used to represent machine-generated artifacts). Usually, LLM-generated code is specified with the use of a prompt (green box—green is used to represent human-generated artifacts), which usually includes the high-level description of the task to be performed (context), some examples of input and output[1] and the user's current input.

---

[1] Note that to improve the LLM performance the model can be fine-tuned through a training phase using a set of examples. Fine-tuned models typically perform better than just prompting [23].

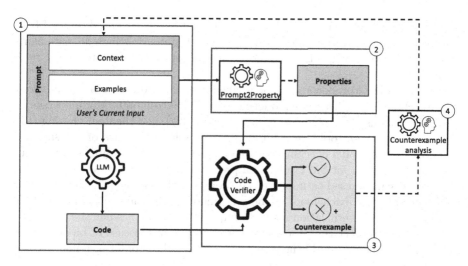

Fig. 1. (Part of the) Vision on the use of Formal Requirements and Formal Methods for and with LLMs (Color figure online)

To address challenges *C1* and *C3*, we propose to verify the code against properties that formally express the user input and ethical requirements (Sect. 3.1). As shown in Fig. 1, box 3, the verification tool takes as input the specified properties that the code has to satisfy and the generated code and checks that the properties hold in the code, and, in case they do not, it generates a counterexample (in yellow in box 3). Given the need to specify properties, use formal tools, and interpret the results of these tools, to address challenges *C4* and *C5*, as we plan to leverage LLMs to make these activities accessible (Sect. 3.2).

Finally, since the probabilistic nature of LLMs does not guarantee consistency and sometimes generates "hallucinations" (*C2*), to address this issue, we propose the use of probabilistic and stochastic verification techniques (Sect. 3.3).

3.1 Formal Requirements and Verification of LLM-Generated Code

> **Vision:** (a) Using formal languages to specify the requirements that the code has to guarantee; (b) leverage existing verification tools to check that the code satisfies the requirements.

To give some guarantees about LLM-generated code, we propose to derive and formalize the requirements that the code has to satisfy from the user's prompt. The prompt includes a description of the user's expectations and thus naturally represents the system's requirements. These requirements can be expressed in different languages. Still, to take full advantage of existing verification tools, we envision they get formalized using some form of temporal logic, such as LTL [25]. In addition to these properties, we need to consider ethical requirements and check the fairness of the generated code. These requirements are necessary to guarantee that the bias contained in the data used by the LLMs does not propagate to the code. Examples of such properties include statements about the independence of the code decisions from specific parameters (e.g., race, inclusivity).

The formalized properties can be verified over the code using program verification techniques, such as code model checking [8], Satisfiability Modulo Theories (SMT) solvers used for bounded model checking [22], and theorem proving [13]. The output of these tools varies: a counterexample in the case of not satisfied properties when model checking is used (as pictured in Fig. 1 box 3); an exemplar of satisfiability for SMT solvers; a proof for theorem provers). In any case, this output can be used to improve the prompt and consequently the code. We represent this "feedback loop" in box 4, where we envision a human-machine support to use the verification result to improve the prompt when the generated code does not respect the properties. For example, in the case of model checking, an idea could be to add the counterexample to the prompt as a non-allowed behavior.

Notice that this approach does not explicitly address concerns related to the interpretability of the generated code, but the verification of the properties provides some output that can be leveraged to understand the code better. For example, in the case of model checking, the verification output either provides guarantees on the code or counterexamples showing why the program does not respect the properties that can help to understand the code and identify why it violates the properties.

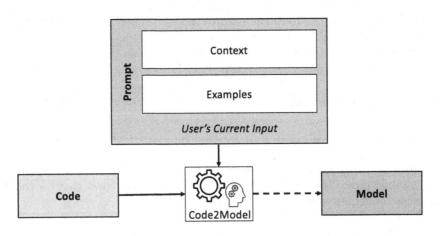

Fig. 2. Model generation from the code. (Color figure online)

The existing code-based verification techniques are less studied than the equivalent techniques that use a model of the systems instead of code. This suggests the need to produce a model of the code by leveraging LLMs fine-tuned for this task as shown in Fig. 2 where we envision that a combination of machine and human support helps to create a model (in the red box—red represents arti-facts that are created as a combination of human and machine effort) starting from the code and leveraging the prompt(s) used to generate it. In addition, having a model of the system could help understand the results of the verification step better and consequently improve the prompt (to improve the final code).

Roadmap to the Proposed Vision

- Develop guidelines to extract the user's requirements from the prompt and formalize them.
- Identify a set of fairness requirements in a parametric form and understand how to contextualize them for a given prompt.
- Test the effectiveness of existing tools in verifying the code against the properties and identify areas of improvement.
- Provide guidelines on how to add the output generated by the verification tools to the prompt to improve the quality of the generated code.
- Integrate this automated process into the software development life cycle to identify and address potential issues systematically.
- Analyze techniques to extract a model of the system from the code and repeat the step done for code verification (Fig. 1, box 4)

3.2 Using LLMs to Democratize Formal Specifications and Methods

Vision: Using fine-tuned LLMs to support the formalization of the properties to be verified and to support the verification process through explanations.

The correct specification and interpretation of LTL formulae require a solid mathematical background and highly trained experts to be done correctly [3,10]. For this reason, there has been work proposing to translate NL to LTL by using conventional neural network models [15,24] and more recently employing LLMs [6,21]. A common obstacle encountered in these studies is the limited availability of training data. Despite the significant advancements demonstrated by LLMs in translation tasks, their performance tends to degrade as the complexity of the formula to generate increases, making it challenging to generalize the approach. We envision a solution to this problem through a combination of LLM-based and human support that help produce the formalized properties, as represented in box 3 in Fig. 1.

Conversely, to complement these approaches and support the understanding of the produced formulae, Cherukuri et al. [7] focus on the translation from LTL to NL. Their results are still in an early stage, likely due to the inherent difficulty in learning representations of high-level abstractions of LTL. The proposed vision includes improving this direction of translation so that the software engineer can be supported in understanding existing properties. This approach helps enhance the understanding between humans and systems.

Another way to democratize formal RE and make it more accessible is to have support to interpret the results provided by model checkers and SMT solvers.

Model checkers give a positive answer when a property is satisfied, and a counterexample shows how the property is violated when it is not. On the contrary, SMT solvers provide a negative response when a property is not satisfiable and an example of how it can be satisfied if it is. These exemplars provided by the verification tools are fundamental to better understanding the quality of the code or model under analysis and improving it. Still, they are often difficult to interpret and thus exploit. LLMs capabilities might offer support to explain their meaning better and include them in the prompt to generate better code.

Finally, LLMs can help in the use of theorem provers. Theorem provers (e.g., [27]) support their users in generating correctness proofs by providing support in solving part of the proof (or the proof itself, in case of decidable problems). Because of the mathematical nature and the complexity of the problem, interacting with a theorem prover is a challenging task, especially for novices [20]. LLMs could provide an opportunity to produce NL explanations of the generated proofs, thus supporting their understanding and the next steps of the proof.

Roadmap to the Proposed Vision

- Generate LLMs support to automatically produce the formalization of the properties from the prompt (Fig. 1, dotted line 2).
- Explore the use of LLMs to explain and generalize counterexamples and make them part of the prompt to be able to correct the resulting code.
- Explore using LLMs to generate explanations and recommendations using theorem provers.

3.3   Verification of LLMs Generation Consistency

> **Vision:** Use probabilistic and stochastic verification tools to improve the consistency of LLMs generated results.

The inherently probabilistic nature of LLMs affects their ability to consistently produce the same output, and sometimes results in the generation of incorrect output and hallucinations [29]. In the context of software development and to make this vision realizable, it is important to take care of this problem. Some preliminary works have been published that address this issue [28]. Given that the root cause of these behaviors comes from the probabilistic behavior of LLMs, we plan to approach this issue by trying to explore how probabilistic [19] and stochastic verification [5] approaches can assist with this issue. Probabilistic and stochastic model checking are formal techniques for analyzing systems that exhibit probabilistic behavior. They can be used to address the probabilistic nature of LLMs and produce consistent output. We envision a process made of three steps, namely modeling, specification, and checking.

- **Modelling:** the LLM is modeled as a probabilistic or stochastic system. This involves defining the states of the system (i.e., the different states of the LLM), the transitions between these states (which could be the different operations or actions your LLM can take), and the probabilities associated with these transitions. The states of the LLM can be defined as the activations of the neurons at each layer. Each state in the neural network is associated with a specific configuration of these activations, and the transitions between states are determined by the weights and biases of the network, as well as the input data. Given a state (i.e., a specific configuration of neuron activations) and an input, the network computes the next state by applying the weighted sums and activation functions. The probabilities of these transitions can be thought of as the likelihood that the network will move from one state to another given the current state and input. The stochastic nature of the LLM comes from the fact that the input data can have inherent randomness, and the network's output (i.e., the next state) is a function of this random input. Moreover, during training, techniques like dropout or noise injection can introduce additional randomness.

- **Specification:** the properties to check are specified in a suitable probabilistic or stochastic logic. These properties could be about the behavior of the LLM, such as "the probability that the output changes by more than a threshold amount given a small perturbation to the input is less than 0.01".
- **Checking:** the probabilistic or stochastic model checker verifies whether the model of the LLM satisfies the specified properties. The model checker does this by systematically exploring the state space of the model and checking the properties against it.

Roadmap to the Proposed Vision
- Identify and formalize properties that express consistency and accuracy of LLMs output.
- Explore ways to automatically create stochastic models from the code of LLMs.
- Explore ways to compress very large stochastic models so that their model checking is feasible.
- Explore using different probabilistic and stochastic verification approaches to better understand the reliability of LLM's output.
- Consider alternative approaches, e.g., extract the model based on repeated observation of the output, and adjust the input to obtain consistent output.

4   Conclusion

The proposed roadmap envisions a transformative journey to make the use of LLMs in code generation reliable by using and rethinking formal RE. Indeed, while formal RE becomes an essential step in LLM-supported software development, our vision redesigns its use through the support of LLMs, which have the potential to make formal languages and formal tools accessible. In our vision, formal tools are also explored to analyze the consistency of LLM's output. The successful implementation of this roadmap has the potential to address the concerns associated with LLM-generated code and lays the foundation for a paradigm of responsible and trustable AI. In this vision, developers, guided by formal RE, work with LLMs to create innovative, reliable, and ethical solutions.

References

1. Bibel, W.: Automated Theorem Proving. Springer, Wiesbaden (2013). https://doi.org/10.1007/978-3-322-90102-6
2. Brun, Y., Meliou, A.: Software fairness. In: Proceedings of the 2018 26th ACM Joint Meeting on European Software Engineering Conference and Symposium on the Foundations of Software Engineering, pp. 754–759 (2018)
3. Brunello, A., Montanari, A., Reynolds, M.: Synthesis of LTL formulas from natural language texts: state of the art and research directions. In: 26th International Symposium on Temporal Representation and Reasoning (TIME 2019). Schloss Dagstuhl-Leibniz-Zentrum fuer Informatik (2019)
4. Chen, M., et al.: Evaluating large language models trained on code. arXiv preprint arXiv:2107.03374 (2021)
5. Chen, T., Forejt, V., Kwiatkowska, M., Parker, D., Simaitis, A.: Automatic verification of competitive stochastic systems. Formal Methods Syst. Des. **43**, 61–92 (2013)
6. Chen, Y., Gandhi, R., Zhang, Y., Fan, C.: NL2TL: transforming natural languages to temporal logics using large language models. arXiv preprint arXiv:2305.07766 (2023)
7. Cherukuri, H., Ferrari, A., Spoletini, P.: Towards explainable formal methods: from LTL to natural language with neural machine translation. In: Gervasi, V., Vogelsang, A. (eds.) REFSQ 2022. LNCS, vol. 13216, pp. 79–86. Springer, Cham (2022). https://doi.org/10.1007/978-3-030-98464-9_7
8. Chong, N., et al.: Code-level model checking in the software development workflow. In: Proceedings of the ACM/IEEE 42nd International Conference on Software Engineering: Software Engineering in Practice (2020)
9. Clarke, E.M., Henzinger, T.A., Veith, H., Bloem, R., et al.: Handbook of Model Checking, vol. 10. Springer, Cham (2018). https://doi.org/10.1007/978-3-319-10575-8
10. Czepa, C., Zdun, U.: On the understandability of temporal properties formalized in linear temporal logic, property specification patterns and event processing language. IEEE Trans. Softw. Eng. **46**(1), 100–112 (2018)
11. De Moura, L., Bjørner, N.: Z3: an efficient SMT solver. In: Ramakrishnan, C.R., Rehof, J. (eds.) TACAS 2008. LNCS, vol. 4963, pp. 337–340. Springer, Heidelberg (2008). https://doi.org/10.1007/978-3-540-78800-3_24
12. Dietsch, D., Langenfeld, V., Westphal, B.: Formal requirements in an informal world. In: 2020 IEEE Workshop on Formal Requirements (FORMREQ), pp. 14–20. IEEE (2020)
13. D'silva, V., Kroening, D., Weissenbacher, G.: A survey of automated techniques for formal software verification. IEEE Trans. Comput. Aided Des. Integr. Circ. Syst. **27**(7), 1165–1178 (2008)
14. Fan, A., et al.: Large language models for software engineering: survey and open problems. arXiv preprint arXiv:2310.03533 (2023)
15. Gopalan, N., Arumugam, D., Wong, L.L., Tellex, S.: Sequence-to-sequence language grounding of non-Markovian task specifications. In: Robotics: Science and Systems, vol. 2018 (2018)
16. Greenspan, S., Mylopoulos, J., Borgida, A.: On formal requirements modeling languages: RML revisited. In: Proceedings of 16th International Conference on Software Engineering, pp. 135–147. IEEE (1994)

17. Hou, X., et al.: Large language models for software engineering: a systematic literature review. arXiv preprint arXiv:2308.10620 (2023)
18. Huang, D., Bu, Q., Zhang, J., Xie, X., Chen, J., Cui, H.: Bias testing and mitigation in LLM-based code generation (2024)
19. Katoen, J.P.: The probabilistic model checking landscape. In: Proceedings of the 31st Annual ACM/IEEE Symposium on Logic in Computer Science (2016)
20. Knobelsdorf, M., Frede, C., Böhne, S., Kreitz, C.: Theorem provers as a learning tool in theory of computation. In: Proceedings of the 2017 ACM Conference on International Computing Education Research, pp. 83–92 (2017)
21. Liu, J.X., et al.: Lang2LTL: translating natural language commands to temporal robot task specification. arXiv preprint arXiv:2302.11649 (2023)
22. Liu, T., Nagel, M., Taghdiri, M.: Bounded program verification using an SMT solver: a case study. In: 2012 IEEE Fifth International Conference on Software Testing, Verification and Validation, pp. 101–110. IEEE (2012)
23. Min, B., et al.: Recent advances in natural language processing via large pre-trained language models: a survey. ACM Comput. Surv. **56**(2), 1–40 (2023)
24. Patel, R., Pavlick, E., Tellex, S.: Grounding language to non-Markovian tasks with no supervision of task specifications. In: Robotics: Science and Systems (2020)
25. Pnueli, A.: The temporal logic of programs. In: 18th Annual Symposium on Foundations of Computer Science (SFCS 1977), pp. 46–57. IEEE (1977)
26. Shah, D., Osiński, B., Levine, S., et al.: LM-NAV: robotic navigation with large pretrained models of language, vision, and action. In: Conference on Robot Learning, pp. 492–504. PMLR (2023)
27. Coq Development Team: he Coq proof assistant (1989–2023). http://coq.inria.fr
28. Wang, R.E., Durmus, E., Goodman, N., Hashimoto, T.: Language modeling via stochastic processes. arXiv preprint arXiv:2203.11370 (2022)
29. Ye, H., Liu, T., Zhang, A., Hua, W., Jia, W.: Cognitive mirage: a review of hallucinations in large language models. arXiv preprint arXiv:2309.06794 (2023)
30. Zan, D., et al.: Large language models meet NL2Code: a survey. In: Proceedings of the 61st Annual Meeting of the Association for Computational Linguistics (Volume 1: Long Papers), pp. 7443–7464 (2023)
31. Zhao, W.X., et al.: A survey of large language models. arXiv preprint arXiv:2303.18223 (2023)

Author Index

D. Mendez and A. Moreira (Eds.): REFSQ 2024, LNCS 14588, pp. 355–356, 2024.
https://doi.org/10.1007/978-3-031-57327-9

Printed in the United States
by Baker & Taylor Publisher Services